Studies in Computational Intelligence

Volume 792

Series editor

Janusz Kacprzyk, Polish Academy of Sciences, Warsaw, Poland
e-mail: kacprzyk@ibspan.waw.pl

The series "Studies in Computational Intelligence" (SCI) publishes new developments and advances in the various areas of computational intelligence—quickly and with a high quality. The intent is to cover the theory, applications, and design methods of computational intelligence, as embedded in the fields of engineering, computer science, physics and life sciences, as well as the methodologies behind them. The series contains monographs, lecture notes and edited volumes in computational intelligence spanning the areas of neural networks, connectionist systems, genetic algorithms, evolutionary computation, artificial intelligence, cellular automata, self-organizing systems, soft computing, fuzzy systems, and hybrid intelligent systems. Of particular value to both the contributors and the readership are the short publication timeframe and the world-wide distribution, which enable both wide and rapid dissemination of research output.

More information about this series at http://www.springer.com/series/7092

Juan Julian Merelo · Fernando Melício
José M. Cadenas · António Dourado
Kurosh Madani · António Ruano
Joaquim Filipe
Editors

Computational Intelligence

International Joint Conference, IJCCI 2016
Porto, Portugal, November 9–11, 2016
Revised Selected Papers

 Springer

Editors
Juan Julian Merelo
Department of Computer Architecture
  and Computer Technology
Universidad de Granada
Granada, Spain

Fernando Melício
ISEL-Instituto Politécnico de Lisboa
Lisboa, Portugal

José M. Cadenas
Facultad de Informática,
Department of Information and
  Communications Engineering
University of Murcia
Murcia, Spain

António Dourado
University of Coimbra
Coimbra, Portugal

Kurosh Madani
Université Paris-Est Créteil (UPEC)
Créteil, France

António Ruano
University of Algarve
Faro, Portugal

Joaquim Filipe
INSTICC
Setúbal, Portugal

ISSN 1860-949X   ISSN 1860-9503 (electronic)
Studies in Computational Intelligence
ISBN 978-3-030-07585-9   ISBN 978-3-319-99283-9 (eBook)
https://doi.org/10.1007/978-3-319-99283-9

This Springer imprint is published by the registered company Springer Nature Switzerland AG
The registered company address is: Gewerbestrasse 11, 6330 Cham, Switzerland

Organization

Conference Chair

Joaquim Filipe, Polytechnic Institute of Setúbal/INSTICC, Portugal

Program Co-chairs

ECTA
Juan Julian Merelo, University of Granada, Spain
Fernando Melício, Instituto Superior de Engenharia de Lisboa, Portugal

FCTA
José M. Cadenas, University of Murcia, Spain
António Dourado, University of Coimbra, Portugal

NCTA
Kurosh Madani, University Paris-Est Créteil (UPEC), France
António Ruano, University of Algarve, Portugal

ECTA Program Committee

Arturo Hernandez Aguirre, Centre for Research in Mathematics, Mexico
Chang Wook Ahn, Sungkyunkwan University, Korea, Republic of
Richard Allmendinger, University of Manchester, UK
Meghna Babbar-Sebens, Oregon State University, USA
Helio J. C. Barbosa, Laboratorio Nacional de Computaçao Cientifica, Brazil
Michal Bidlo, Brno University of Technology, Faculty of Information Technology, Czech Republic
Tim Blackwell, University of London, UK
William R. Buckley, California Evolution Institute, USA
Mauro Castelli, Universidade Nova de Lisboa, Portugal
Pedro A. Castillo, University of Granada, Spain

Sung-Bae Cho, Yonsei University, Korea, Republic of
Chi-Yin Chow, City University of Hong Kong, Hong Kong
Vic Ciesielski, RMIT University, Australia
Carlos Cotta, Universidad de Málaga, Spain
Kyriakos Deliparaschos, Cyprus University of Technology, Cyprus
Rolf Dornberger, University of Applied Sciences and Arts Northwestern Switzerland, Switzerland
Peter Duerr, Sony Corporation, Japan
Marc Ebner, Ernst-Moritz-Arndt-Universität Greifswald, Germany
Fabio Fassetti, DIMES, University of Calabria, Italy
Carlos M. Fernandes, University of Lisbon, Portugal
Eliseo Ferrante, KU Leuven and Universite' de Technologie de Compiegne, Belgium
Stefka Fidanova, Bulgarian Academy of Sciences, Bulgaria
Valeria Fionda, University of Calabria, Italy
Dalila B. M. M. Fontes, Faculdade de Economia and LIAAD-INESC TEC, Universidade do Porto, Portugal
Ewa Gajda-Zagórska, Institute of Science and Technology Austria, Austria
Marian Gheorghe, University of Bradford, UK
Alexandros Giagkos, Aberystwyth University, UK
Crina Grosan, Brunel University London, UK
Jörg Hähner, Universität Augsburg, Germany
Lutz Hamel, University of Rhode Island, USA
Thomas Hanne, University of Applied Arts and Sciences Northwestern Switzerland, Switzerland
J. Ignacio Hidalgo, Universidad Complutense de Madrid, Spain
Wei-Chiang Hong, Jiangsu Normal University, China
Katsunobu Imai, Hiroshima University, Japan
Hesam Izakian, Alberta Centre for Child, Family and Community Research, Canada
Colin Johnson, University of Kent, UK
Iwona Karcz-Duleba, Wroclaw University of Science Technology, Poland
Wali Khan, Kohat University of Science and Technology (KUST), Kohat, Pakistan
Mario Köppen, Kyushu Institute of Technology, Japan
Ondrej Krejcar, University of Hradec Kralove, Czech Republic
Rajeev Kumar, Indian Institute of Technology, India
Halina Kwasnicka, Wroclaw University of Technology, Poland
Nuno Leite, Instituto Superior de Engenharia de Lisboa, Portugal
Diego Pérez Liébana, Queen Mary University of London, UK
Piotr Lipinski, University of Wroclaw, Poland
Shih-Hsi Liu, California State University, Fresno, USA
Wenjian Luo, University of Science and Technology of China, China
Penousal Machado, University of Coimbra, Portugal
Stephen Majercik, Bowdoin College, USA
Simon Martin, University of Stirling, UK

FCTA Program Committee

Ajith Abraham, Machine Intelligence Research Labs (MIR Labs), USA
Salmiah Ahmad, International Islamic University Malaysia, Malaysia
Rafael Alcala, University of Granada, Spain
Jesús Alcalá-Fdez, University of Granada, Spain
Ismail H. Altas, Karadeniz Technical University, Turkey
Michela Antonelli, University of Pisa, Italy
Sansanee Auephanwiriyakul, Chiang Mai University, Thailand
Bernard De Baets, Ghent University, Belgium
Rosangela Ballini, Unicamp, Brazil
Alper Basturk, Erciyes University, Melikgazi, Turkey
Mokhtar Beldjehem, University of Ottawa, Canada
Faouzi Bouslama, Dubai Men's College/Higher Colleges of Technology, UAE
Ahmed Bufardi, NA, Switzerland
Humberto Bustince, Public University of Navarra, Spain
José M. Cadenas, University of Murcia, Spain
Jinhai Cai, University of South Australia, Australia
Daniel Antonio Callegari, PUCRS—Pontificia Universidade Catolica do Rio Grande do Sul, Brazil
Heloisa Camargo, UFSCar, Brazil
Rahul Caprihan, Dayalbagh Educational Institute, India
Pablo Carmona, University of Extremadura, Spain
João Paulo Carvalho, INESC-ID/Instituto Superior Técnico, Portugal
Fabio Casciati, Università degli Studi di Pavia, Italy
Giovanna Castellano, University of Bari, Italy
Kit Yan Chan, Curtin University, Australia
Wen-Jer Chang, National Taiwan Ocean University, Taiwan
Chun-Hao Chen, Tamkang University, Taiwan
Jianhua Chen, Louisiana State University and A&M College, USA
France Cheong, RMIT University, Australia
Jyh-Horng Chou, National Kaohsiung University of Applied Sciences, Taiwan, Taiwan
Hung-Yuan Chung, National Central University, Taiwan
Mikael Collan, Lappeenranta University of Technology, Finland
Keeley Crockett, Manchester Metropolitan University, UK
Valerie Cross, Miami University, USA
Martina Dankova, University of Ostrava, Czech Republic
Bijan Davvaz, Yazd University, Iran, Islamic Republic of
Irene Diaz, University of Oviedo, Spain
József Dombi, University of Szeged, Institute of Informatics, Hungary
António Dourado, University of Coimbra, Portugal
Pietro Ducange, University of Pisa, Italy
Francesc Esteva, Artificial Intelligence Research Institute (IIIA-CSIC), Spain
Mario Fedrizzi, University of Trento, Italy

Yoshikazu Fukuyama, Meiji University, Japan
Angel Garrido, Universidad Nacional de Educacion a Distancia (UNED), Facultad de Ciencias, Spain
Alexander Gegov, University of Portsmouth, UK
Antonio Gonzalez, University of Granada, Spain
Sarah Greenfield, De Montfort University, UK
Masafumi Hagiwara, Keio University, Japan
Susana Muñoz Hernández, Universidad Politécnica de Madrid (UPM), Spain
Katsuhiro Honda, Osaka Prefecture University, Japan
Chih-Cheng Hung, Kennesaw State University, USA
Lazaros S. Iliadis, Democritus University of Thrace, Greece
Chia-Feng Juang, National Chung Hsing University, Taiwan
Cengiz Kahraman, Istanbul Technical University, Turkey
Etienne Kerre, Ghent University, Belgium
Shubhalaxmi Kher, Arkansas State University, USA
Frank Klawonn, Ostfalia University of Applied Sciences, Germany
Laszlo T. Koczy, Szechenyi Istvan University, Hungary
Vladik Kreinovich, University of Texas at El Paso, USA
Yau-Hwang Kuo, National Cheng Kung University, Taiwan
Anne Laurent, Lirmm, University of Montpellier, France
Huey-Ming Lee, Chinese Culture University, Taiwan
Kang Li, Queen's University Belfast, UK
Lvzhou Li, Sun Yat-sen University, China
Chin-Teng Lin, National Chiao Tung University, Taiwan
José Manuel Molina López, Universidad Carlos III de Madrid, Spain
Edwin Lughofer, Johannes Kepler University, Austria
Francisco Gallego Lupianez, Universidad Complutense de Madrid, Spain
Francesco Marcelloni, University of Pisa, Italy
Agnese Marchini, Pavia University, Italy
Christophe Marsala, LIP6, France
Corrado Mencar, University of Bari, Italy
José M. Merigó, University of Chile, Chile
Araceli Sanchis de Miguel, University Carlos III, Spain
Ludmil Mikhailov, University of Manchester, UK
Valeri Mladenov, Technical University Sofia, Bulgaria
Javier Montero, Complutense University of Madrid, Spain
Alejandro Carrasco Muñoz, University of Seville, Spain
Hiroshi Nakajima, Omron Corporation, Japan
Maria do Carmo Nicoletti, Universidade Federal de São Carlos, Brazil
Vesa Niskanen, University of Helsinki, Finland
Yusuke Nojima, Osaka Prefecture University, Japan
Vilém Novák, University of Ostrava, Czech Republic
David A. Pelta, University of Granada, Spain
Parag Pendharkar, Pennsylvania State University, USA
Irina Perfilieva, University of Ostrava, Czech Republic

Radu-Emil Precup, Politehnica University of Timisoara, Romania
Daowen Qiu, Sun Yat-sen University, China
Leonid Reznik, Rochester Institute of Technology, USA
Antonello Rizzi, Università di Roma "La Sapienza", Italy
Reza Saatchi, Sheffield Hallam University, UK
Indrajit Saha, National Institute of Technical Teachers' Training and Research, India
Daniel Sánchez, University of Granada, Spain
Jurek Sasiadek, Carleton University, Canada
Daniel Schwartz, Florida State University, USA
Rudolf Seising, Deutsches Museum München, Germany
Qiang Shen, Aberystwyth University, UK
Giorgos Stamou, National Technical University of Athens, Greece
Hooman Tahayori, Ryerson University, Canada
C. W. Tao, National Ilan University, Taiwan
Mohammad Teshnehlab, K. N. Toosi University of Technology, Iran, Islamic Republic of
Dat Tran, University of Canberra, Australia
George Tsihrintzis, University of Piraeus, Greece
Leonilde Varela, University of Minho, School of Engineering, Portugal
Jasmin Velagic, Faculty of Electrical Engineering, Bosnia and Herzegovina
Susana M. Vieira, University of Lisbon, Portugal
Wen-June Wang, National Central University, Taiwan
Junzo Watada, Universiti Teknologi PETRONAS, Malaysia
Ronald R. Yager, Iona College, USA
Takeshi Yamakawa, Fuzzy Logic Systems Institute (FLSI), Japan
Chung-Hsing Yeh, Monash University, Australia
Jianqiang Yi, Institute of Automation, Chinese Academy of Sciences, China
Slawomir Zadrozny, Polish Academy of Sciences, Poland
Hans-Jürgen Zimmermann, European Laboratory for Intelligent Techniques Engineering (ELITE), Germany

NCTA Program Committee

Francisco Martínez Álvarez, Pablo de Olavide University of Seville, Spain
Veronique Amarger, University Paris-Est Créteil (UPEC), France
Anang Hudaya Muhamad Amin, Multimedia University, Malaysia
William Armitage, University of South Florida, USA
Vijayan Asari, University of Dayton, USA
Peter M. Atkinson, University of Southampton, UK
Gilles Bernard, Paris 8 University, France
Yevgeniy Bodyanskiy, Kharkiv National University of Radio Electronics, Ukraine
Ivo Bukovsky, Czech Technical University in Prague, Faculty of Mechanical Engineering, Czech Republic
Javier Fernandez de Canete, University of Malaga, Spain

Yuriy L. Orlov, Novosibirsk State University and Institute of Cytology and Genetics SB RAS, Russian Federation
Mourad Oussalah, University of Oulu, Finland
Mark Oxley, Air Force Institute of Technology, USA
Tim C. Pearce, University of Leicester, UK
Parag Pendharkar, Pennsylvania State University, USA
Amaryllis Raouzaiou, National Technical University of Athens, Greece
Neil Rowe, Naval Postgraduate School, USA
Christophe Sabourin, IUT Sénart, University Paris-Est Créteil (UPEC), France
Virgilijus Sakalauskas, Vilnius University, Lithuania
Gerald Schaefer, Loughborough University, UK
Christoph Schommer, University Luxembourg, Campus Belval, Maison du Nombre, Luxembourg
Emilio Soria-Olivas, University of Valencia, Spain
Lambert Spaanenburg, Lund University, Lund Institute of Technology, Sweden
Jochen Steil, Bielefeld University, Germany
Ruedi Stoop, Universität Zürich/ETH Zürich, Switzerland
Catherine Stringfellow, Midwestern State University, USA
Mu-Chun Su, National Central University, Taiwan
Johan Suykens, KU Leuven, Belgium
Norikazu Takahashi, Okayama University, Japan
Antonio J. Tallón-Ballesteros, Universidade de Lisboa, Portugal
Yi Tang, Yunnan University of Nationalities, China
Juan-Manuel Torres-Moreno, Ecole Polytechnique de Montréal, Canada
Carlos M. Travieso, University of Las Palmas de Gran Canaria, Spain
Jessica Turner, Georgia State University, USA
Andrei Utkin, INOV INESC Inovação, Portugal
Alfredo Vellido, Universitat Politècnica de Catalunya, Spain
Salvatore Vitabile, University of Palermo, Italy
Shuai Wan, Northwestern Polytechnical University, China
Jingyu Wang, Northwestern Polytechnical University (NPU), China
Guanghui Wen, Southeast University, China
Bo Yang, University of Jinan, China
Chung-Hsing Yeh, Monash University, Australia
Wenwu Yu, Southeast University, China
Cleber Zanchettin, Federal University of Pernambuco, Brazil
Huiyu Zhou, Queen's University Belfast, UK

Invited Speakers

Thomas Stützle, Université Libre de Bruxelles, Belgium
Bernadette Bouchon-Meunier, University Pierre et Marie Curie-Paris 6, France
Una-May O'Reilly, MIT Computer Science and Artificial Intelligence Laboratory, USA
Juan Julian Merelo, University of Granada, Spain
Daniel Chirtes, EASME—Executive Agency for SMEs—European Commission, Romania

Preface

The present book includes extended and revised versions of a set of selected papers from the 8th International Joint Conference on Computational Intelligence (IJCCI 2016), held in Porto, Portugal, from 9 to 11 November 2016.

IJCCI 2016 received 74 paper submissions from 33 countries, of which 19% were included in this book. The papers were selected by the event chairs, and their selection is based on a number of criteria that include the classifications and comments provided by the program committee members, the session chairs' assessment, and also the program chairs' global view of all papers included in the technical program. The authors of selected papers were then invited to submit a revised and extended version of their papers having at least 30% innovative material.

The purpose of IJCCI is to bring together researchers, engineers, and practitioners on the areas of fuzzy computation, evolutionary computation, and neural computation. IJCCI is composed of three co-located conferences, each specialized in at least one of the aforementioned main knowledge areas.

The papers selected to be included in this book contribute to the understanding of relevant topics under the aforementioned areas. Seven of the selected papers are closely related to evolutionary computation, discussing aspects such as evolution strategies and the evolution of complex cellular automata, genetic algorithms, swarm/collective intelligence, artificial life, game theory and applications. Another four papers are addressing topics under the fuzzy computation area, from fuzzy modeling, control, and prediction in human–robot systems, to pattern recognition, including fuzzy clustering and classifiers as well as fuzzy inference. Finally, three papers were selected from the area of neural computation, two of them addressing in particular computational neuroscience, neuroinformatics, and bioinformatics and one addressing an application in particular, namely an e-mail spam filter based on unsupervised neural architectures and thematic categories.

We would like to thank all the authors for their contributions and also to the reviewers who have helped ensuring the quality of this publication.

Granada, Spain   Juan Julian Merelo
Lisboa, Portugal   Fernando Melício
Murcia, Spain   José M. Cadenas
Coimbra, Portugal   António Dourado
Créteil, France   Kurosh Madani
Faro, Portugal   António Ruano
Setúbal, Portugal   Joaquim Filipe
November 2016

Contents

Part I
Evolutionary Computation Theory and Applications

An Adapting Quantum Field Surrogate for Evolution Strategies

Jörg Bremer and Sebastian Lehnhoff

Abstract Black-box optimization largely suffers from the absence of analytical objective forms. Thus, no derivatives or other structure information can be harnessed for guiding optimization algorithms. But, even with analytic form, high-dimensionality and multi-modality often hinder the search for a global optimum. Heuristics based on populations or evolution strategies are a proven and effective means to, at least partly, overcome these problems. Evolution strategies have thus been successfully applied to optimization problems with rugged, multi-modal fitness landscapes from numerous applications, to nonlinear problems, and to derivative free optimization. One obstacle in heuristics in general is the occurrence of premature convergence, meaning a solution population converges too early and gets stuck in a local optimum. In this paper, we present an approach that harnesses the adapting quantum potential field determined by the spatial distribution of elitist solutions as guidance for the next generation. The potential field evolves to a smoother surface leveling local optima but keeping the global structure what in turn allows for a faster convergence of the solution set. On the other hand, the likelihood of premature convergence decreases. We demonstrate the applicability and the competitiveness of our approach compared with particle swarm optimization and the well-established evolution strategy CMA-ES.

Keywords Evolution strategies · Global optimization · Surrogate optimization · Premature convergence · Quantum potential

J. Bremer (✉) · S. Lehnhoff
University of Oldenburg, 26129 Oldenburg, Germany
e-mail: joerg.bremer@uni-oldenburg.de

S. Lehnhoff
e-mail: sebastian.lehnhoff@uni-oldenburg.de

© Springer Nature Switzerland AG 2019
J. J. Merelo et al. (eds.), *Computational Intelligence*,
Studies in Computational Intelligence 792,
https://doi.org/10.1007/978-3-319-99283-9_1

1 Introduction

Global optimization comprises many problems in practice as well as in the scientific community. These problems are often hallmarked by presence of a rugged fitness landscape with many local optima and non-linearity. Thus optimization algorithms are likely to become stuck in local optima and guaranteeing the exact optimum is often intractable.

Evolution Strategies [1] have shown excellent performance in global optimization especially when it comes to complex multi-modal, high dimensional, real valued problems [2, 3]. A major drawback of population based algorithms is the large number of objective function evaluations. Real world problems often face computational efforts for fitness evaluations; e. g. in Smart Grid load planning scenarios, fitness evaluation involves simulating a large number of energy resources and their behaviour [4].

Another problem is known as premature convergence [5–7], when a heuristics converges too early towards a sub-optimal solution and then gets stuck in this local optimum. This might for instance happen if an adaption strategy decreases the mutation range and thus the range of the currently searched sub-region and possible ways out are no longer scrutinized.

In [8] a surrogate approach that harnesses a continuously updating quantum potential field determined by elitist solutions has been proposed. On the one hand side, the quantum field exploits global information by aggregating over scattered fitness information similar to scale space approaches [9, 10], on the other side by continuously adapting to elitist solutions, the quantum field surface quickly flattens to a smooth surrogate for guiding further sampling directions. To achieve this, the surrogate is a result of Schrödinger's equation of which a probability function is derived that determines the potential function after a clustering approach (cf. [9]). Here, we extend the work from [8] with additional results.

The Schrödinger equation [11] is one of the fundamental concepts of quantum mechanics [12]. After the Copenhagen interpretation [13], the equation delivers the likelihood of finding a quantum mechanical particle at a certain point. Forces causing a certain deflection from particles' ground state are given by a potential field in the equation. Usually the field is given by a known wave function operator and the likely positions of particles are sought. But one can put it also the other way round and determine the field that might have been the root cause for a given certain deflection; for instance by given solution positions.

We associate solutions and their position with quantum mechanical particles and associate minima of the potential field, created in denser regions of good solutions' positions with areas of interest for further investigation. Thus, offspring solutions are generated with a trend in descending the potential field. By harnessing the quantum potential as a surrogate, we achieve a faster convergence with less objective function calls compared with using the objective function alone. In lieu thereof the potential field has to be evaluated at selected point. Although a fine grained computation of the potential field would be a computationally hard task in higher dimensions [9], we achieve a better overall performance because we need to calculate the field only at isolated data points.

The paper starts with a review of using quantum mechanics in computational intelligence and in particular in evolutionary algorithms; we briefly recap the quantum potential approach for clustering and present the adaption for integration into evolution strategies as introduced in [8]. We extend the work from [8] by additionally scrutinizing the effect of larger sample sizes. We conclude with an evaluation of the approach with the help of several well-known benchmark test functions and demonstrate the competitiveness to two well established algorithms: particle swarm optimization (PSO) and co-variance matrix adaption evolution strategy (CMA-ES).

2 Related Work

Optimization approaches in general can roughly be classified into deterministic and probabilistic methods. Deterministic approaches like interval methods [14], Cutting Plane methods [15], or Lipschitzian methods [16] often suffer from intractability of the problem or getting stuck in local optima [17].

Several evolutionary approaches have been introduced to solve nonlinear complex optimization problems with multi-modal, rugged fitness landscapes [18, 19]. Each of these methods has its own characteristics, strengths and weaknesses. A common characteristics in all evolutionary algorithms is the generation of an offspring solution set in order to explore the characteristics of the objective function in the neighbourhood of existing solutions. When the solution space is hard to explore or objective evaluations are costly, computational effort is a common drawback for all population-based schemes. Many efforts have already been spent to accelerate convergence of these methods. Example techniques are: improved population initialization [20], adaptive populations sizes [21] or exploiting sub-populations [22].

Sometimes a surrogate model is used in case of computational expensive objective functions [23] to substitute a share of objective function evaluations with cheap surrogate model evaluations. The surrogate model represents a learned model of the original objective function. Recent approaches use Radial Basis Functions, Polynomial Regression, Support Vector Regression, Artificial Neural Network or Kriging [24]; each approach exhibiting individual advantages and drawbacks. The famous CMA-ES for example learns a second order model of the objective function by adapting the co-variance matrix and thus the distribution of solutions according to past good steps in evolution for future solution sampling [25].

At the same time, quantum mechanics has inspired several fields of computational intelligence such as data mining, pattern recognition or optimization. In [9] a quantum mechanics based method for clustering has been introduced. Quantum clustering extends the ideas of Scale Space algorithms and Support Vector Clustering [26, 27] by representing an operator in Hilbert space by a scaled Schrödinger equation that yields a probability function as result. The inherent potential function of the equation that can be analytically derived from the probability function is used to identify barycenters of data cluster by associating minima with centers. In [28] this approach has been extended to a dynamic approach that uses the fully fledged time dependant variant of the Schrödinger equation to allow for a interactive visual data mining especially for large data sets [29].

The authors in [8] adapted and extended the quantum field part of the clustering approach to optimization and use the potential function to associate already found solutions from the objective domain with feature vectors in Hilbert space; but with keeping an emphasis on the total sum (cf. [9]) and thus with keeping in mind all improvements of the ongoing optimum search.

Reference [30] used a quantum potential approach derived from quantum clustering to detect abnormal events in multidimensional data streams. Reference [31] used quantum clustering for weighing linear programming support vector regression. In this work, we derive a sampling method for a $(\mu + \lambda)$-ES from the quantum potential approach originally used for clustering by [32].

A quantum mechanical extension to particle swarm optimization has been presented e.g. in [33, 34]. Here particles move according to quantum mechanical behavior in contrast to the classical mechanics ruled movement of particles in standard PSO. Usually a harmonic oscillator is used. In [35] both methods quantum clustering and quantum PSO have been combined by deriving good particle starting positions from the clustering method first. For the simulated Annealing (SA) approach also a quantum extension has been developed [36]. Whereas in classical SA local minima are escaped by leaping over the barrier with a thermal jump, quantum SA introduces the quantum mechanical tunneling effect for such escapes.

We integrated the quantum concept into evolution strategies; but by using a different approach: we harness the information in the quantum field about the so far gained success as a surrogate for generating the offspring generation. By using the potential field, information from all samples at the same time is condensed into a directed generation of the next generation.

3   The Schrödinger Potential

We start with a brief recap of the Schrödinger potential and describe the concept following [9, 32, 37]. Let

$$H\psi \equiv \left(-\frac{\sigma_{\text{pot}}^2}{2}\nabla^2 + V(x) \right)\psi(x) = E\psi(x) \tag{1}$$

be the Schrödinger equation rescaled to a single free parameter σ_{pot} and eigenstate $\psi(x)$. H denotes the Hamiltonian operator corresponding to the total energy E of the system. ψ is the wave function of the quantum system and ∇^2 denotes the Laplacian differential operator. V corresponds to the potential energy in the system. In case of a single point at x_0 Eq. (1) results in

$$V = \frac{1}{2}\sigma_{\text{pot}}^2(x - x_0)^2 \tag{2}$$

$$E = \frac{d}{2} \tag{3}$$

with d denoting the dimension of the field. In this case ψ is the ground state of a harmonic oscillator. Given an arbitrary set of points, the potential at a point x can be expressed by

$$
\begin{aligned}
V(x) &= E + \frac{\frac{\sigma_{\text{pot}}^2}{2}\nabla^2\psi}{\psi} \\
&= E - \frac{d}{2} + \frac{1}{2\sigma_{\text{pot}}^2\psi}\sum_i(x - x_i)^2 e^{-\frac{(x-x_i)^2}{2\sigma_{\text{pot}}^2}} .
\end{aligned}
\tag{4}
$$

In Eq. (4) the Gaussian wave function

$$\psi(x) = \sum_i(x - x_i)^2 e^{-\frac{(x-x_i)^2}{2\sigma_{\text{pot}}^2}} \tag{5}$$

is associated to each point and summed up. Please note that the bandwidth parameter (usually named σ, cf. [32]) has been denoted σ_{pot} to discriminate the bandwidth of the wave function and the variance σ in the mutation used later in the evolution strategy. In quantum mechanics, usually the potential $V(x)$ is given and solutions or eigenfunctions $\psi(x)$ are sought. In our approach we are already given $\psi(x)$ determined by a set of data points. The set of data points is given by elitist solutions. We then look for the potential $V(x)$ whose solution is $\psi(x)$.

The wave function part corresponds with the Parzen window estimator approach for data clustering [38] or with scale-space clustering [10] that interprets this wave function as the density function that could have generated the underlying data set. The maxima of this density function correspond therefore with data centers.

In quantum clustering and by requiring ψ to be the ground state of the Hamiltonian H the potential field V establishes a surface that shows more pronounced minima than other kernel estimators [37]. V is unique up to a constant factor. By setting the minimum of V to zero it follows that

$$E = -min\frac{\frac{\sigma_{\text{pot}}^2}{2}\nabla^2\psi}{\psi} . \tag{6}$$

With this convention V is determined uniquely with $0 \leq E \leq \frac{d}{2}$. In this case, E is the lowest eigenvalue of the operator H and thus describes the ground state.

V is expected to exhibit one or more minima within some region around the data set and grow quadratically on the outside [32]. In quantum clustering, these minima are associated with cluster centers. We will interpret them as balance points or nuclei where the minimum of an associated function f lies if the set of data points that defines V is a selection of good points (in the sense of a good fitness according to f).

4 The Algorithm

We start with a general description of the idea. In our approach we generate the quantum potential field of an elitist selection of samples. Out of a sample of λ solutions the best μ are selected according to the objective function. These μ solutions then define a quantum potential field that exhibits troughs at the barycenters of good solution taking into account all good solutions at the same time. In the next step this potential field is used to guide the sampling of the next generation of λ offspring solutions from which the next generation is selected that defines the new field. In this way, the potential field continuously adapts in each iteration to the so far found best solutions.

The advantage of using the potential field results from its good performance in identifying the barycenters of data points. Horn and Gottlieb [9] demonstrated the superior performance compared with density based approaches like Scale Space or Parzen Window approaches [38, 39]. Transfered to optimization this means the quantum potential allows for a better identification of local optima. As they can be explored faster they can be neglected earlier which in turn leads to a faster convergence of the potential field towards the global optimum (cf. Fig. 1).

Figure 1 gives an impression of the adaption process that transforms the quantum field into an easier searchable function. Each column shows the situation after 1, 3 and 8 iterations (top to bottom) for different 2-dimensional objective functions. The top row thus displays the original objective function; from top to bottom the evolving potential field is displayed together with the respective offspring solutions that represent the so far best. The minimum of the potential field evolves towards the minimum of the objective function (or towards more than one optimum if applicable).

Figure 2 shows the approach formally. Starting from an initially generated sample $\mathcal{X}$ equally distributed across the whole search domain defined by a box constraint in each dimension. Next, the offspring is generated by sampling λ points normally distributed around the μ solutions from the old generation with an at each time randomly chosen parent solution as expectation and with variance σ^2 that decreases with each generation. We use a rejection sampling approach with the metropolis criterion [40] for acceptance applied to the difference in the potential field between a new candidate solution and the old solution. The new sample is accepted with probability

$$p_a = \min(1, e^{\Delta V}). \tag{7}$$

$\Delta V = V(x_{old}) - V(x_{new})$ denotes the level difference in the quantum field. A descent within the potential field is always accepted. A (temporary) degradation in quantum potential level is accepted with a probability P_a (cf. Eq. (7)) determined by the level of degradation.

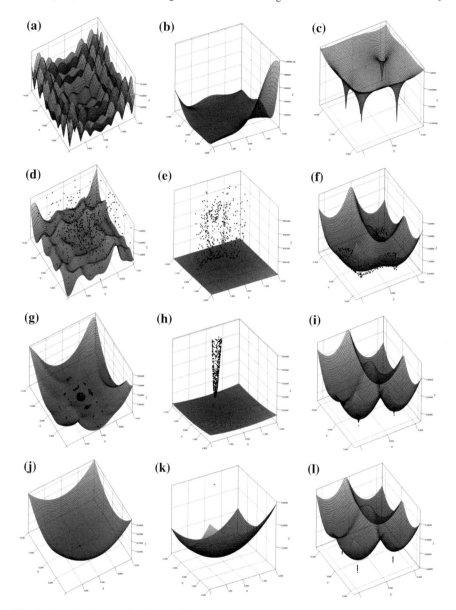

Fig. 1 Function (top row) and exemplary evolution (after 1, 3 and 8 iterations) of the quantum potential (from to to the bottom) that guides the search for minima for the test functions: Alpine, Goldstein-Price and Himmelblau (from left to right); Extracted from [8]

As long as there exists at least one pair $\{x_1, x_2\} \subset \mathcal{S}$ with $x_1 \neq x_2$, the potential field has a minimum at x' with $x' \neq x_1 \wedge x' \neq x_1$. Thus, the sampling will find new candidates. The sample variance σ^2 is decreased in each iteration by a rate ω.

$\mathcal{X} := \{\boldsymbol{x}_i \sim \mathcal{U}(x_{lo}, x_{up})^d\},\ 1 \leqslant i \leqslant n$ // initialize solution population randomly uniform
distributed over the domain
repeat
   $\mathcal{S} \leftarrow \varnothing$ // start with empty offspring set
   repeat
      $\boldsymbol{x}_z := \boldsymbol{x}_i \in \mathcal{X},\ i \sim \mathcal{U}(1, |\mathcal{X}|)$ ˜   // selct random solution from parents
      $\boldsymbol{s} \sim \mathcal{N}(\boldsymbol{x}_z, \sigma^2)$ // randomly mutate with variance σ
      if $p \leqslant e^{V(\boldsymbol{x}_z) - V(\boldsymbol{s})},\ p \sim \mathcal{U}(0, 1)$ **then**
         $\mathcal{S} \leftarrow \mathcal{S} \cup \boldsymbol{s}$ // keep if Metropolis criterion met
      end if
   until $|\mathcal{S}| == \lambda$
   $V \leftarrow V(\mathcal{S}, \sigma_{\text{pot}})$ // update potential field
   $\mathcal{X} \leftarrow \text{select}(\mathcal{S}, f, \mu)$ // select elitist solutions
   $\sigma \leftarrow \sigma \cdot \omega$ // adapt step size
until $\|f(x_{best}) - f(x^*)\| \leqslant \epsilon$ // until stopping criterion met

Fig. 2 Basic scheme for the Quantum Sampling ES Algorithm

Finally, for the next iteration, the solution set $\mathcal{X}$ is updated by selecting the μ best
from offspring $\mathcal{S}$.

The process is repeated until any stopping criterion is met; apart from having
come near enough the minimum we regularly used an upper bound for the maximum
number of iterations (or rather: number of fitness evaluations respectively).

5   Results

We evaluated our evolution strategy with a set of well known test functions developed
for benchmarking optimization heuristics. We used the following established benchmark functions: Alpine, Goldstein-Price, Himmelblau, Bohachevsky 1, generalized
Rosenbrock, Griewank, Sphere, Booth, Chichinadze and Zakharov [3, 21, 41, 42];
see also appendix A. These functions represent a mix of multi-modal, 2-dimensional
and multi-dimensional functions, partly with a huge number of local minima and
steep as well as shoal surroundings of the global optimum and broad variations in
characteristics and domain sizes.

Figure 1 shows some of the used functions (left column) together with the respective evolution of the quantum potential field that guides the search towards the minimum at $(0, 0)$ for the Alpine function and $(0, -1)$ for Goldstein-Price; the Himmelblau function has four global minima which are all found. The figure also shows the
evolution of the solution population. In the next step, we tested the performance of
our approach against competitive approaches.

First, we tested the effect of using the quantum field as adaptive surrogate compared with the same update strategy working directly on the fitness landscape of the
objective function. Figure 3 shows the convergence of the error on the Alpine test
function for both cases. Although the approach with surrogate converges slightly

Fig. 3 Comparing the convergence of using the quantum field as surrogate with the plain approach working on the objective function directly. For testing, the 2-dimensional Alpine function has been used. Extracted from [8]

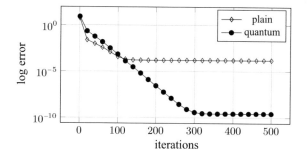

Fig. 4 Convergence of using the quantum field as surrogate compared to the plain approach. Depicted are the means of 100 runs on the 20-dimensional Alpine function. Extracted from [8]

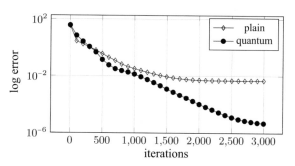

slower in the beginning it clearly outperforms the plain approach without quantum surrogate. Figure 4 shows the same effect for the 20-dimensional case. Both results show the convergence of the mean error for 100 runs each.

In a next step, we compared our approach with two well-known heuristics: particle swarm optimization (PSO) from [43] and the covariance matrix adaption evolution strategy (CMA-ES) by [44]. Both strategies are well-known, established, and have been applied to wide range of optimization problems. We used readily available and evaluated implementations from Jswarm-PSO (http://jswarm-pso.sourceforge.net) and commons math (http://commons.apache.org/proper/commons-math).

All algorithms have an individual, strategy specific set of parameters that usually can be tweaked to some degree for a problem specific adaption. Nevertheless, default values that are applicable for a wide range of functions are usually available. For our experiments, we used the following default settings. For the CMA-ES, the (external) strategy parameters are λ, μ, $w_{i=1...\mu}$, controlling selection and recombination; c_σ and d_σ for step size control and c_c and μ_{cov} controlling the covariance matrix adaption. We have chosen to set these values after [44].

$$\lambda = 4 + \lfloor 3 \ln n \rfloor, \quad \mu = \left[\frac{\lambda}{2}\right], \tag{8}$$

$$w_i = \frac{\ln(\frac{\lambda}{2} + 0.5) - \ln i}{\sum_{\mu}^{j=1} \frac{\lambda}{2} + 0.5) - \ln i}, \quad i = 1, \ldots, \mu \tag{9}$$

Fig. 5 Comparing the convergence (means of 100 runs) of CMA-ES and the quantum approach on the 2-dimensional Booth function. Extracted from [8]

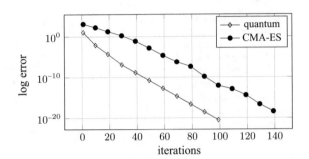

$$C_c = \frac{4}{n+4}, \quad \mu_{\text{cov}} = \mu_e f f, \tag{10}$$

$$C_{\text{cov}} = \frac{1}{\mu_{\text{cov}}} \frac{2}{(n+\sqrt{2})^2} + \left(1 - \frac{1}{\mu_{\text{cov}}}\right) \min\left(1, \frac{2\mu_{\text{cov}} - 1}{(n+2)^2 + \mu_{\text{cov}}}\right), \tag{11}$$

These settings are specific to the dimension N of the objective function. An in-depth discussion of these parameters is also given in [45].

For the PSO, we used values of 0.9 for the weights and 1 for the inertia parameter as default setting as recommended for example in [46].

For the quantum field strategy, we empirically found the following values as a useful setting for a wide range of objective functions. The initial mutation variance has been set to $\sigma = d/10$ for an initial diameter d of the search space (domain of the objective function). The shrinking rate of the variance has been set to $\omega = 0.98$ and the bandwidth in the potential equation (4) has been set to $\sigma_{\text{pot}} = 0.4$. For the population size we chose $\mu = 10$ and $\lambda = 50$ if not otherwise stated.

First, we compared the convergence of the quantum strategy with the CMA-ES. Figure 5 shows a first result for the 2-dimensional Booth function (with a minimum of zero). The quantum approach has been stopped at errors below 1×10^{-21} to avoid numerical instabilities. The used CMA-ES implementation has a similar condition integrated into its code. Comparing iterations, the quantum approach converges faster than CMA-ES. Figure 5 shows the result for the 20-dimensional Griewank function with a search domain of $[-2049, 2048]^{20}$. Here, the quantum approach achieves about the same result as the CMA-ES within less iterations. Comparing iterations does not yet shed light on performance.

As the performance is determined by the number of operations that have to be conducted in each iteration, the following experiments consider the number of function evaluation calls rather than iterations. Table 1 shows the results for a bunch of 2-dimensional test functions. For each test function and each algorithm the achieved mean (averaged over 100 runs) solution quality and the needed number of function evaluations is displayed. The solution quality is expressed as the error in terms of

Table 1 Results for comparing CMA-ES, PSO and the quantum approach with a set of 2-dimensional test functions. The error denotes the remaining difference to the known optimum objective value; evaluations refer to the number of conducted function evaluations and in case of the quantum approach to the sum of function and quantum surrogate evaluations (the share of objective evaluations is for the quantum case given in brackets). Extracted from [8]

Problem	Algorithm	Error	Evaluations
Alpine	CMA-ES	$3.954 \times 10^{-12} \pm 4.797 \times 10^{-12}$	746.38 ± 90.81
	PSO	$5.543 \times 10^{-9} \pm 3.971 \times 10^{-8}$	500000.00 ± 0.00
	Quantum	$8.356 \times 10^{-16} \pm 4.135 \times 10^{-16}$	$198996.53 \pm 132200.50(99363)$
Griewank	CMA-ES	$2.821 \times 10^{2} \pm 2.254 \times 10^{2}$	174.22 ± 137.63
	PSO	$4.192 \times 10^{-4} \pm 1.683 \times 10^{-3}$	500000.00 ± 0.00
	Quantum	$6.577 \times 10^{-3} \pm 5.175 \times 10^{-3}$	$472348.75 \pm 94420.25(201361)$
Goldstein price	CMA-ES	$0.459 \times 10^{1} \pm 1.194 \times 10^{2}$	613.96 ± 203.03
	PSO	$1.698 \times 10^{0} \pm 1.196 \times 10^{1}$	500000.00 ± 0.00
	Quantum	$1.130 \times 10^{0} \pm 1.255 \times 10^{1}$	$500012.75 \pm 3.48(250000)$
Bohachevsky 1	CMA-ES	$3.301 \times 10^{-2} \pm 1.094 \times 10^{-1}$	672.70 ± 59.91
	PSO	$1.626 \times 10^{-10} \pm 1.568 \times 10^{-9}$	500000.00 ± 0.00
	Quantum	$4.707 \times 10^{-16} \pm 3.246 \times 10^{-16}$	$87548.03 \pm 3377.00(39641)$
Booth	CMA-ES	$4.826 \times 10^{-17} \pm 1.124 \times 10^{-16}$	605.68 ± 49.55
	PSO	$5.985 \times 10^{-14} \pm 3.743 \times 10^{-13}$	500000.00 ± 0.00
	Quantum	$5.695 \times 10^{-16} \pm 2.957 \times 10^{-16}$	$68283.18 \pm 2711.51(33696)$
Chichinadze	CMA-ES	$0.953 \times 10^{1} \pm 4.125 \times 10^{1}$	698.44 ± 200.31
	PSO	$0.226 \times 10^{0} \pm 2.247 \times 10^{-1}$	500000.00 ± 0.00
	Quantum	$1.327 \times 10^{1} \pm 8.376 \times 10^{0}$	$180.68 \pm 12.63(50)$

Fig. 6 Comparing the
convergence of CMA-ES
and the quantum approach
on the 20-dimensional
Griewank function on the
domain $[-2048, 2048]^{20}$.
Both algorithms have been
stopped after 1500 iterations.
Extracted from [8]

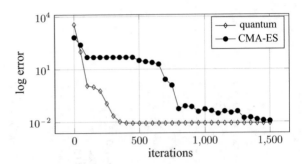

remaining difference to the known global optimum. As stopping criterion this time each algorithm has been equipped with two conditions: error below 5×10^{-17} and a given budget of at most 5×10^7 evaluations. For the quantum approach two counts of evaluation functions are given, because due to the nature of surrogate approaches a share of function evaluations is substituted by surrogate evaluations. Thus, total evaluations refers to the sum of function and surrogate evaluations. The number of mere objective evaluations is given in brackets (Fig. 6).

The results show that CMA-ES is in general unbeatable in terms of function evaluations whereas the quantum approach in half of the cases gains the more accurate result. Nevertheless, CMA-ES also needs additional calculations for eigen decomposition and update of the co-variance matrix. The PSO succeeds for the Griewank and the Chichinadze function. For the 20-dimensional cases in Table 2 the quantum approach gains the most accurate result in most of the cases. The CMA-ES winning margin of a low number of evaluations decreases compared with the quantum approach, but is still prominent. Nevertheless, the number of necessary function evaluations for the quantum approach can still be reduced when using a lower population size. But, such tuning is subject to the problem at hand. On the other hand, notwithstanding the low number of objective evaluations, the CMA-ES needs higher processing time for high-dimensional problems due to the fact that CMA-ES needs – among others – to conduct eigenvalue decompositions of its co-variance matrix ($\mathcal{O}(n^3)$) with number of dimensions n [47]. Table 7 gives an impression on necessary computation (CPU-) times (Java 8, 2.7 GHz Quadcore) for the 100-dimensional Sphere function for CMA-ES and a quantum approach with reduced populations size ($\mu = 4$, $\lambda = 12$).

All in all, the quantum approach is competitive to the established algorithms and in some cases even superior.

Tables 3, 4, 5 and 6 show some results for lower dimensional problems. In this experiment we explored the effect of population size. Thus we set up all three algorithms with the same number of solutions per iteration and thus the same population size (Table 7).

All algorithms are compared by the following indicators: the success rate denotes the share of successful optimizations runs (here, an optimization run was considered successful if it terminates with an residual error less than 10^{-24}).

Table 2 Results for comparing CMA-ES, PSO and the quantum approach with a set of 20-dimensional test functions with the same setting as in Table 1. Extracted from [8]

Problem	Algorithm	Error	Evaluations
Rosenbrock	CMA-ES	$1.594 \times 10^7 \pm 6.524 \times 10^7$	10678.60 ± 6514.47
	PSO	$1.629 \times 10^{10} \pm 1.759 \times 10^{10}$	50000000 ± 0.00
	Quantum	$3.884 \times 10^7 \pm 1.470 \times 10^8$	$50001008.27 \pm 557.53(23291850)$
Griewank	CMA-ES	$5.878 \times 10^1 \pm 1.082 \times 10^2$	8734.48 ± 7556.50
	PSO	$1.429 \times 10^2 \pm 3.539 \times 10^2$	50000000 ± 0.00
	Quantum	$2.267 \times 10^{-3} \pm 3.973 \times 10^{-3}$	$17269617.9 \pm 1949920.8(6929090)$
Zakharov	CMA-ES	$9.184 \times 10^{-16} \pm 1.068 \times 10^{-15}$	8902.24 ± 845.05
	PSO	$7.711 \times 10^1 \pm 5.961 \times 10^1$	50000000 ± 0.00
	Quantum	$8.978 \times 10^{-17} \pm 9.645 \times 10^{-18}$	$2962737.19 \pm 10126.34(1021370)$
Spherical	CMA-ES	$1.200 \times 10^{-15} \pm 1.258 \times 10^{-15}$	8678.80 ± 912.13
	PSO	$2.637 \times 10^0 \pm 7.517 \times 10^0$	50000000 ± 0.00
	Quantum	$8.943 \times 10^{-17} \pm 1.036 \times 10^{-17}$	$2674716.70 \pm 8844.04(973750)$
Alpine	CMA-ES	$9.490 \times 10^{-12} \pm 3.331 \times 10^{-11}$	15196.48 ± 570.76
	PSO	$4.021 \times 10^0 \pm 2.442 \times 10^0$	50000000 ± 0.0
	Quantum	$9.272 \times 10^{-17} \pm 5.857 \times 10^{-18}$	$4771580.5 \pm 8390.4(1867830)$

Table 3 Results for 3-dimensional test function. Best values for each category, algorithm and indicator are highlighted. The success rate of zero for the CMA-ES results from an algorithm specific premature abort to avoid numerical instabilities resulting in an incomparable small number of evaluations. The error denotes the mean difference of found and true optimum. All results are the mean of 50 runs

Problem	Algorithm	Succ. rate	Error
Rosenbrock	CMA-ES	0.000	$5.552 \times 10^{-19} \pm 5.758 \times 10^{-19}$
	PSO	**0.980**	$8.330 \times 10^{-25} \pm 5.890 \times 10^{-24}$
	Quantum	**0.980**	$\mathbf{1.233 \times 10^{-33} \pm 8.716 \times 10^{-33}}$
Griewank	CMA-ES	0.040	$2.395 \times 10^{-2} \pm 1.589 \times 10^{-2}$
	PSO	**1.000**	$\mathbf{0.000 \times 10^{0} \pm 0.000 \times 10^{0}}$
	Quantum	**1.000**	$\mathbf{0.000 \times 10^{0} \pm 0.000 \times 10^{0}}$
Zakharov	CMA-ES	0.000	$8.941 \times 10^{-19} \pm 1.150 \times 10^{-18}$
	PSO	0.740	$1.922 \times 10^{-16} \pm 1.359 \times 10^{-15}$
	Quantum	**1.000**	$\mathbf{0.000 \times 10^{0} \pm 0.000 \times 10^{0}}$

Table 4 Results for 3-dimensional test functions. Best values for each category, algorithm and indicator are highlighted. The distance denotes the Euclidean distance to the known global optimum, the number of evaluations for the quantum case again includes surrogate calls. All results are the mean of 50 runs

Problem	Algorithm	Distance	# Evaluations
Rosenbrock	CMA-ES	$6.528 \times 10^{-10} \pm 5.382 \times 10^{-10}$	$9.992 \times 10^{4} \pm 3.811 \times 10^{3}$
	PSO	$2.634 \times 10^{-13} \pm 1.862 \times 10^{-12}$	$6.560 \times 10^{5} \pm 2.331 \times 10^{5}$
	Quantum	$\mathbf{1.018 \times 10^{-17} \pm 7.195 \times 10^{-17}}$	$\mathbf{3.834 \times 10^{5} \pm 3.488 \times 10^{5}}$
Griewank	CMA-ES	$9.044 \times 10^{0} \pm 3.756 \times 10^{0}$	$8.130 \times 10^{4} \pm 2.577 \times 10^{4}$
	PSO	$1.530 \times 10^{-8} \pm 4.252 \times 10^{-9}$	$\mathbf{1.405 \times 10^{5} \pm 1.016 \times 10^{5}}$
	Quantum	$\mathbf{1.521 \times 10^{-8} \pm 4.405 \times 10^{-9}}$	$1.792 \times 10^{5} \pm 1.547 \times 10^{5}$
Zakharov	CMA-ES	$1.732 \times 10^{0} \pm 3.046 \times 10^{-10}$	$8.615 \times 10^{4} \pm 2.433 \times 10^{3}$
	PSO	$1.630 \times 10^{-9} \pm 1.152 \times 10^{-8}$	$1.777 \times 10^{6} \pm 4.331 \times 10^{5}$
	Quantum	$\mathbf{0.000 \times 10^{0} \pm 0.000 \times 10^{0}}$	$\mathbf{7.731 \times 10^{5} \pm 1.217 \times 10^{3}}$

In Tables 3 and 5, the error denotes the mean difference of the found minimum value to the real minimum, whereas the distance in Tables 4 and 6 denotes the mean L^2 distance of a found solution to the minimum point. Please note that finding the exact minimum value not necessarily coincides with finding the exact minimum point; e. g. in case of a shallow objective functions. In this case a mean residual distance between real and found optimum point is also present for 100% success in finding the optimum value. All results are means of 50 runs each. The population size λ has been set to 1000 for 3-dimensional and 5000 for the 8-dimensional case. Accordingly, the population sizes of CMA-ES an for the PSO have been set to the same values.

In these experiments the population size was set to large values. In [48] population sizes up to $4 \cdot n^2$ (256 in the 8 dimensional case) have been tested. Table 2 shows success rate and mean residual error for three benchmark functions and all

Table 5 Results for 8-dimensional test functions. The number of entities (particles) for the PSO has been massively increased (ten times more then quantum) to reach a sufficient high success rate. The same setting as in Table 3 has been used

Problem	Algorithm	Succ. rate	Error
Rosenbrock	CMA-ES	0.000	$1.366 \times 10^{-16} \pm 5.229 \times 10^{-17}$
	PSO	**1.000**	$\mathbf{0.000 \times 10^{0} \pm 0.000 \times 10^{0}}$
	Quantum	**1.000**	$\mathbf{0.000 \times 10^{0} \pm 0.000 \times 10^{0}}$
Griewank	CMA-ES	0.040	$6.784 \times 10^{-2} \pm 3.237 \times 10^{-2}$
	PSO	**1.000**	$\mathbf{0.000 \times 10^{0} \pm 0.000 \times 10^{0}}$
	Quantum	**1.000**	$\mathbf{0.000 \times 10^{0} \pm 0.000 \times 10^{0}}$
Zakharov	CMA-ES	0.020	$6.612 \times 10^{-2} \pm 2.881 \times 10^{-2}$
	PSO	0.800	$3.543 \times 10^{-157} \pm 2.505 \times 10^{-156}$
	Quantum	**1.000**	$\mathbf{0.000 \times 10^{0} \pm 0.000 \times 10^{0}}$

Table 6 Results for 8-dimensional test functions. The number of entities (particles) for the PSO has been massively increased (ten times more then quantum) to reach a sufficient high success rate. The same setting as in Table 4 has been used

Problem	Algorithm	Distance	# Evaluations
Rosenbrock	CMA-ES	$7.191 \times 10^{-9} \pm 7.051 \times 10^{-9}$	$4.900 \times 10^{5} \pm 5.774 \times 10^{3}$
	PSO	$4.879 \times 10^{-15} \pm 3.040 \times 10^{-16}$	$4.406 \times 10^{8} \pm 1.795 \times 10^{8}$
	Quantum	$\mathbf{0.000 \times 10^{0} \pm 0.000 \times 10^{0}}$	$\mathbf{5.808 \times 10^{6} \pm 2.683 \times 10^{4}}$
Griewank	CMA-ES	$1.548 \times 10^{1} \pm 5.642 \times 10^{0}$	$2.592 \times 10^{5} \pm 5.034 \times 10^{4}$
	PSO	$\mathbf{3.535 \times 10^{-8} \pm 6.730 \times 10^{-9}}$	$9.442 \times 10^{6} \pm 4.674 \times 10^{6}$
	Quantum	$3.672 \times 10^{-8} \pm 5.180 \times 10^{-9}$	$\mathbf{1.370 \times 10^{6} \pm 1.107 \times 10^{6}}$
Zakharov	CMA-ES	$1.551 \times 10^{1} \pm 4.867 \times 10^{0}$	$2.648 \times 10^{5} \pm 5.234 \times 10^{4}$
	PSO	$8.390 \times 10^{-80} \pm 5.932 \times 10^{-79}$	$1.086 \times 10^{7} \pm 7.143 \times 10^{6}$
	Quantum	$\mathbf{0.000 \times 10^{0} \pm 0.000 \times 10^{0}}$	$\mathbf{5.244 \times 10^{6} \pm 2.603 \times 10^{3}}$

Table 7 Comparing computational performace of CMA-ES and quantum approach with the 100-dimensional Sphere function

Algorithm	Error	Total evaluations	CPU time/nsec.
CMA-ES	$1.593 \times 10^{-16} \pm 1.379 \times 10^{-16}$	129204.4 ± 18336.1	$5.189 \times 10^{10} \pm 6.918 \times 10^{9}$
Quantum	$9.861 \times 10^{-17} \pm 1.503 \times 10^{-18}$	161172.8 ± 626.8	$4.029 \times 10^{9} \pm 1.040 \times 10^{8}$

three algorithms. As can be seen, particle swarm and the quantum approach benefit significantly from growing population sizes as both of them find a solution below the threshold in almost every case. CMA-ES on the other hand seems to get stuck in some local optimum almost always, but this effect is also due to an internal mechanism of the implementation that we used leading to a premature abort to prevent insta-bilities when calculating the covariance matrix as soon as the population diameter becomes too small. Regarding the gained result qualities in terms of residual error, the CMA-ES is competitive although the found solution positions differ a bit more from the real optimum position than the ones found by PSO and quantum ES. It is nevertheless unbeaten in terms of function evaluations. The same holds true for the results in Tables 5 and 6 for the 8-dimensional case. The quantum approach several times reaches an optimal result with an error of zero if using such large population sizes. An error of zero in this case has to be seen as lower than machine epsilon of the used Java programming language rather than exactly zero.

On the face of it, the CMA-ES seems to dominate the results in Tables 1 and 2 with an unbeaten low number of fitness evaluations. But, the reason for the low number lies in the fact that the used CMA-ES implementation almost every time stopped before reaching the optimum (expressed by the low success rate). This is due to additional stopping criteria that are used internally to prevent ill-conditioned covariance matrices (condition number) and mathematical illnesses for eigenvalue decompositions. Unfortunately, finding the exact optimum is also preluded. In order to rate the low evaluations number we conducted an additional experiment and compared with the quantum approach. The quantum approach needs for the 8-dimensional Zakharov function (lowermost row in Table 2) $2.58 \times 10^5 \pm 2.74 \times 10^4$ evaluations to reach the same suboptimal result ($f(x) \approx 1 \times 10^{-16}$ and a distance to the optimum of $d \approx 1 \times 10^{-8}$). The number of evaluations is competitive in this case. However, due to the fact that CMA-ES needs – among others – to conduct eigenvalue decompositions of its co-variance matrix ($\mathcal{O}(n^3)$ with number of dimensions n [47]), the quantum approach is faster considering computation time for high-dimensional cases.

The success rate of the PSO is competitive but far more fitness evaluations are used. All algorithms use the same number of entities, i.e. number of particles or samples for a comparison of this feature. This time, a meta-optimization of the algorithm parameters was done using a PSO.

Finally, we compared our approach with CMA-ES performing at high-dimensional problems with low number of individuals. We used the sphere function with dimension 100 and a population size of $\mu = 2$ and $\lambda = 10$. Figure 7 shows the result. Depicted are merely the number of objective evaluations as both CMA-ES and the

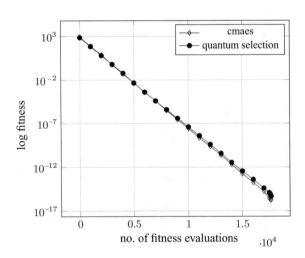

Fig. 7 Comparison of CMA-ES and quantum ES for the 100 dimensional sphere function with $\mu = 2$ and $\lambda = 10$, $f(x^*) = 0$

quantum approach do additional calculations for updating the respectively used surrogates. Again, our method is competitive with CMA-ES in terms of function evaluations, but performs faster due to the high dimensionality of the problem.

6 Conclusion

We introduced a novel evolution strategy for global optimization that uses the quantum potential field defined by elitist solutions for generating the offspring solution set. The quantum field exploits global information by aggregating over scattered fitness information similar to scale space approaches and on the other side continuously adapts to elitist solutions. In this way, the quantum field surface quickly flattens to a smooth surrogate for guiding further sampling directions.

By using the quantum potential, information about the fitness landscape of scattered points is condensed into a smoother surrogate for guiding further sampling instead of looking at different single solutions; one at a time. In this way, the quantum surrogate tries to fit the search distribution to the shape of the objective function like CMA-ES [49]. The quantum surrogate adapts continuously as the optimization process zooms into areas of interest.

Compared with a population based solver and CMA-ES as established evolution strategy, we achieved a competitive and sometimes faster convergence with less objective function calls and less susceptibility to premature convergence. We tested our method on ill-conditioned problems as well as on simple problems finding it performing equally good on both.

Appendix

For evaluation we used a wide variety of objective functions with different dimensionality and characteristics taken from [3, 21, 41, 42, 50].

Alpine:

$$f_1(\boldsymbol{x}) = \sum_{i=1}^{n} |x_i \sin(x_i) + 0.1 x_i|, \tag{12}$$

$-10 \le x_i \le 10$ with $\boldsymbol{x}^* = (0, \ldots, 0)$ and $f_1(\boldsymbol{x}^*) = 0$.

Goldstein-Price:

$$\begin{aligned}
f_2(\boldsymbol{x}) =& (1 + (x_1 + x_2 + 1)^2 \cdot \\
& (19 - 14x_1 + 2x_1^2 - 14x_2 + 6x_2 x_2 + 3x_2^2)) \cdot \\
& (30 + (2x_1 - 3x_2)^2 \cdot \\
& (18 - 32x_1 + 12x_1^2 + 48x_2 - 36x1x2 + 27x2^2)),
\end{aligned} \tag{13}$$

$-2 \leq x_1, x_2 \leq 2$ with $x^* = (0, -1)$ and $f_2(x^*) = 3$.

Himmelblau:

$$f_3(x) = (x_1^2 + x_2 - 11)^2 + (x_1 + x_2^2 - 7)^2 \tag{14}$$

$-10 \leq x_1, x_2 \leq 10$ with $f_3(x^*) = 0$ at four identical local minima.

Bohachevsky 1:

$$f_4(x) = x_1^2 + 2x_2^2 - 0.3\cos(2\pi x_1) - 0.4\cos(4\pi x_2) + 0.7, \tag{15}$$

$-100 \leq x_1, x_2 \leq 100$ with $x^* = (0, 0)$ and $f_4(x^*) = 0$.

Generalized Rosenbrock:

$$f_5(x) = \sum_{i=1}^{n-1} ((1 - x_i)^2 + 100(x_{i+1} - x_i^2)^2), \tag{16}$$

$-2048 \leq x_i \leq 2048$ with $x^* = (1, \ldots, 1)$ $f_5(x^*) = 0$.

Griewank:

$$f_6(x) = 1 + \frac{1}{200}\sum_{i=1}^{n} x_i^2 - \prod_{i=0}^{n} \cos(\frac{x_i}{\sqrt{i}}), \tag{17}$$

$-100 \leq x_i \leq 100$ with $x^* = (0, \ldots, 0)$ $f_6(x^*) = 0$.

Zakharov:

$$f_7(x) = \sum_{i=1}^{n} x_i^2 + (\sum_{i=1}^{n} 0.5ix_i)^2 + (\sum_{i=1}^{n} 0.5ix_i)^4, \tag{18}$$

$-5 \leq x_i \leq 10$ with $x^* = (0, \ldots, 0)$ $f_7(x^*) = 0$.

Sphere:

$$f_8(x) = \sum_{i=1}^{n} x_i^2, \tag{19}$$

$-5 \leq x_i \leq 5$ with $x^* = (0, \ldots, 0)$ $f_8(x^*) = 0$.

Chichinadze:

$$f_9(x) = x_1^2 - 12x_1 + 11 + 10\cos(\pi x_1/2)$$
$$+ 8\sin(5\pi x_1) - (1/5)^{0.5}e^{-0.5(x_2-0.5)^2}, \tag{20}$$

$-30 \leq x_1, x_2 \leq 30$ with $x^* = (5.90133, 0.5)$ $f_9(x^*) = -43.3159$.

Booth:

$$f_10(x) = (x_1 + 2x_2 - 7)^2(2x_1 + x_2 - 5)^2, \tag{21}$$

$-20 \leq x_1, x_2 \leq 20$ with $x^* = (1, 3)$ $f_9(x^*) = 0$.

References

1. Rechenberg, I.: Cybernetic solution path of an experimental problem. Technical report, Royal Air Force Establishment (1965)
2. Kramer, O.: A review of constraint-handling techniques for evolution strategies. Appl. Comput. Intell. Soft Comput. **2010**, 1–19 (2010)
3. Ulmer, H., Streichert, F., Zell, A.: Evolution strategies assisted by gaussian processes with improved pre-selection criterion. In: IEEE Congress on Evolutionary Computation, CEC 2003, pp. 692–699 (2003)
4. Bremer, J., Sonnenschein, M.: Parallel tempering for constrained many criteria optimization in dynamic virtual power plants. In: IEEE Symposium on Computational Intelligence Applications in Smart Grid (CIASG), pp. 1–8 (2014)
5. Leung, Y., Gao, Y., Xu, Z.B.: Degree of population diversity - a perspective on premature convergence in genetic algorithms and its markov chain analysis. IEEE Trans. Neural Netw. **8**, 1165–1176 (1997)
6. Trelea, I.C.: The particle swarm optimization algorithm: convergence analysis and parameter selection. Inf. Process. Lett. **85**, 317–325 (2003)
7. Rudolph, G.: Self-adaptive mutations may lead to premature convergence. IEEE Trans. Evolut. Comput. **5**, 410–414 (2001)
8. Bremer, J., Lehnhoff, S.: A quantum field evolution strategy - an adaptive surrogate approach. In: Merelo, J.J., Melício, F., Cadenas, J.M., Dourado, A., Madani, K., Ruano, A., Filipe, J. (eds.) Proceedings of the 8th International Joint Conference on Computational Intelligence. Volume 1 of ECTA. SCITEPRESS - Science and Technology Publications, Lda., Portugal (2016)
9. Horn, D., Gottlieb, A.: The method of quantum clustering. In: Neural Information Processing Systems, pp. 769–776 (2001)
10. Leung, Y., Zhang, J.S., Xu, Z.B.: Clustering by scale-space filtering. IEEE Trans. Pattern Anal. Mach. Intell. **22**, 1396–1410 (2000)
11. Schleich, W.P., Greenberger, D.M., Kobe, D.H., Scully, M.O.: Schrödinger equation revisited. Proc. Natl. Acad. Sci. **110**, 5374–5379 (2013)
12. Feynman, R.P., Leighton, R.B., Sands, M., Lindsay, R.B.: The Feynman Lectures on Physics, vol. 3: Quantum mechanics (1966)
13. Wimmel, H.: Quantum Physics and Observed Reality: A Critical Interpretation of Quantum Mechanics. World Scientific (1992)
14. Hansen, E.: Global optimization using interval analysis - the multi-dimensional case. Numer. Math. **34**, 247–270 (1980)
15. Tuy, H., Thieu, T., Thai, N.: A conical algorithm for globally minimizing a concave function over a closed convex set. Math. Oper. Res. **10**, 498–514 (1985)
16. Hansen, P., Jaumard, B., Lu, S.H.: Global optimization of univariate lipschitz functions ii: new algorithms and computational comparison. Math. Program. **55**, 273–292 (1992)
17. Simon, D.: Evolutionary Optimization Algorithms. Wiley (2013)
18. Bäck, T., Fogel, D.B., Michalewicz, Z. (eds.): Handbook of Evolutionary Computation, 1st edn. IOP Publishing Ltd., UK (1997)
19. Horst, R., Pardalos, P.M. (eds.): Handbook of Global Optimization. Kluwer Academic Publishers, Netherlands (1995)
20. Rahnamayan, S., Tizhoosh, H.R., Salama, M.M.: A novel population initialization method for accelerating evolutionary algorithms. Comput. Math. Appl. **53**, 1605–1614 (2007)
21. Ahrari, A., Shariat-Panahi, M.: An improved evolution strategy with adaptive population size. Optimization **64**, 1–20 (2013)
22. Rigling, B.D., Moore, F.W.: Exploitation of sub-populations in evolution strategies for improved numerical optimization. Ann. Arbor. **1001**, 48105 (1999)
23. Loshchilov, I., Schoenauer, M., Sebag, M.: Self-adaptive surrogate-assisted covariance matrix adaptation evolution strategy. CoRR abs/1204.2356 (2012)

24. Gano, S.E., Kim, H., Brown II, D.E.: Comparison of three surrogate modeling techniques: Datascape, kriging, and second order regression. In: Proceedings of the 11th AIAA/ISSMO Multidisciplinary Analysis and Optimization Conference, AIAA-2006-7048. Portsmouth, Virginia (2006)
25. CMA-ES: A Function Value Free Second Order Optimization Method. In: PGMO COPI 2014. Paris, France (2014)
26. Ben-Hur, A., Siegelmann, H.T., Horn, D., Vapnik, V.: Support vector clustering. J. Mach. Learn. Res. **2**, 125–137 (2001)
27. Bremer, J., Rapp, B., Sonnenschein, M.: Support vector based encoding of distributed energy resources' feasible load spaces. In: IEEE PES Conference on Innovative Smart Grid Technologies Europe, Chalmers Lindholmen, Gothenburg, Sweden (2010)
28. Weinstein, M., Horn, D.: Dynamic quantum clustering: a method for visual exploration of structures in data. Phys. Rev. E **80**, 066117 (2009)
29. Weinstein, M., Meirer, F., Hume, A., Sciau, P., Shaked, G., Hofstetter, R., Persi, E., Mehta, A., Horn, D.: Analyzing big data with dynamic quantum clustering. CoRR abs/1310.2700 (2013)
30. Rapp, B., Bremer, J.: Design of an event engine for next generation cemis: A use case. In: Arndt, H.-K., Gerlinde Knetsch, W.P.E. (eds.) EnviroInfo 2012 – 26th International Conference on Informatics for Environmental Protection, pp. 753–760. Shaker Verlag (2012). ISBN 978-3-8440-1248-4
31. Yu, Y., Qian, F., Liu, H.: Quantum clustering-based weighted linear programming support vector regression for multivariable nonlinear problem. Soft Comput. **14**, 921–929 (2010)
32. Horn, D., Gottlieb, A.: Algorithm for data clustering in pattern recognition problems based on quantum mechanics. Phys. Rev. Lett. **88** (2002)
33. Sun, J., Feng, B., Xu, W.: Particle swarm optimization with particles having quantum behavior. In: Congress on Evolutionary Computation, 2004. CEC2004, vol. 1, pp. 325–331 (2004)
34. Feng, B., Xu, W.: Quantum oscillator model of particle swarm system. In: ICARCV, IEEE (2004) 1454–1459
35. Loo, C.K., Mastorakis, N.E.: Quantum potential swarm optimization of pd controller for cargo ship steering. In: Proceedings of the 11th WSEAS International Conference on Applied Mathematics, Dallas, USA (2007)
36. Suzuki, S., Nishimori, H.: Quantum annealing by transverse ferromagnetic interaction. In: Pietronero, L., Loreto, V., Zapperi, S. (eds.) Abstract Book of the XXIII IUPAP International Conference on Statistical Physics. Genova, Italy (2007)
37. Weinstein, M., Horn, D.: Dynamic quantum clustering: a method for visual exploration of structures in data. Computing Research Repository abs/0908.2 (2009)
38. Parzen, E.: On estimation of a probability density function and mode. Ann. Math. Stat. **33**, 1065–1076 (1962)
39. Roberts, S.: Parametric and non-parametric unsupervised cluster analysis. Pattern Recogn. **30**, 261–272 (1997)
40. Metropolis, N., Rosenbluth, A.W., Rosenbluth, M.N., Teller, A.H., Teller, E.: Equation of state calculations by fast computing machines. J. Chem. Phys. **21**, 1087–1092 (1953)
41. Himmelblau, D.M.: Applied Nonlinear Programming [by] David M. Himmelblau, McGraw-Hill New York (1972)
42. Yao, X., Liu, Y., Lin, G.: Evolutionary programming made faster. IEEE Trans. Evolut. Comput. **3**, 82–102 (1999)
43. Kennedy, J., Eberhart, R.: Particle swarm optimization. In: Proceedings of the IEEE International Conference on Neural Networks, vol. 4, pp. 1942–1948. IEEE (1995)
44. Hansen, N.: The CMA Evolution Strategy: A Tutorial. Technical report (2011)
45. Hansen, N., Ostermeier, A.: Completely derandomized self-adaptation in evolution strategies. Evolut. Comput. **9**, 159–195 (2001)
46. Shi, Y., Eberhart, R.: A modified particle swarm optimizer. In: International Conference on Evolutionary Computation (1998)
47. Knight, J.N., Lunacek, M.: Reducing the space-time complexity of the CMA-ES. In: Genetic and Evolutionary Computation Conference, pp. 658–665 (2007)

48. Müller, S.D., Hansen, N., Koumoutsakos, P.: Increasing the Serial and the Parallel Performance of the CMA-Evolution Strategy with Large Populations, pp. 422–431. Springer, Berlin Heidelberg, Berlin, Heidelberg (2002)

49. Hansen, N.: The CMA evolution strategy: a comparing review. In: Lozano, J., Larranaga, P., Inza, I., Bengoetxea, E. (eds.) Towards a New Evolutionary Computation. Advances on Estimation of Distribution Algorithms, pp. 75–102. Springer, Berlin (2006)

50. Mishra, S.: Some new test functions for global optimization and performance of repulsive particle swarm method. Technical report (2006)

DynaGrow: Next Generation Software for Multi-Objective and Energy Cost-Efficient Control of Supplemental Light in Greenhouses

Jan Corfixen Sørensen, Katrine Heinsvig Kjaer,
Carl-Otto Ottosen and Bo Nørregaard Jørgensen

Abstract It is not possible for growers to compromise product quality by saving energy but the increasing electricity prices challenge the growers economically. Optimization of such multiple conflicting goals requires advanced strategies that are currently not supported in existing greenhouse climate control systems. DynaGrow is built on top of the existing climate control computers and utilizes the existing hardware. By integrating with exiting hardware it is possibly to support advanced multi-objective optimization of climate parameters without investing in new hardware. Furthermore, DynaGrow integrates with local climate data, electricity price forecasts and outdoor weather forecasts, in order to formulate advanced control objectives. In September 2014 and February 2015 two greenhouse experiments were run to evaluate the effects of DynaGrow. By applying multi-objective optimization, it was possible to produce a number of different cultivars and save energy without compromising quality. The best energy savings were achieved in the February 2015 experiment where the contribution from natural light was limited.

Keywords Multi-objective optimization

J. C. Sørensen (✉) · B. N. Jørgensen
The Maersk Mc-Kinney Moller Institute, University of Southern Denmark,
Campusvej 55, 5230 Odense M, Denmark
e-mail: jcs@mmmi.sdu.dk

B. N. Jørgensen
e-mail: bnj@mmmi.sdu.dk

K. H. Kjaer · C.-O. Ottosen
Department of Horticulture, Aarhus University,
Kirstinebjergvej 10, 5792 Aarslev, Denmark
e-mail: katrine.kjaer@agrsci.dk

C.-O. Ottosen
e-mail: co.ottosen@agrsci.dk

© Springer Nature Switzerland AG 2019
J. J. Merelo et al. (eds.), *Computational Intelligence*,
Studies in Computational Intelligence 792,
https://doi.org/10.1007/978-3-319-99283-9_2

1 Introduction

In 2009, Danish horticulture industry consumed 0.8% of the total national electricity and 75% of this consumed energy was estimated to come from supplemental light alone.

Several countries are at present time in a transition towards non-fossil renewable energy sources such as wind turbines. The large contribution of energy from renewable energy sources results in irregular electricity production, that leads to fluctuation in the electricity prices. An increased electricity price is a challenge for the horticulture industry that is positioned in a highly competitive market. A spot-market electricity price structure has been introduced in the Scandinavian countries to provide an incentive for industry to utilize energy in cost effective hours when the supply of energy is plentiful [1].

Existing fixed rate supplemental light control strategies are in contradiction to the flexible price structure. Fixed rate light plans often consume energy at hours that are costly. The fixed rate strategies only plan according to the fixed time periods and do not take price structures into account. However, changing the lighting patterns may have severe effects on how the cultivar reacts. For example, negative effects could be bud dormancy, delayed leaf development, stem elongation, late seed germination and early flower initiation [2]. Hence, there is a need to optimize the utility of supplemental light to ensure that the light plans represent the cheapest electricity prices and promote a high quality of the produced cultivar.

This paper proposes a system, DynaGrow, that integrates local climate data, electricity price and outdoor weather forecasts to formulate advanced control objectives. The core of the system is a customized Multi-Objective Evolutionary Algorithm (MOEA) that searches for coordinated Pareto-optimal light plans that fulfil the specified climate control objectives. DynaGrow is a feature-oriented software system divided into a number of features. Each feature encompasses an individual unit of software functionality and is implemented as loosely coupled plug-in modules. The feature-oriented separation of DynaGrow makes it relatively strait forward to integrate into existing hardware devices and configure the system to optimize different objectives. Each set of objectives are strictly separated and can be configured to fulfil the specific climate control requirements given by the problem domain. At the present time, DynaGrow supports two different climate control hardware platforms and support 37 different objectives [3].

Related work is shortly summarized in Sect. 2. Section 3 describes the different elements of DynaGrow and how they are connected. How the DynaGrow core is implemented as a genetic MOEA is described in Sect. 4. Section 6 describes the experiments that evaluate DynaGrow by growing different cultivars based on three different climate control settings over two different periods. The results of the experiments are described in Sect. 7. Next, the discussion reflects on how well DynaGrow optimized the identified objectives of the climate control problem. Last, Sect. 9 summarizes the article.

2   Related Work

Research literature describes independent models that can contribute to an optimized greenhouse production and cut the energy consumption through development of intelligent control strategies. Aaslyng et al. created the foundation for a component-based climate control system, IntelliGrow, that optimizes the greenhouse climate [4]. The results showed that it was possible to reduce energy consumption by more than 20%. Subsequently, there have been several projects in which the models and control strategies have been optimized [5, 6]. The IntelliGrow concept is documented by Aaslyng et al. in [7]. Kjaer et al. developed the DynaLight system that provides a search-based approach to find the most cost-efficient use of supplemental light, based on a predefined setpoint of Daily Photosynthesis Integral (DPI), forecasted solar light and the spot-market electricity price [8–10]. The DynaLight algorithm is tightly integrated with the weather and electricity forecast services and does not support optimization of multiple objectives.

None of the mentioned approaches support Pareto optimization of multiple independently developed objectives, to generate a coordinated set of setpoints that can be effectuated by the greenhouse climate control system. DynaGrow is the only approach that supports optimization of multiple objectives that are based on weather and electricity price forecasts.

3   DynaGrow

DynaGrow is designed to control climate-related growth factors by sensing and manipulating the greenhouse climate through the use of sensors and actuators. The physical setting of DynaGrow is a combination of a control machine, a number of connection domains and a controlled domain (Fig. 1). The *control machine* consists of the DynaGrow software running on a PC that is connected to a set of climate controllers. The climate controllers are connected to sensors and actuators (Connection Domains) that interact with the indoor climate of the greenhouse (Controlled Domain). A *connection domain* can act as a sensor or an actuator. Sensors provide measured input information m in inform of input variables i to the *control machine*. Contrary, actuators influence the physical phenomenon c in the *controlled domain* according to output variables o provided by the machine.

The output of DynaGrow is determined by the system's *control objectives* that typically incorporate *models* of the physical environment, e.g. a model of the photosynthesis process of a given plant in a greenhouse. Each control objective can be formulated either as a minimization of a cost function or as a constraint. The objective function evaluates options proposed by the MOEA during the optimization process. The *option* argument is passed over by the MOEA and is a data structure that contains a set of input variables i and output variables o that connect the *control machine* to the *connection domains*. The control objectives are evaluated continuously in cycles

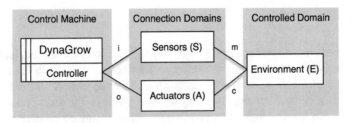

Fig. 1 DynaGrow system overview. Figure extracted from [11]

by the machine, and is integrated into the *control process*. Each cycle is triggered at specific time intervals. For each control cycle, the system perceives the environment through its sensor variable i, optimizes the control objectives and changes the environment through its actuators to obtain the desired objective of the system. The result of a control cycle is a set of *output variables o* (*setpoints*) that are written to the actuators.

The control objectives are optimized using the CONTROLEUM-GA to guarantee a Pareto optimal trade-off between the multiple control objectives within computational tractable time [12]. The CONTROLEUM-GA is a multi-objective genetic algorithm that incorporates domain specific variables and operators to solve dynamic optimization problems. The support for domain specific variables and operators, enable the algorithm to converge fast enough to optimize a major number of objectives within each optimization cycle of DynaGrow.

4 Implementation

The CONTROLEUM-GA function shows the pseudo code for the genetic algorithm implementation used by the core of DynaGrow. Note that the line numbers break and continue in places where sub-functions are called from CONTROLEUM-GA ().

CONTROLEUM- GA(*time, old Pop*)

```
 1  if old Pop.isNotEmpty
 2     for each old Solution ∈ old Pop
 3         ADD-NONDOMSOLUTION(
                              COPY(time, solution))
 4  for i = 0 to POPSIZE
 5     ADD-NONDOMSOLUTION(D-INIT(time))
17  while isNotTerminated
18     for i = 0 to i ≤ POPSIZE
19         if RANDOM-DOUBLE() < MUTATIONRATE
20             child = S-MUTATE(RANDOM(pop))
27         else
28             child = S-CROSSOVER(RANDOM(pop),
                                   RANDOM(pop))
34         ADD-NONDOMSOLUTION(child)
```

The CONTROLEUM-GA function has two arguments: (1) a time stamp *time* for when the algorithm is executed, and (2) the population *oldPop* from previous executions. The time stamp *time* is used for dynamic optimization problems that use the start time of the optimization. The algorithm is separated into the following phases: Initialization, Ranking, Mutation, Crossover and Termination.

Initialization: A population consists of a number of non-dominated Pareto optimal solutions. Each solution is represented by a data structure that has a collection of objective results *solution.objectives* and decision variables *solution.variables*. A solution can have multiple different types of domain specific variables; e.g., temperature, CO_2 and light plan. In this work we focus on the light plan variable. Line 1 checks if the previous population *oldPop* is empty. The population *oldPop* is empty the first time the algorithm is executed. If the population *oldPop* exists from previous executions, it is copied into the new non-dominated population *pop* (Line 3). A domain-specific initialization operator D- INIT is implemented for each type of decision variable (Line 5). For example, the supplemental light plan is initialized by the domain specific initialization operator that is an implementation of the D- INIT function. For example, the different time resolutions of light plans are encoded in the light plan variable.

CONTROLEUM-GA (Line 17) test if the evolution should terminate. Evolution is terminated after a specified time limit, after a number of generations or when the population is stable.

Ranking: The function ADD- NONDOMSOLUTION sorts all solutions in the population *pop* according to the Pareto dominance relation (Line 8). That is, the objectives are ranked given the proposed decision variables *solution.variables*. The results of the evaluations are assigned to the objective values *solution.objectives* for each proposed solution. Only non-dominated solutions are added to the population *pop*.

ADD- NONDOMSOLUTION(*newSolutionA*)

```
 6  for each oldSolutionB ∈ pop
 7      flag =
 8          PARETO- COMPARE(newSolutionA, oldSolutionB)
 9      if flag == ADOMINATESB
10          REMOVE(oldSolutionB, pop)
11      elseif flag == BDOMINATESA
12          return false
13      elseif DISTANCE(newSolutionA, oldSolutionB) < EPS
14          return false
15  ADD(newSolutionA, pop)
16  return true
```

The function PARETO- COMPARE compares if a solution *newSolutionA* dominates a solution *oldSolutionB* or vice versa. If solution *newSolutionA* dominates solution *oldSolutionB* then solution *oldSolutionB* is removed from the population *pop* (Line 10). Contrary, if solution *oldSolutionB* dominates solution *newSolutionA* then it is not added to the population *pop* (Line 12). Two solutions are defined as the same, if the Euclidean DISTANCE between two solutions in the objective space is less than the level of significance defined by constant EPS. In case *newSolutionA* is the same as *oldSolutionB* then it is not added to the population *pop* (Line 14).

Mutation: For each generation, solutions are randomly selected a number of times for mutation. The number of mutations is determined by the constants MUTATIONRATE and POPSIZE. For example, if POPSIZE is 100 and MUTATIONRATE is 50% then 50 randomly selected solutions are mutated. If a mutation results in a non-dominated solution then it is added to the population *pop*.

S- MUTATE(*solution*)

21 $i =$ RANDOM- INT()
22 *solution.variables*[i] $=$ D- MUTATE(*solution.variables*[i])

Mutation is applied at solution level and domain variable level. At solution level a random decision variable is selected for mutation in (Line 21) using a generic uniform mutation (UM) operator. Each decision variable has its own domain-specific mutation operator D- MUTATE. The D- MUTATE operator is applied on the randomly selected variable (Line 22). D- MUTATE-LIGHT shows the implementation of D- MUTATE for a light plan variable. The selected light plan variable is copied and a randomly selected index in the plan is negated. That is, if the light state for the selected index was ON, then after mutation it will be set to OFF.

D- MUTATE- LIGHT(*lightPlan*)

23 $lp =$ COPY(*lightPlan*, TIMEINTERVAL)
24 $i =$ RANDOM- INT($lp.size$)
25 $lp[i] = \neg lp[i]$
26 **return** lp

Implementations of D- MUTATION operators incorporate domain knowledge to ensure that the values of the decision variables are always viable. In case of the light plan variable, the D- MUTATION-LIGHT incorporates knowledge about time-resolution and which index of the light plan that is viable for change (Lines 23 and 24). For example, if a light interval only can change once for a given period of time, or if a light state is always fixed, then it is implemented in the implementation of the D- MUTATE function for the given variable. Each domain mutation operator defines a range for a specific type of decision variable (temperature, energy, CO_2, etc.). The intersection of these ranges defines the viability space of the decision variable.

Crossover: Solutions are randomly selected for crossover for a number of iterations. The number of crossover iterations is determined by the constant POPSIZE. Crossover is applied at solution and decision variable level. The solution level crossover function S- CROSSOVER is called in CONTROLEUM-GA (Line 28).

S- CROSSOVER($solutionA$, $solutionB$)

```
29   i = RANDOM- INT(solutionA.variables.size)
30   solutionA.variables[i] =
         D- CROSSOVER(solutionA.variables[i],
                        solutionB.variables[i])
31   for j = i + 1 to j < solutionA.variables.size − 1
32        solutionA.variables[j] = solutionB.variables[j]
33   return solutionA
```

Random variables from two solutions *solutionA* and *solutionB* are selected for crossover at decision variable level using a generic one-point crossover operator (Lines 29 and 30). The other variables from the selected index $i + 1$ till the last index from *solutionB* is copied to *solutionA*.

The domain-specific D- CROSSOVER operator is applied on the selected decision variables (Line 30). The D- CROSSOVER function is the domain specific crossover operator for two light plans *lightPlanA* and *lightPlanB*. In the DynaGrow experiment, the two light plans are crossed by a standard one-point crossover operator but more complex outputs require more knowledge encoded into the output data structure.

5   Objectives

SPARBAL(option)

```
1   lightPlanSum = CALCPARSUM(LightPlan, LampPAR)
2   balance =
         5 × ParDLI − (ParHist + ParFuture + lightPlanSum)
3   return balance
```

The PAR Light Sum Balance objective (SPARBAL) minimizes a Photosynthetically Active Radiation (PAR) sum balance (Line 2). PAR designates the spectral wave band of solar radiation from 400 to 700 nanometers that photosynthetic organisms are able to utilize for photosynthesis. The *balance* is calculated over a five-days time window defined by the current day, two days in the past and two days in the future. The PAR Integral Today and Past Two Days) (*ParHist*) input is derived from historical data stored by DynaGrow. Data for the Expected Natural PAR Sum Remaining Day and Future Two days (*ParFuture*) input is provided by Conwx Intelligent Forecast Systems [13]. The PAR Day Light Integral (*ParDLI*) input specifies the average goal to be achieved over the five-day period and is provided by the grower. The total light plan PAR sum (*lightPlanSum*) variable is calculated based on the Installed Lamp PAR(*LampPAR*) input and the number of suggested light intervals in the *LightPlan* output. The *balance* is then calculated as the difference between the provided goal and the total of *ParHist*, *ParFuture* and the *lightPlanSum*. At the end of each day, the *ParHist* input will be updated with data from the past days. Similarly, the *ParFuture* input will be updated with data from next day. That is, the balance is calculated based on a five-day sliding window.

SCHEAPLIGHT(option)

1   $cost = \sum_{i=1}^{n} LightPlan.Switch_i \times ElForecasts.Price_i \times$
        $(TotalLoad \times LpTimeslot_i)$

2   **return** $cost$

The Prefer Cheap Light objective (SCHEAPLIGHT) is specified as a cost function that minimizes the price of the Light Plan (*LightPlan*) based on El. Spot and Prognosis Prices (*ElCompPrices*). The electricity spot-market price forecast is provided by Nord Pool and the three day prognosis is provided by Energi Danmark [1, 14]. The Total Lamp Load (*TotalLoad*) is calculated as the Installed Lamp Effect (*InstLampEffect*) multiplied by the Greenhouse Size (*GreenhouseSize*). The index i is the time slot index of the *LightPlan*. For each light time slot i, the sub-cost is calculated as the Total Lamp Load (*TotalLoad*) multiplied by the light plan time slot interval T_i and the electricity price *ElForecasts.Price_i*. The total cost of the Light Plan (*LightPlan*) is then the sum of all the sub-costs for each of the light intervals (*LightPlan.Switch_i*). The *LightPlan.Switch_i* is zero for light switched off and one for when the supplemental light is lit.

6   Experiments

In autumn 2014 and spring 2015 two greenhouse experiments were executed to evaluate the cost effectiveness and the qualities of the DynaGrow software. The experiments were executed as three treatments in three identical greenhouse compartments for each of the experiment periods. To illustrate the effect of the systems we have selected a period for each of the two experiments. The period evaluated for the 2014 experiment is September 22–28 and the period for the 2015 experiment is February 6–12.

S SON-T 14 and 15: The S SON-T 14 and 15 treatments are executed by a standard control system with SON-T lamps and a fixed day length of 18 hours.

Each compartment is 76 m². Each SON-T compartment has 16 SON-T lamps installed and each has an effect of 600 Watt. That is, the installed effect for the SON-T compartments is $(16 \times 600 \, W)/76 \, m^2 = 126 \, W/m^2$.

DG SON-T 14 and 15: The DG SON-T 14 and 15 compartments are equipped with SON-T lamps but controlled by DynaGrow. The granularity of the SON-T light plans is set to one hour due the physical properties of the lamps. A SON-T lamp cannot tolerate to be switched ON/OFF too often due to heating. The granularity of the LED light plans is set to 15 min as LED lamps tolerate to be lit in short intervals.

DG LED 14 and 15: The DG LED 14 and 15 treatments are equipped with LED lamps and are controlled by DynaGrow. A LED lamp has an effect of 190 W and the LED compartment has 38 LED lamps installed. The total installed effect for the LED compartment is then $(38 \times 190 \, W)/76 \, m^2 = 95 \, W/m^2$.

Common Settings: The energy cost used for supplemental light is calculated based on when the light is lit, and the price of electricity for that specific time. The electricity prices are for the west part of the Danish electricity grid (DK1).

The PAR lamp intensity for both LED and SON-T lamps is theoretically set to 100 μmol m^{-2} s^{-1}. That is, if the lamps are lit for a day (86400 s), they will contribute with 100 μmol m^{-2} s^{-1} $\times$ 86400 s, which is equal to 8.6 mol m^{-2} d^{-1}.

For the DynaGrow compartments, the control cycle is set to 5 min. The DynaGrow experiments (DG) included optimization of the PAR Light Sum Balance (SPARBAL) and Prefer Cheap Light (SCHEAPLIGHT) objectives.

The daylight integral (DLI) goal for all of the evaluated control strategies in the 2014 experiments is set to 12 mol m^{-2} d^{-1}. The DLI goal for all of the evaluated control strategies in the 2015 experiments is set to 8 mol m^{-2} d^{-1}.

The indoor light is derived from the outdoor light and is measured in *klux* but is converted into μmol m^{-2} s^{-1} by multiplying by a factor 18. Similarly, the DLI is measured in *kluxh* but is converted to mol m^{-2} d^{-1} by multiplying by factor 0.07 [15].

The CO_2, heating, ventilation, and screen setpoints for all experiments were controlled by a standard Day/Night threshold control strategy. The Day/Night CO_2 limits were set to 350 PPM and 800 PPM, respectively. The Day/Night Heatinglimits were both set to 20°C. The Day/Night Screenlimits were configured to 0% (no screens during day) and 100% (screens during night), respectively. The Day/Night Windowslimits were both set to 30°C.

Cultivars: A number of different plant species and batches were grown in the three treatments over the two seasons to evaluate the effect of the different light plans. The cultivars were harvested at maturation, meaning that plants from the different treatments were sometimes harvested at different times. Vegetative refers to the production of leaves and stems, and generative refers to the production of flowers, fruits and seeds (Table 1).

7 Results

2014: Figure 2a visualizes the light plans for each of the 2014 experiments together with the electricity prices [1]. The S SON-T 14 light plans are regular as expected given the standard interval control strategy and do not compensate for the expensive light hours, see Fig. 2. In contrast, Figs. 2b, c illustrate the light plans from the DG SON-T 14 and DG LED 14 treatments that are irregular but compensate for the expensive light hours. The DG SON-T 14 and DG LED 14 light plans are similar but the DG LED 14 light plans are more fine-grained because of the smaller 15 min light intervals.

Figure 3 shows the accumulated light plan hours from a sample period of the experiment, September 22-28 2014. The S SON-T 14 treatment accumulates light hours at regular intervals and do not compensate for the expensive hours. Figure 3a illustrates that light is lit during expensive hours around the evening of September 23.

Table 1 A summary of some of the plant species grown in the three treatments

Specie	Cultivar	Sown	1. Harvest	2. Harvest	Harvest diff.	Veg/Gen
Tomato	Arvento RZ	18.09.14	8.10.14	28.10.14	16	Veg
		31.10.14	24.11.14	09.12.14	15	Veg
Rose	Alaska, Felicita, Apatche	16.9.14	30.09.14	21–27.10.14	21–27	Gen
	Alaska, Felicita, Apatche	21.10.14	03.11.14	26.11–09.12.14	23–36	Gen
	Alaska, Felicita, Apatche	20.01.15	03.02.15	02.03–08.03.15	27–32	Gen
	Alaska, Felicita, Apatche	05.03.15	20.03.15	13.04–23.04.15	26–34	Gen
Kalanchoe	Simone, Molly	19.09.14	17.10.14	21.11.14	24	Veg
	Jackie, Simone	17.10.14	14.11.14	19.12.14	30	Veg
	Jackie, Simone	17.10.14	14.11.14	19.11–09.12.14	5-25	Gen
	Evita, Simone	23.01.15	23.01.15	21.02–27.03.15	29–63	Veg/Gen
Chili	Macho, Hot Fajita, Hot Burito, Snack Orange	10.02.15	03.03.15	17.03–06.05.15	13–63	Gen

Figures 3b, c illustrate the accumulated light plan hours for the DG SON-T 14 and DG LED 14 experiments. The SParBal and SCheapLight objectives are optimized in the DynaGrow experiments and for that reason the two figures are very similar.

Table 2 summarizes the supplemental light hour results. An average DLI of 8 mol m^{-2} d^{-1} over the experiment period was from natural light. The S SON-T 14 result is the reference and contributes with 386 hours of supplemental light and an average of 12.5 supplemental light hour per day. The reference DLI from natural and supplemental light is 12.5 mol m^{-2} d^{-1}. The DG SON-T 14 result has 5% lower DLI and 8% less supplemental light hours compared to the reference. The DG LED 14 result has the same DLI and 3% less supplemental light hours compared to the reference.

Table 3 compares the energy consumption and cost for each of the 2014 experiments. The S SON-T 14 experiment is the reference and consumed 3.7 MWh ($386.8h \times 600W \times 16$ lamps). In comparison to the reference, the DG SON-T 14 and DG LED 14 experiments consumed 8 and 27% less energy, respectively.

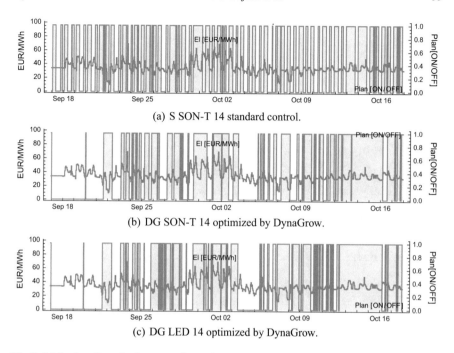

(a) S SON-T 14 standard control.

(b) DG SON-T 14 optimized by DynaGrow.

(c) DG LED 14 optimized by DynaGrow.

Fig. 2 Light plans from the three experimental compartments optimized by different control strategies from September 17 to October 18 2014

Table 2 Light hours and DLIs of the experiment September 17 to October 18 2014

Experiment	Aver. DLI [mol m^{-2} d^{-1}]		Supp. light [h]		Aver. Supp. light [h day^{-1}]
S SON-T 14	12.4	100%	386.8	100%	12.5
DG SON-T 14	11.9	95%	355.1	92%	11.5
DG LED 14	12.5	101%	374.7	97%	12.1

Table 3 Energy results of the experiment September 17 to October 18 2014

Experiment	Energy [KWh]		Cost [€]	
S SON-T 14	3703.5	100%	135.88	100%
DG SON-T 14	3400.3	92%	116.99	86%
DG LED 14	2705.1	73%	93.29	69%

The costs of the light plans are calculated based on the hourly electricity prices and the energy consumption [1]. The reference cost of the S SON-T 14 light plan is 135.88 €. In comparison, the cost of the DG SON-T 14 light plan is 14% lower than the reference and the DG LED 14 light plan cost is 31% lower than the reference. *2015:* Figure 4 illustrates the light plans effectuated for each of the compartments during the 2015 experiment period, together with the electricity prices provided from

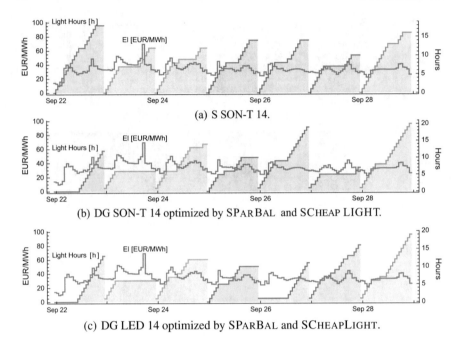

(a) S SON-T 14.

(b) DG SON-T 14 optimized by SPARBAL and SCHEAP LIGHT.

(c) DG LED 14 optimized by SPARBAL and SCHEAPLIGHT.

Fig. 3 Comparison of daily accumulated light hours optimized in September 2014 by different objectives from the S SON-T 14, DG SON-T 14 and DG LED 14 experiments

the Nord Pool power spot-market [1]. The light plan for the S SON-T 15 compartment has fixed rates as expected from the standard fixed rate light control strategy. Figure 4a shows that the fixed-rate light plan clearly requires supplemental light at hours when the electricity price is high.

In contrast, the DG SON-T 15 light plan is quite different to the S SON-T 15 light plan. Figure 4b illustrates the DG SON-T 15 light plan for the same experiment period. The DG SON-T 15 light plan has the same one-hour granularity as the S SON-T 15 light plan but expensive supplemental light hours are avoided. That is, the CONTROLEUM-GA has clearly optimized the SCHEAPLIGHT objective in order to generate the DG SON-T 15 light plan.

The DG LED 15 light plan is similar to the DG SON-T 15 light plan as both light plans are optimized by the same MOEA with the same objectives (Fig. 4c). The differences between the DG SON-T 15 and DG LED 15 light plans are the granularity of the generated light plans and the energy consumed by the different type of lamps. The finer granularity (15 min) of the DG LED 15 light plans can be observed in February 20–23 (Fig. 4c).

Figure 5 shows how the light plan hours accumulate for each of the treatments for a sample period within the experiment period. For the S SON-T 15 experiment, the light plan hours are accumulated continuously over the day or during the morning and evening (Fig. 5a). Figures 5b, c illustrate that the accumulated hours from the

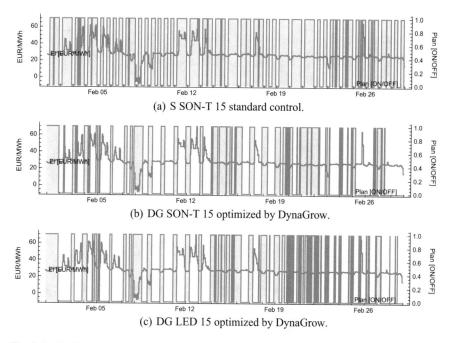

(a) S SON-T 15 standard control.

(b) DG SON-T 15 optimized by DynaGrow.

(c) DG LED 15 optimized by DynaGrow.

Fig. 4 Light plans from the three experimental compartments optimized by different control strategies in February 2015. Figure extracted from [11]

DG SON-T 15 and DG LED 15 light plans are similar as the DG SON-T 15 and DG LED 15 are optimized by the same SCHEAPLIGHT and SPARBAL objectives.

Table 4 shows the experimental DLI results for the three compartments in February 2015. Natural light contributed with 3.5 mol m^{-2} d^{-1} on average. For the reference S SON-T 15 treatment, supplemental light was lit in 432.3 h. In comparison, supplemental light was lit for 259.3 and 253.3 in the DG SON-T 15 and DG LED 15 treatments, respectively. That is, the DG SON-T 15 and DG LED 15 treatments had 40 and 41% less supplemental light hours than the reference, respectively. The DLI goal for each treatment was 8 mol m^{-2} d^{-1}. Additionally, the average DLI for the S SON-T 15 treatment was higher than the DG SON-T 15 and DG LED 15 treatments. The two DynaGrow treatments obtained approximately the same average DLI of 6.8 mol m^{-2} d^{-1} which is 26% lower than the reference and 1.2 mol m^{-2} d^{-1} lower than the DLI goal.

Table 5 provides a summary of the experimental energy results for February 2015. The total energy consumed by the S SON-T treatment (control) was 4149.6 KWh and was set as the reference. The DG SON-T 15 the DG LED 15 treatments consumed 40 and 56% less energy compared to the reference, respectively. Furthermore, the total cost of S SON-T 15 light plan (reference) was 128.9 €. The DG SON-T 15 and DG LED 15 treatments was 52 and 65% cheaper than the reference, respectively.

Cultivars: The properties of the cultivars are evaluated by measuring the relative growth rate (RGR), relative dry weight (RDW) and the number of flowers. In general

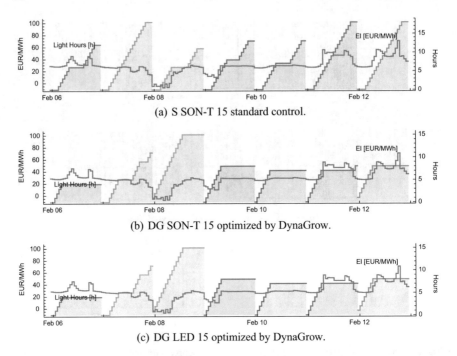

(a) S SON-T 15 standard control.

(b) DG SON-T 15 optimized by DynaGrow.

(c) DG LED 15 optimized by DynaGrow.

Fig. 5 Comparison of daily accumulated light hours optimized in February 2015 by different objectives from the S SON-T 15, DG SON-T 15 and DG LED 15 experiments

Table 4 Light hours and DLIs of the experiment February 2015

Experiment	Average DLI [mol m^{-2} d^{-1}]		Supp. Light [h]		Aver. Supp. Light [h day^{-1}]
S SON-T 15	9.2	100%	432.3	100%	15.4
DG SON-T 15	6.8	74%	259.3	60%	9.3
DG LED 15	6.8	74%	253.3	59%	9.0

Table 5 Energy results of the experiment February 2015

Control	Energy [KWh]		Cost [€]	
S SON-T 15	4149.6	100%	128.95	100%
DG SON-T 15	2488.8	60%	62.46	48%
DG LED 15	1828.5	44%	44.61	35%

the results are reflecting that plant growth was related to the climate conditions of the treatments with species and genotype-specific differences. All the plants grew well in the three climates and reached maturation within acceptable time. Figure 6 provides pictures of Kalanchoe (Simone), Felicitas, Apache, Hot Burrito, Hot Fajita, Macho and Alaska grown under the S SON-T (left), DG SON-T (Middle) and DG

(a) Felicitas.

(b) Apache.

(c) Hot Burrito.

(d) Hot Fajita.

(e) Kalanchoe (Simone).

(f) Macho.

(g) Alaska.

Fig. 6 Cultivar samples resulting from the S SON-T 15 (left), DG SON-T 15 (Middle) and DG LED 15 (Right) treatments

LED (Right) treatments. Further details about the results of growing the different cultivars in irregular light conditions can be found in work by Ottosen et al. [8, 9, 16]. Last, is was concluded that all the grown cultivars was in a sales-ready quality.

8 Discussion

2014: The average DLI from natural light was 8.0 mol m^{-2} d^{-1} during the 2014 experiment period. The lamps can in best case deliver 8.6 mol m^{-2} d^{-1} supplemental light per day. That is, the maximum natural and supplemental light per day is 16.6 mol m^{-2} d^{-1}.

The large contribution of the DLI from natural light results in less supplemental light and less potential for cost optimization. Despite the small optimization potential, the DG LED 14 resulted in the same DLI as the reference with 3% less light hours. The DG SON-T 14 obtained a 5% lower DLI than reference but with 8% less supplemental light hours. The explanation of how it is possible to obtain almost the same DLI with less hours, can be seen in the distribution of the light plans in Fig. 2. The S SON-T 14 light plan is regular at a daily basis due to the fixed light interval strategy. That is, the DLI goal has to be obtained per day. The DynaGrow strategies allow for more flexibility and do not require the same DLI on a daily basis but requires an average DLI spread over several days. The flexibility allows light hours to be postponed till more cost-effective intervals. Some cultivars, like tomatoes, cannot tolerate this form of flexibility. For that reason, the SCHEAPLIGHT and SPARBAL optimizations have to be effectuated with caution. The optimization works well with various long-day ornamental cultivars, like Kalanchoe and Campanula [16]. It requires more research to identify which cultivars can tolerate the irregular light intervals.

The DG SON-T 14 and DG LED 14 treatments were 14 and 31% cheaper than the reference, respectively (Table 3). In euro, that is a saving of 42.59 € per month in the best case for a compartment of 76 m^2. The electricity cost savings are not impressive and is related to the fact that September and October 2014 had a large amount of solar irradiation.

The DG SON-T 14 and DG LED 14 accumulated light hours have small variations. For example, September 26 DG SON-T 14 accumulates more light than DG LED 14. In contrast, DG LED 14 accumulates more light in September 27 than DG SON-T 14. This small variation can be explained by the Prefer Cheap Light (SCHEAPLIGHT) optimization. The maximum electricity prices for September 26 and 27 are similar and the price for the two light plans will be the same weather the light is lit more on September 26 and less on September 27 or vice versa.

2015: The lower DG SON-T 15 and DG LED 15 DLIs were not expected as similar effect (light intensity) should have been installed in the S SON-T 15 treatment. The equal DG SON-T 15 and DG LED 15 DLIs indicate that the optimization of PAR Light Sum Balance objective has been achieved but with a too low goal. A reason for the lower DLI can be explained by the natural light forecast provided by the external

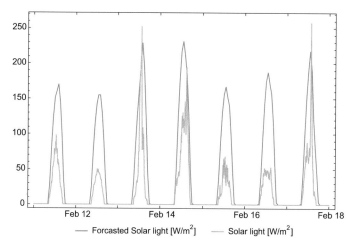

Fig. 7 Optimistic outdoor light forecast (bell curve) and the actual outdoor light data for February 11–18. Figure extracted from [11]

service. The cost calculation in the SPARBAL objective depends on a precise outdoor light forecast, see SPARBAL Sect. 3.

Figure 7 illustrates sample time series of the outdoor light forecast and the actual outdoor light. The time-series reveal that the outdoor light forecast has a tendency to be too optimistic. An optimistic light forecast will influence the sliding window balance in the SPARBAL objective as the *ParFuture* will promise more light than is the actual light. If the *ParFuture* is optimistic over several days, like illustrated in the time-series in Fig. 7 (February 11–18), then the average of the *ParDLI* will never be achieved within the sliding window. That is, the result will be a lower average DLI as indicated by Table 4.

The DG SON-T 15 and DG LED 15 treatments were 52 and 65% cheaper than the reference, respectively (Table 5). In euro, that is a saving of 84.34 € per month in the best case for a compartment of 76 m^2. In Denmark, some industrial growers have greenhouse facilities that is more than 65000 m^2 and the documented energy saving can have a huge potential economical impact. It is important to emphasize that the price calculations do not include rates for Public Service Obligations (PSO), but is based purely on the spot-market prices and the theoretically installed effect in each compartment.

The 2015 savings are impressive and should be considered in relation to the low average DLI from natural light of 3.5 mol m^{-2} d^{-1}. A low natural DLI results in a better potential for electricity cost optimization.

Validity: The focus of this work has been on dynamic light control. To be valid in a production setting, other climate control parameters, e.g., humidity, CO_2, temperature, have to be considered. That is, the light-control objectives have to be optimized simultaneously with the other parameters.

The experiments assume that the same lamp light intensity is installed for each of the greenhouse compartments. In reality, the installed lamps did not provide the same light intensity. To compare the result, the lamp intensity has been theoretically calculated to be the same for each of the compartments. It is assumed that the supplemental light can be added with the natural light. That is, when supplemental light is lit then 100 μmol m^{-2} s^{-1} is added to the given natural light at that time.

As a result of the differences of the installed lamps in each compartment, it was not possible to compare the measured electricity consumption. Instead, the energy results have been calculated and compared assuming theoretically installed lamp effects, i.e., 95 W/m^2 for LED and 126 W/m^2 for SON-T lamps.

9 Conclusion

Increasing energy prices are a challenge for growers and they need to apply supplemental light when electricity prices are low. It is important that the energy savings do not compromise the quality and growth of the crops. Multi-objective optimization was applied to find Pareto optimal trade-offs between achieving the DLIs and minimizing the electricity price of the light plans.

The S SON-T 14, DG SON-T 14 and DG LED 14 experiments resulted in comparable average DLIs of approximately 12 mol m^{-2} d^{-1} from September 17 to October 18. The S SON-T 14, DG SON-T 14 and DG LED 14 had 387, 355, 375 supplemental light hours, respectively. In conclusion, it is possible to achieve comparable average DLIs over a longer period with less supplemental light. That is, supplemental light is applied at intervals that contribute more to the average DLI over the period and at the same time minimize the total electricity price. Due to the relatively large contribution of natural light during September 2014 the energy savings were much less than in the February 2015 experiment. In comparison to the reference, the DG SON-T 14 and DG LED 14 experiments consumed 8 and 27% less energy, respectively.

February 2015 had less natural light and the energy cost savings were considerable but the provided DLIs were lower. The two DynaGrow treatments obtained approximately the same average DLI of 6.8 mol m^{-2} d^{-1} which is 26% lower than the reference. The DG SON-T 15 the DG LED 15 treatments consumed 40 and 56% less energy compared to the reference, respectively. Furthermore, the DG SON-T 15 and DG LED 15 treatments were 52 and 65% cheaper than the reference, respectively.

In general, the result of the experiments demonstrates that DynaGrow utilizes supplemental light at low electricity prices without compromising the growth and quality of the crop compared to standard fixed-rate supplemental light control. It was possible to produce a number of different cultivars where the supplemental light (SON-T or LED) were controlled by the DynaGrow software. The energy savings are achieved in relation to a control treatment with a fixed day length, but only if the DLIs are comparable between the treatments.

In Denmark, DynaGrow will have a high impact on cost in the beginning and end of the growing season, when there is a huge potential for optimizing the supplemental light.

There is an unexplored potential to optimize the utilization of supplemental light, temperature, CO_2, humidity and other climate variables simultaneously by formulating multiple advanced control objectives based on models already available from the extensive horticultural literature.

The results clearly demonstrate that DynaGrow supports a dynamic climate control strategy by optimizing multiple control objectives that result in a cost-effective control of the greenhouse climate.

References

1. Nord Pool: Nord pool spot market electricity prices (2016). http://www.nordpoolspot.com
2. Thomas, B., Vince-Prue, D.: Photoperiodism in Plants. Elsevier Science Direct (1997)
3. Sørensen, J.C., Jørgensen, B.N., Klein, M., Demazeau, Y.: An agent-based extensible climate control system for sustainable greenhouse production. In: 14th International Conference on Agents in Principle, Agents in Practice, vol. 7047, pp. 218–233. PRIMA: Wollongong, Australia (2011)
4. Aaslyng, J.M., Ehler, N., Karlsen, P., Høgh-Schmidt, K., Rosenqvist, E.: Decreasing the environmental load by a photosynthetic based system for greenhouse climate control. Acta Hortic. Int. Symp. Plant Prod. Closed Ecosyst. (ISHS) **1**, 105–110 (1996)
5. Aaslyng, J.M., Ehler, N., Karlsen, P., Rosenqvist, E.: IntelliGrow: a component-based Climate control System for Decreasing the Greenhouse Energy Consumption. Acta Hortic. 35–41 (1999)
6. Lund, J.B., Rosenqvist, E., Ottosen, C.O., Aaslyng, J.M.: Effect of a dynamic climate on energy consumption and production of hibiscus rosa-sinensis L in greenhouses. HortScience **41**, 384–388 (2006)
7. Aaslyng, J.M., Ehler, N., Jakobsen, L.: Climate control software integration with a greenhouse environmental control computer. Environ. Model. Softw. **20**(20), 521–527 (2005)
8. Kjaer, K.H., Ottosen, C.O.: Growth of chrysanthemum in response to supplemental light provided by irregularlight breaks during the night. J. Am. Soc. Hortic. Sci. **136**, 3–9 (2011)
9. Kjaer, K.H., Ottosen, C.O., Jørgensen, B.N.: Timing growth and development of Campanula by daily light integral and supplemental light level in a cost-efficient light control system. Scientia Horticulturae (2012)
10. Clausen, A., Sørensen, J.C., Jørgensen, B.N., Kjaer, K.H., Ottosen, C.O.: Integrating commercial greenhouses in the smart grid with demand response based control of supplemental lighting. In: International Conference on Agriculture, Environment, Energy and Biotechnology(ICAEEB 2014) (2012)
11. Sørensen, J., Kjær, K., Ottosen, C.O., Nørregaard Jørgensen, B.: Dynagrow – multi-objective optimization for energy cost-efficient control of supplemental light in greenhouses. In: Proceedings fra ECTA 35th Annual Conference, pp. 1–9 (2016)
12. Ghoreishi, S.N., Sørensen, J.C., Jørgensen, B.N.: Enhancing State-of-the-art Multi-objective optimization algorithms by applying domain specific operators. In: IEEE Symposium on Computational Intelligence in Dynamic and Uncertain Environments (IEEE CIDUE'15), p. 8. IEEE Computational Intelligence Society (CIS), South Africa (2015)

13. Conwx: Conwx intelligent forecasting systems (2016). http://www.conwx.com
14. Energi Danmark: Energi danmark electricity forecasts (2016). http://www.energidanmark.dk
15. Enoch, H., Kimball, B.: Carbon Dioxide Enrichment of Greenhouse Crops. CRC Press Inc., FL (1986)
16. Kjaer, K.H., Ottosen, C.O., Jørgensen, B.N.: Cost-efficient light control for production of two campanula species. Scientia Hortic. **129**, 825–831 (2011)

A System for Evolving Art Using Supervised Learning and Aesthetic Analogies

Aidan Breen and Colm O'Riordan

Abstract Aesthetic experience is an important aspect of creativity and our perception of the world around us. Analogy is a tool we use as part of the creative process to translate our perceptions into creative works of art. In this paper we present our research on the development of an artificially intelligent system for the creation of art in the form of real-time visual displays to accompany a given music piece. The presented system achieves this by using Grammatical Evolution, a form of Evolutionary Computation, to evolve Mapping Expressions. These expressions form part of a conceptual structure, described herein, which allows aesthetic data to be gathered and analogies to be made between music and visuals. The system then uses the evolved mapping expressions to generate visuals in real-time, given some musical input. The output is a novel visual display, similar to concert or stage lighting which is reactive to input from a performer.

Keywords Genetic algorithms · Evolutionary art and design · Genetic programming · Hybrid systems · Computational analogy · Aesthetics

1 Introduction

Analogy is the comparison of separate domains. The process of analogy has strong applications in communication, logical reasoning, and creativity. A human artist will often take some source material as inspiration and create an equivalent, or related art piece in their chosen artistic domain. This process of metaphor is the equivalent of making an artistic analogy and has been used successfully in a literal form by artists like Klee [1], Kandinsky [2] and more recently Snibbe [3]. Similar approaches

A. Breen (✉) · C. O'Riordan
National University of Ireland, Galway, Ireland
e-mail: a.breen2@nuigalway.ie
URL: http://www3.it.nuigalway.ie/cirg/

C. O'Riordan
e-mail: c.oriordan@nuigalway.ie

© Springer Nature Switzerland AG 2019
J. J. Merelo et al. (eds.), *Computational Intelligence*,
Studies in Computational Intelligence 792,
https://doi.org/10.1007/978-3-319-99283-9_3

are often taken in a less direct form by stage lighting designers or film soundtrack composers.

Our aim is to make computational analogies between the domains of music and visuals by making use of aesthetic models, computational analogy, and grammatical evolution.

This work has direct practical applications for live performance and stage lighting design. The work in this paper may also have less direct applications in user interface and user experience design with particular use in the automatic generation of user interfaces and subconscious feedback mechanisms. Beyond these application domains, our research motivation also includes gaining insight into aesthetics and analogical reasoning.

1.1 Creating Aesthetic Analogies

One of the major challenges of computational art is to understand what makes an art piece *good*. Indeed the cultural and contextual influences of an art piece may define what makes it emotive, such as Duchamp's Fountain [4] or René Magritte's The Treachery of Images [5], but beyond that we rely on the aesthetics of an object to decide if it is pleasurable to perceive. Aesthetics provide an objective description of this perception. We use this objective description as a tool upon which to build our analogies.

Every domain has its own aesthetic measures—musical harmony, visual symmetry, rhythm and combinations thereof. In some cases, these measures can be used to describe objects in more than one domain. Symmetry, for example, can describe both a visual image, and a phrase of music. The example we demonstrate in this paper is harmony. Musical harmony can be measured by the consonance or dissonance of musical notes. Visual harmony can be measured directly as the harmony of colours.

The analogy we are hoping to create is described as follows: given some musical input with harmony value x, a *mapping expression* can be created to generate a visual output with a harmony value y such that $x \simeq y$. Furthermore, we posit that when performed together, both input music and output visuals will create a pleasing experience. In other words, can we take music and create a visual with a similar harmony and will they go together?

For this simple example, it is clear that a suitable expression could be created by hand with some knowledge of music and colour theory. However, if we extend the system to include more aesthetic measures, such as symmetry, intensity, contrast or granularity, defining an analogy by use of a *mapping expression* becomes far more complex. While developing a system to capture more complex mappings is beyond the scope of this paper, we aim to build the system such that it may be extended to do so.

1.2 Grammatical Evolution and Mapping Expressions

A genetic algorithm (GA) provides a useful method of traversing an artistic search space, as demonstrated by Boden and Edmonds in their 2009 review [6]. Grammatical evolution (GE) [7], in particular allows us to provide a simple grammar which defines the structure of *mapping expressions* which can be evolved using a GA. This allows us to flexibly incorporate aesthetic data, operators and constants while producing human readable output. Importantly, we make no assumptions about the relationships between input and output. This approach does not restrict the output to any rigid pattern; potentially allowing the creation of novel and interesting relationships between any two domains, music and visuals or otherwise.

No single set of *mapping expressions* would be capable of creating pleasing output in every circumstance. In this respect, we intend to find suitable expressions for a specific input, such as a verse, chorus or phrase. Expressions may then be used in real-time when required and would handle improvisation or unexpected performance variations. Expressions produced by Grammatical Evolution are naturally well suited to this task as they can be stored or loaded when necessary, and evaluated in real-time.

1.3 Contributions and Layout

The main contribution of this work is an implementation of Grammatical Evolution using music and empirically developed aesthetic models to produce novel visual displays. Secondary contributions include a structural framework for aesthetic analogies used to guide the gathering of data and development of evolutionary art using *mapping expressions*, and preliminary results produced by our implementation of the system.

The layout of this paper is as follows. Section 2 outlines related work in the areas of computational analogy, computational aesthetics, and computational art. Section 3 introduces our proposed method including a general description of our analogy structure, aesthetic models and the structure of our evolutionary system. Section 4 presents the details of our implementation in two distinct phases, the evolutionary phase (Sect. 4.1) and the evaluation phase (Sect. 4.2). Our results are presented in Sect. 5 followed by our conclusion in Sect. 7 including a brief discussion of future work (Sect. 6.3).

2 Related Work

"Analogy underpins language, art, music, invention and science" [8]. In particular, Computational Analogy (CA) combines computer science and psychology. CA aims to gain some insight into analogy making through computational experimentation.

As a research domain, it has been active since the late 1960s, accelerated in the 1980s and continues today. Computational analogy systems historically fall into three main categories: symbolic systems, connectionist systems and hybrid systems. Symbolic systems make use of symbolic logic, means-ends analysis and search heuristics. Connectionist systems make use of networks with spreading activation and back-propagation techniques. Hybrid systems often use agent based systems taking aspects of both symbolic and connectionist systems. For further reading, see [9, 10].

Birkhoff is often cited as one of the first to consider aesthetics from a scientific point of view. His simplistic 'aesthetic measure' formula, $M = O/C$, was simply the ratio of order (O) to complexity (C) [11]. Of course, this is over simplified and abstract, but it did begin a long running discussion on aesthetics and how we can use aesthetics to learn about the higher functions of human cognition.

More recently, the discussion has been reignited by Ramachandran who has outlined a set of 8 'laws of artistic experience' [12]. In this paper a number of factors are outlined which may influence how the human brain perceives art. Some of these factors are measurable, such as contrast and symmetry, but others remain more abstract such as the grouping of figures. Nonetheless, it has inspired further discussion [13–16].

Within specific domains, heuristics can be formalized and used to generate derivative pieces in a particular style or to solve particular challenges in the creation of the art itself. GAs in particular have proven to be quite effective due to their ability to traverse a large search space. In music for example, GAs have been used to piece together particular musical phrases [17], generate complex rhythmic patterns [18] or even generate entire music pieces [19]. Similar systems have also been used to create visuals [20, 21], sculpture [22] and even poetry [23].

Indeed the use of aesthetic measures in combination with GAs has also been reviewed [24] and an approach has been outlined to demonstrate the potential application of Multi-Objective Optimization to combine these measures [25]. While the system does produce computational art that may be described as aesthetic, it is also limited strictly by the aesthetic measures used, without any artistic context.

It is clear that computational systems can work in tandem with aesthetics to generate art and explore the possible applications of computational intelligence. Up to this point however, popular approaches have been remarkably rigid. Our work aims to explore a more flexible approach and perhaps discover a more natural artistic framework through analogy.

3   Proposed Method

3.1   Analogy Structure

We make use of a conceptual structure to provide a basis for our aesthetic data and analogies. The structure is shown in Fig. 1 with measurable aesthetic attributes in separate domains which are connected by a set of *mapping expressions* which may

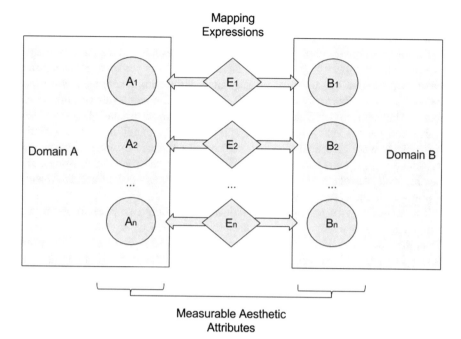

Fig. 1 Analogy structure overview. The *Mapping Expressions*, E_1 to E_n, are encoded as chromosomes and evolved using the genetic algorithm. Extracted from [26]

be evolved using grammatical evolution. The implementation in this paper uses a single attribute in each domain, however, the structure is not restricted to a bijective mapping.

Some aesthetic attributes may be more suitable than others for use in a structure as described in Fig. 1. Harmony is selected for use in this paper as it has a strong impact on the overall aesthetic quality of an art piece, and can be measured quite easily in separate domains. In music, the harmony of notes being played is often referred to as the consonance—or conversely, dissonance—of those notes. While the timbre of notes has an impact on consonance, an estimate can be obtained from pitch alone. In the visual domain, colour harmony, or how pleasing a set of colours are in combination, can be measured as a function of the positions of those colours in some colour space, such as RGB (red, green and blue dimensions), CMYK (cyan, magenta, yellow and black dimensions), HSV (hue, saturation and value dimensions) or LAB (one lightness dimension and two opposing colour dimensions of red/green and yellow/blue). The conceptual similarities between musical consonance and visual harmony provide a convenient and understandable starting point.

Consonance values for any two notes have been measured [27–29] and numerous methods have been proposed that suggest a consonance value can be obtained for larger sets of notes [30–33]. The simplest general approach is to sum the consonances for all pairs of notes in a set. This provides a good estimation for chords with the same number of notes and can be normalized to account for chords of different cardinalities.

For this preliminary implementation, we enforce a number of restrictions. Firstly, we restrict the number of inputs to two musical notes at any one time. This simplifies the grammar and allows us to more easily analyse the output *mapping expressions*. Secondly, musical harmony is calculated using just 12 note classes within a single octave. This helps to avoid consonance variations for lower frequencies.

The consonance values for each note classes were gathered in a study whereby test subjects were presented with two note pairs and asked to select the note pair that sounded more consonant. By using this *two alternative forced choice* approach, together with a novel ranking algorithm, a number of issues which affected the results of previous studies were avoided.

The subjectivity of responses was reduced by forcing two pairs to be directly compared. Contextual issues, where the order in which pairs were ranked might affect their actual ranking, were also avoided in this way. The ranking algorithm used a graph based approach which allowed a full ranking of all 12 note classes to be found with a minimal number of comparisons. This was particularly important as subject fatigue had a large impact on the quality of responses and can increase in a very short space of time. The performance of the ranking algorithm has been further analysed and compared to other ranking approaches demonstrating the effectiveness of the algorithm for this application [34].

The observed consonance values were then compared to previous studies and historical observations, showing a high correlation. Figure 2 shows the consonance values for note pairs used based on this study [29].

Similarly, colour harmony values can be measured and modelled [35–37]. While the harmony of more than 2 colours may be obtained with a similar approach to music

Fig. 2 Consonance Values for musical intervals used to calculate musical Harmony Values. Extracted from [26]

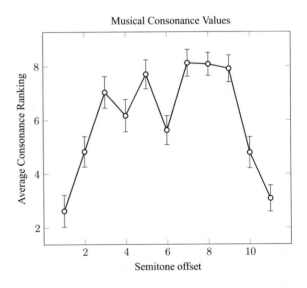

Fig. 3 Average Colour
Harmony values. Extracted
from [26]

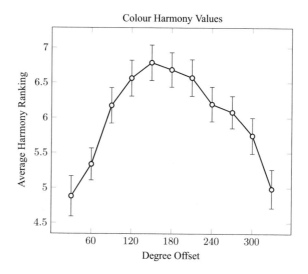

chords, the pattern in which colours are displayed adds an extra level of complexity. To combat this, we assume our visual display is not a strict two dimensional image, but rather a pair of lights emitting coloured light into some space. For example, a pair of LED stage lights for a small musical performance.

The harmony values for colour pairs were obtained using the same methodology used to obtain musical data. A *two alternative forced choice* style study using the same ranking algorithm was employed to prevent subjectivity and contextual pitfalls and to reduce subject fatigue. The results were compared, once again, to previous studies and historical observations. Figure 3 shows the harmony values obtained in this study and used in the work presented here.

3.2 Evolutionary System

Our evolutionary approach is based upon Grammatical Evolution [7]. We use a Genetic Algorithm (GA) to evolve *mapping expressions* based upon a given grammar. The evolved expression allows us to create a real time system rather than an offline visual output. In this way, any particular performance is not limited to a strict musical input thereby allowing improvisation, timing and phrasing variation, and handling of human error.

Another advantage of this particular GA approach is the flexibility by which we can incorporate aesthetic data. By using a grammar, we are decoupling the structure of the output of the GA from the actual process of evolution. To extend the current implementation to include intensity as an aesthetic attribute, for example, we simply edit the grammar to accommodate this. We can also extend the grammar to include other operators, such as predefined functions if we feel their inclusion may improve the performance of the system.

Further, the human readability of the output expression is extremely valuable, not only as a sanity check to debug unexpected behaviour, but also to analyse the output of a particular evolutionary run. The goal of this work is not simply to create analogies, but to understand the process by which they are created. The readability of output expressions at each stage of the evolutionary process is therefore a huge advantage over other *black box* AI techniques which produce results without any way of understanding how or why they were produced.

Mapping Expressions. *Mapping expressions* are created by using an individual chromosome to guide the construction of a symbolic expression by use of the given grammar. The following is an example of a symbolic expression representing a nested list of operators (addition and multiplication) and parameters (2, 8 and 5) using prefix notation. This is the same structure used by *mapping expressions*.

```
(+  2  (*  8  5))
```

The symbolic expression is evaluated from the inside out, by evaluating nested expressions and replacing them with their result. In this example, the nested expression multiplying 8 by 5 will be evaluated producing the following intermediate expression.

```
(+  2  40)
```

Once all nested expressions are evaluated, the final expression can return a result, in this case, 42.

Grammar. The grammar defines the structure of an expression using terminal and non-terminal lexical operators. Terminals are literal symbols that may appear within the expression, such as +, * and the integers in the example expression above. Non-terminals are symbols that can be replaced. Non-terminals often represent a class of symbols such as operators of a specific cardinality, other non-terminals, or specific terminals. In the example expression above, the + and * symbols might be represented by a single *operator* non-terminal that accepts two parameters. Parameters may be an integer, or a sub expression. Finally, the integers in the expression above could be replaced by an integer non-terminal. This leaves us with the simple grammar shown in Table 1 which is capable of representing not only the example expression, but also any other expression, of any size, that adds and multiplies integers.

Given a defined grammar, such as the example grammar in Table 1, an expression can be built from a chromosome using the following approach. Beginning with a starting non-terminal, each value in the chromosome is used in series as the index of the next legal terminal or non-terminal. This mapping continues until either the expression requires no more arguments, or a size limit is reached. If the chromosome is not long enough to complete the expression, we simply begin reading from the start of the chromosome again.

Table 1 Example grammar

Non-terminals	Possible replacements
Parameter	Integer, (Operator Parameter Parameter)
Operator	+,*
Integer	Any integer
Terminals	
+	
*	
Any Integer	

Evolution. The evolution process is straightforward thanks to the simple structure of chromosomes. Each *mapping expression* becomes an individual within out population possessing a single chromosome. Beginning with a population of individuals with randomly generated chromosomes, known as Generation 0, successive generations are produced using Selection, Crossover and Mutation genetic operators.

Tournament selection wth elitism is used in this implementation. At generation G_n, the fitness of each individual is calculated. A small number of the highest fitness individuals are moved to generation G_{n+1}. This elitism is used to prevent the maximum fitness of the population from declining. The selection phase now begins to population the rest of generation G_{n+1}. Two individuals are selected at random. The individual with more favourable fitness is placed into generation G_{n+1}. The process continues until the desired size of generation G_{n+1} is reached.

When selection is complete, crossover begins using both single point and double point crossover. Two individuals $P1$ and $P2$ are selection at random from the newly selected generation G_{n+1}. If single point crossover is to be used, a random crossover point is selected and two children, $C1$ and $C2$ are produced. These children take the place of their parents in the new generation.

$C1$ will receive a chromosome made up of the $P1$ codons to the left of the crossover point, and the $P2$ codons to the right of the crossover point. $C2$ will receive the alternative, that is, a chromosome made up of the $P1$ codons to the right of the crossover point, and the $P2$ codons to the left of the crossover point. If double point crossover is to be used, the same steps are taken, but twice. This results in one child receiving outer codons from one parent and inner codons from the other.

The effect of double point crossover is quite strong when using the described implementation of Grammatical Evolution. Due to the tree-like, recursive way in which an expression is built, single point crossover invariably has large effects on the final expression. Large portions of the expression are likely to be completely different, even to the parent expression, as previous codons determine the meaning of following codons. Double point crossover alternatively, can operate similar to subtree substitution, affecting only a small portion of the final expression and retaining the overall structure of the expression tree.

Finally, once generation G_{n+1} is complete, a mutation operator is applied. Mutation consists of setting a very small number of codons to new values. Both the mutated codon and the new value are randomly selected.

This process relies heavily on the fitness function. Calculating the fitness of any *mapping expression* without some guidelines would be extremely subjective. In our implementation we take a heuristic approach that rewards solutions that produce outputs with a similar normalised aesthetic value as inputs. An in-depth description of the implemented fitness function is presented in Sect. 4.1.

4 Implementation

We now discuss how the structure introduced above together with the data gathered has been implemented. We demonstrate how the following system has been used to evolve *mapping expressions* that generate a visual output when given a musical input. The system may be used to generate visuals in time with music by use of a time synchronized subsystem utilizing a music synthesizer and visualization server.

4.1 Evolution Phase

Figure 4 shows the structure of the Evolution Phase. This phase is centred about the GE algorithm. In our implementation we use a population of 50 chromosomes. Chromosomes are stored as 8 bit integer arrays, with values between 0 and 255. A chromosome length of 60 integer values was used in the work presented in this paper.

Musical input is taken in the form of *Musical Instrument Digital Interface* (MIDI) data. The MIDI protocol represents digital music signals, originally designed as a transmission protocol to allow musical signals to be sent between instruments and synthesizers. Musical notes are sent as packet pairs (*note on* and *note off*) containing the note pitch and the 'velocity', or strength of the note which is often translated to volume. The MIDI protocol also allows data to be stored as a file with each packet

Fig. 4 Evolution Phase overview. Extracted from [26]

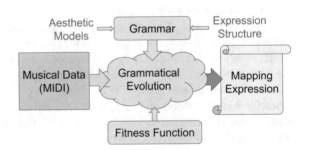

Table 2 Grammar terminal operators

Expression	Arguments
Plus 90°	1
Plus 180°	1
Sin	1
Cos	1
Log	1
Addition	2
Subtraction	2
Multiplication	2
Division	2
Music harmony constant	2
Visual harmony constant	2
Ternary conditional operator	3

Table 3 Grammar terminal values

Expression	Range
Constant integer value	0–255
Musical input 1	0–255
Musical input 2	0–255

containing a timing value. We use a file to store a sample musical input using this format and determine which notes are being played using the timing value.

The implemented grammar contains a list of operators, and values (variables and constants) which are presented in Tables 2 and 3. Of note here are the aesthetic values for music and visuals which can be inserted directly into an expression as constants, or read at run-time as variables in the case of musical input. The aesthetic models use normalised values based on the values shown in Figs. 2 and 3. Aesthetic constant expressions such as *Musical Harmony Constant* and *Visual Harmony Constant*, accept two arguments representing two music notes or two colour hues. The expression returns the aesthetic value of those two notes or colours.

Our fitness function, as introduced above, aims to maximise the similarity between input and output harmony. The fitness for any n pairs of input musical notes is calculated as follows, where M is a function representing the musical harmony of a pair of notes, and V is a function representing the visual harmony of a pair of colour hues.

$$fitness = \frac{1}{n} \sum_{i=1}^{n} 255 - |M(input) - V(output)| \tag{1}$$

Both M and V are normalised between 0 and 255, which produces a fitness range of 0–255.

Tournament selection is carried out to select individuals for evolution. A combination of single point and double point crossover is used to build a succeeding generation. Elitism is used to maintain the maximum fitness of the population by promoting the best performing individuals to the next generation without crossover or mutation.

Mutation is applied at the gene level. A gene is mutated by randomly resetting its value. The mutation rate is the probability with which a gene will be mutated. The mutation rate is varied based on the number of generations since a new peak fitness has been reached. This allows us to optimise locally for a period, and introduce hyper-mutation after an appropriate number of generations without any increase in peak fitness. We call this the Mutation Threshold. The standard mutation rate (Mut_1) is calculated as:

$$Mut_1 = \left(\frac{0.02}{70}\alpha\right) + 0.01 \tag{2}$$

where α represents the number of generations since a new peak fitness was reached.

After the Mutation Threshold is reached, indicating a local optima, hyper-mutation (Mut_2) is introduced to encourage further exploration of the fitness landscape.

$$Mut_2 = 1.0 \tag{3}$$

If a fitter solution is discovered, mutation is again reduced to Mut_1 to allow smaller variations to occur.

At each generation, the *mapping expressions* represented by chromosomes are stored. This allows us to monitor the structure and size of the expressions as they are created. We can also use the stored expressions to compare earlier generation to later generations.

Evolution is halted after a Halting Threshold has been reached. The Halting Threshold is measured as the number of generations without an increase of peak fitness. Details of the parameters used can be found in Table 4.

The output of this process is a set of *mapping expressions* representing the final generation. From this set, the fittest expression is selected for evaluation and comparison to the fittest expression from previous generations.

Table 4 Genetic algorithm parameters

Parameter	Value
Population size	50
Chromosome length	60
Crossover rate	0.8
Standard mutation Rate (Mut_1)	see Eq. (2)
Hyper-Mutation rate (Mut_2)	1.0
Mutation threshold	100
Halting threshold	200

4.2 Evaluation Phase

In order to evaluate the performance of an evolved *mapping expression*, we must play both music and visuals together. To this end, we have built the evaluation system as outlined in Fig. 5. While the evolved expression is capable of generating visual output in realtime, the evaluation for this work is conducted offline, using a pre-calculated file containing the original musical data, and the generated visual data.

In order to perform music in synchrony with generated visuals, an *extended MIDI player* subsystem is required. Musical data (MIDI) and visual data are combined in an *extended MIDI file* (MIDIX). The *extended MIDI player* then parses this file and uses an internal time synchronisation process to send MIDI signals to two separate systems: a music synthesizer and a visualization server. Both systems are capable of reading MIDI signals and producing output in real time.

The music synthesizer is a common tool in music creation and performance. Historically, synthesizers existed as hardware devices connected to some digital instrument producing MIDI signals. The synthesizer would listen for input from a physical MIDI cable, and produce a signal which could be sent to an amplifier and speakers to create audio. Modern synthesizers are typically digital systems that listen to digital MIDI ports for message packets and produce audio using modern audio interfaces. While hardware based synthesizers are highly sought after by electronic musicians, software based synthesizers are indistinguishable in most circumstances.

We make use of digital synthesizers in this work due to their flexibility and cost effectiveness. We use a standard MIDI port to send and receive musical data to an open source software synthesizer, part of the Reaper digital audio workstation [38]. The signals received by the synthesizer are used to produce realistic sounding music.

The visualization server is a software system created specifically for this implementation. The server works in a similar fashion to a music synthesizer; however, rather than using a MIDI port, the server uses HTTP and websockets. With this approach, a webserver accepts HTTP messages sent from the *extended MIDI player* and uses websockets to update a javascript application running in a web browser. This technology was chosen due to the cross platform support for javascript applications and also due to the ease of rendering a visual display using standard web technologies. When the javascript application receives an update through the websocket, the display is updated accordingly. This ensures the visuals remain synchronized with the audio being played.

Fig. 5 Evaluation Phase overview. Extracted from [26]

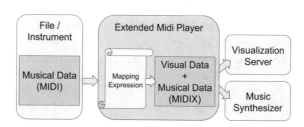

4.3   Supervised Fitness

Using the evaluation system outlined above, we can interactively evaluate the performance of a particular *mapping expression*. We can either use a static MIDI file to compare individual *expressions* or we can use a live MIDI instrument to send live MIDI signals to evaluate how it performs with improvised and varying input.

Expressions that are deemed fit by human supervision may then be reintroduced to the evolution phase to continue the process. This step is independent of the fitness function in order to capture aesthetic results beyond its capabilities.

5   Results

5.1   Evolutionary Phase

Using the approach outlined above we successfully evolved *mapping expressions* capable of mapping musical input to visual output.

Many of the random seed expressions such as the following example simply produced constant values:

```
['plus180',['plus90',['sin', 215]]]
```

In later generations however, we see more complex expressions producing better fitting results:

```
['add',56, ['musicalHarmony',94,['cos','mus2']]]
```

Here we see the expression makes use of the input variable mus2 and the musical harmony constant musicalHarmony which produces a dynamic output. The example chosen here is one of the smallest expressions created.

Further visual analysis of the smallest evolved expressions shows regular use of dynamic input variables. A script was used to search through the entire set of final expressions from all evolutionary runs. The smallest expressions were selected for manual evaluation and visual inspection. The sampled expressions achieved high fitness scores, above 200 in all cases, while using less than 10 sub-expressions. Leaf nodes of these sampled expression trees consist of at least one input variable in all cases. All sampled expressions also made use of musical or visual harmony constants at least once.

Figure 6 shows the distribution of fitness values for randomly generated expressions versus evolved expressions. We see the distribution for random expressions is heavily skewed towards the minimum value of 100. This is due to the number of expressions which produce a constant output. Evolved expressions however show a much tighter distribution with significantly higher fitness values.

The distribution of intervals in the input M will affect the fitness of the evolved expression. An evolved expression may be directly compared to the target visual

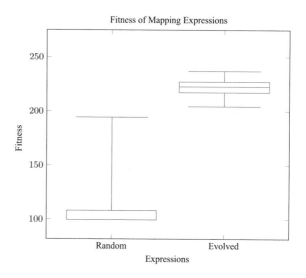

Fig. 6 Fitness of 100 randomly generated *Mapping Expressions* versus Evolved expressions. Extracted from [26]

harmony by using an equally distributed input. Our input is a set of 11 note intervals, 1–11, excluding the unison and octave intervals 0 and 12 respectively. In Fig. 7 we see a demonstration of this comparison. Previous generations are shown as dotted lines with the final fittest individual in solid black. The target output, the output that would produce an optimum fitness value, is shown in red. We see as the generations pass, the output matches the target more closely. Of note here are the horizontal dotted lines indicating older generations producing constant outputs which have been superseded by generations producing closer matching dynamic outputs.

Figure 8 shows the fitness of a single population across a number of generations. In blue we see the maximum fitness of each generation. This value never decreases

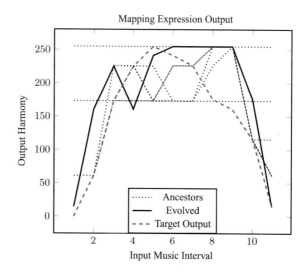

Fig. 7 Output of one evolved expression and its ancestors compared to target visual harmony. Extracted from [26]

Fig. 8 Population fitness for
631 generations of a typical
run. Extracted from [26]

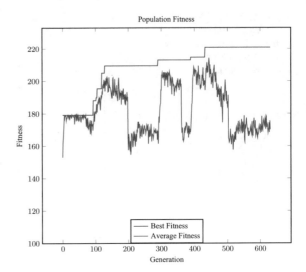

due to the use of elitism. Highly fit individuals are brought from one generation to the next to prevent the population from decaying and losing a possible fit solution.

In red we see the average fitness of the population. During early generations we see dramatic improvements in fitness followed by a series of incremental increases punctuated by dramatic decreases in average fitness. These incremental increases in fitness are an indication of local optima being discovered with low mutation. Hypermutation is then introduced causing the dramatic reduction in average fitness as more individuals mutate in more extreme ways. While this seems to have a negative effect on the population as a whole, it allows us to find fitter solutions and prevents premature population convergence at a local optima.

5.2 Evaluation Phase

Preliminary results have been obtained based on the initial implementation described in Sect. 4.2. These results demonstrate that a visual display, however rudimentary, can be produced based on musical data in real time. The *extended MIDI player* was used to play a file containing a 10 s music piece with musical intervals of varying harmony. Visuals generated by a *mapping expression* were displayed on a computer screen. Visuals were observed to be in time with the synthesized music. An example of the visual display with screenshots taken at 2 s intervals are shown in Fig. 9. The colour pattern used in the visual display was similar to that used to collect colour harmony data. This ensures that there is no variation between observed colour harmony and generated colour harmony.

Our initial observations indicate the the evolved expressions produce an analogy that results in more enjoyable visuals than randomly generated colours. The colours

Fig. 9 Generated visual
display. Extracted from [26]

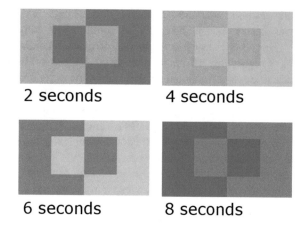

tend to follow a general pattern where similar note pairs, with similar intervals, produce similar colours.

6 Discussion

In this work we aim to lay the foundation for further developments in the use of aesthetic analogy and supervised learning in art. Our work takes a pragmatic approach, with the objective of using these tools to create an output which can be built upon and extended in further studies.

In this regard, the results presented above show a positive outcome. We were successful in defining an analogical structure which was codified such that Grammatical Evolution could be utilised to produce an output that represents an analogy. That analog—or its representation—can then be used to create a visual output where before there was only sound—or the digital representation of sound.

Realistically, of course, the visual output of the system is lacking in a number of ways. It might be a stretch to call the display shown in Fig. 9 a piece of art, but it *is* undeniably a visual display which has been created by a computer system. This system knows only the given parameters of aesthetics, a grammar and a way to compare results in the form of a fitness function. In this respect, the results might be compared to the exploratory play of a toddler as she learns. While the results might not be on par with more advanced or mature systems, what we are shown is really the beginning of what might be possible.

6.1 Analogical Structure

One of the clear drawbacks of the work presented here is the restriction in the structure of the analogy used. Currently, the system is designed to create an analogy using only one aesthetic attribute—harmony. This attribute was chosen specifically because it was clearly present and easily measured in both domains.

Harmony is however, only one of many potential attributes within the context of the aesthetics of music and visuals. The resulting analogy could therefore be described as a fractional aesthetic analogy, or an analogy of harmony alone which may contribute to the overall aesthetic quality. We hope to achieve a full analogy of aesthetics by making use of more than one attribute. To this end, we have highlighted a number of potential candidates for future attributes to be included.

Symmetry, intensity, contrast and complexity have specifically been highlighted as candidates for our future work. All four attributes can be observed in each domain and we believe they all may be measured, to some degree, without a great deal of subjectiveness or computational expense. Contrast is perhaps the easiest attribute to measure in both domains with contrast in visuals being measurable on a per-pixel basis and in music on a per-note basis. Symmetry is similarly directly measurable in visuals, and potentially measurable in music using heuristics such as matching beat patterns or relative pitch changes. Intensity in visuals may be measured as a function of colour contrast and fractal index while in music it may be measured by volume and tempo. Finally, complexity is commonly measured in visuals using a fractal index and may be measured in music simply by measuring the number of notes being played in a period of time.

These attributes may not represent the full aesthetic gamut but we believe that they will provide a strong toolset, beyond harmony alone.

6.2 Evaluation

The evaluation phase, as described above, refers to the evaluation of the output visual display in tandem with the input music. We report a positive preliminary result however, fully evaluating the output of the system in any reasonably objective way would require a great deal of work beyond the scope of this paper.

A study of human subjects would be required to compare randomly generated visual displays, to displays created with *mapping expressions* of various fitness. Due to the subjective nature of human aesthetic response, a reasonably large study would be required. A similar problem was faced when gathering the data used to build the aesthetic models used in this work. Even when tasked with ranking simple musical note pairs—or colour pairs—the participants in the study suffered from fatigue and boredom extremely quickly which may have had strong effects on the data gathered.

In the case of musical note pairs, however, the study could be conducted at a reasonably quick speed as each note pair would play for only a few seconds. To evaluate the output of the system presented in this paper, a longer segment of music must be played, significantly increasing the time it takes to gather information from any individual subject. This has a knock-on effect of increasing fatigue and boredom, reducing the quality of the data gathered. It is possible that any data gathered in a study like this would simply by too strongly affected by subject fatigue and boredom to be useful in any meaningful way.

In practice however, no visual display is created in this manner. A display for a live music performance would typically be created at the discretion of the lighting designer or artist, and a small number of advisors with very little specific feedback from the audience. This observation should serve to guide the development of the system as an artistic tool, rather than a replacement of the artist.

6.3   Future Work

We have shown that *mapping expressions* can be evolved using a fitness function based on empirically developed aesthetic models. However, we have not evaluated the perceived aesthetic differences between expressions of varying fitness. Further research is required to fully evaluate the strength of this correlation.

At present we restrict the number of input musical notes to simplify the grammar and allow analysis of the evolved expressions. This clearly limits the application of this system greatly. Future iterations should accommodate varying musical input lengths.

The results presented were obtained using only one *mapping expression* between musical consonance and colour harmony. We have not explored the possibilities of using multiple mapping expressions incorporating many attributes. We believe this will improve the quality of generated visuals dramatically.

As shown in Sect. 5, the fitness of a population has certain limitations. We hope to improve the speed at which fitness increases and also increase the maximum fitness achievable by any individual by tuning the parameters of the genetic operators.

The *extended MIDI* format has a number of useful applications beyond its use in this implementation. The format may also be useful for predefined visual displays and synchronised performances. With this in mind, we would like to fully define our version of the protocol and make it available to the public.

In a similar vein, the visualization server, which uses the *extended MIDI* format may also be improved. Most immediately, it should be able to handle all of the attributes used by *mapping expressions* to generate varied and immersing visual displays. Also, the server is currently restricted to displaying visual displays on a computer screen, which is not suitable for a live performance. We hope to develop functionality to allow the visualization server to accept an *extended MIDI* signal and control stage lighting hardware using industry standard protocols.

The outlined system is certainly capable of producing some visual output. Whether that output is deemed aesthetically pleasing is still an open question. In order to determine the actual performance of the final output of the system, we hope to conduct a study with human subjects. Our hypothesis here is: *the system produces more pleasing visual displays than random colour changes.*

The proposed study would demonstrate if we are moving in the right direction, however, the overall goal of this research is to create a system that can create art, and perform it. To this end, the success of the system should be evaluated with a live performance.

7 Conclusion

The reported results show the effectiveness of the analogical structure as shown in Fig. 1, which can be used in combination with Grammatical Evolution to produce novel visual displays. This analogical structure successfully guides the collection of data in each aesthetic domain. This structure also allows the creation of a fitness function which serves to allow the generation of effective mapping expressions which maximise the aesthetic similarity between input music and output visuals over the course of an evolutionary run. The use of mapping expressions, evolved using this fitness function, allows the creation of real-time aesthetic analogies based solely on human aesthetic experience, without assuming any structure or weighting between aesthetic values across domains.

Results show mapping expressions that have achieved similarity in harmony between input music and output visuals using this approach. We also show the increase in fitness over time across an evolutionary run, demonstrating the effectiveness of Grammatical Evolution in this application.

Acknowledgements The work presented in this paper was kindly funded by the Hardiman Scholarship, National University of Ireland, Galway. The authors would also like to thank the staff and students of the Computational Intelligence Research Group (CIRG), Department of Information Technology, and the School of Mathematics, Statistics and Applied Mathematics, NUIG for their guidance and support throughout the development of this work.

References

1. Klee, P.: Pedagogical Sketchbook. Praeger Publishers, Washington (1925)
2. Kandinsky, W., Rebay, H.: Point and Line to Plane. Courier Corporation, (1947)
3. Snibbe, S.S., Levin, G.: Interactive dynamic abstraction. In: Proceedings of the 1st International Symposium on Non-photorealistic Animation and Rendering, ACM, 21–29 (2000)
4. Cameld, W.A.: Marcel Duchamp's fountain: Its history and aesthetics in the context of 1917. Artist of the century, Marcel Duchamp (1990)
5. Magritte, R.: The treachery of images. Oil Canvas **231**, 1928–1929 (1928)
6. Boden, M.A., Edmonds, E.A.: What is generative art? Digit. Creat. **20**, 21–46 (2009)
7. O'Neil, M., Ryan, C.: Grammatical evolution. In: Grammatical Evolution. Springer (2003) 33–47
8. Gentner, D., Forbus, K.D.: Computational models of analogy. Wiley Interdiscip. Rev.: Cogn. Sci. **2**(3), 266–276 (2011)
9. French, R.M.: The computational modeling of analogy-making. Trends Cogn. Sci. **6**, 200–205 (2002)
10. Hall, R.P.: Computational approaches to analogical reasoning: a comparative analysis. Artif. Intell. **39**(1), 39–120 (1989)

11. Birkhoff, G.: Aesthetic Measure. Cambridge University Press (1933)
12. Ramachandran, V.S., Hirstein, W.: The science of art: a neurological theory of aesthetic experience. J. Conscious. Stud. **6**, 15–35 (1999)
13. Goguen, J.A.: Art and the brain: editorial introduction. J. Conscious. Stud. **6**, 5–14 (1999)
14. Huang, M.: The neuroscience of art. Stanf. J. Neurosci. **2**, 24–26 (2009)
15. Hagendoorn, I.: The dancing brain. cerebrum: the dana forum on brain. Science **5**, 19–34 (2003)
16. Palmer, S.E., Schloss, K.B., Sammartino, J.: Visual aesthetics and human preference. Ann. Rev. Psychol. **64**, 77–107 (2013)
17. Todd, P.M., Werner, G.M.: Frankensteinian methods for evolutionary music. Musical networks: parallel distributed perception and performace, 313 (1999)
18. Eigenfeldt, A.: The evolution of evolutionary software: intelligent rhythm generation in Kinetic Engine. In: Applications of Evolutionary Computing, pp. 498–507. Springer (2009)
19. Fox, R., Crawford, R.: A hybrid approach to automated music composition. In: Artificial Intelligence Perspectives in Intelligent Systems, pp. 213–223. Springer (2016)
20. Heidarpour, M., Hoseini, S.M.: Generating art tile patterns using genetic algorithm. In: 2015 4th Iranian Joint Congress on Fuzzy and Intelligent Systems (CFIS), pp. 1–4. IEEE (2015)
21. Garcia-Sanchez, P., et al.: Testing the diferences of using RGB and HSV histograms during evolution in evolutionary Art. In: ECTA. (2013)
22. Bergen, S., Ross, B.J.: Aesthetic 3D model evolution. Genet. Program. Evol. Mach. **14**, 339–367 (2013)
23. Yang, W., Cheng, Y., He, J., Hu, W., Lin, X.: Research on community competition and adaptive genetic algorithm for automatic generation of tang poetry. Math. Probl. Eng. 2016 **2016**, (2016)
24. den Heijer, E., Eiben, A.E.: Comparing aesthetic measures for evolutionary art. In: Applications of Evolutionary Computation, pp. 311–320. Springer (2010)
25. den Heijer, E., Eiben, A.E.: Evolving art using multiple aesthetic measures. In: Applications of Evolutionary Computation, pp. 234–243. Springer (2011)
26. Breen, A., O'Riordan, C.: Evolving art using aesthetic analogies: evolutionary supervised learning to generate art with grammatical evolution. In: 8th International Conference on Evolutionary Computation, Theory and Applications. Porto, Portugal, pp. 59–68.(2016)
27. Malmberg, C.F.: The perception of consonance and dissonance. Psychol. Monogr. **25**, 93–133 (1918)
28. Kameoka, A., Kuriyagawa, M.: Consonance theory part I: consonance of dyads. J. Acoust. Soc. Am. **45**, 1451–1459 (1969)
29. Breen, A., O'Riordan, C.: Capturing and ranking perspectives on the consonance and dissonance of dyads. In: Sound and Music Computing Conference, pp. 125–132. Maynooth (2015)
30. Von Helmholtz, H.: On the Sensations of Tone as a Physiological Basis for the Theory of Music. Longmans, Green (1912)
31. Plomp, R., Levelt, W.J.M.: Tonal consonance and critical bandwidth. J. Acoust. Soc. Am. **38**, 548–560 (1965)
32. Hutchinson, W., Knopoff, L.: The acoustic component of Western consonance. J. New Music Res. **7**, 1–29 (1978)
33. Vassilakis, P.N.: Auditory roughness as means of musical expression. Sel. Rep. Ethnomusicol. **12**, 119–144 (2005)
34. Breen, A., O'Riordan, C.: Capturing data in the presence of noise for artificial intelligence systems. In: Irish Conference on Artificial Intelligence and Cognitive Science, pp. 204–216. Dublin (2016)
35. Chuang, M.C., Ou, L.C.: Influence of a holistic color interval on color harmony. COLOR Res. Appl. **26**, 29–39 (2001)
36. Szabó, F., Bodrogi, P., Schanda, J.: Experimental modeling of colour harmony. Color Res. Appl. **35**, 34–49 (2010)
37. Schloss, K.B., Palmer, S.E.: Aesthetic response to color combinations: preference, harmony, and similarity. Atted. Percept. Psychophys. **73**, 551–571 (2011)
38. Cockos: Reaper (2016)

Particle Convergence Expected Time in the Stochastic Model of PSO

Krzysztof Trojanowski and Tomasz Kulpa

Abstract Convergence properties in the model of PSO with inertia weight are a subject of analysis. Particularly, we are interested in estimating the time necessary for a particle to obtain equilibrium state in deterministic and stochastic models. For the deterministic model, an earlier defined upper bound of particle convergence time (*pctb*) is revised and updated. For the stochastic model, four new measures of the expected particle convergence time are proposed: (1) the convergence of the expected location of the particle, (2) the particle location variance convergence and (3)–(4) their respective weak versions. In the experimental part of the research, graphs of recorded expected running time (ERT) values are compared to graphs of upper bound of *pct* from the deterministic model as well as graphs of recorded convergence times of the particle location *pwcet* from the stochastic model.

1 Introduction

A predominance of one optimization method over another may depend on a difference either in the quality of the results for the given same computational cost or in the computational costs for the requested quality of the result. However, in the case of stochastic optimization algorithms, one can not guarantee that the optimal solution is found even in the finite amount of time. Therefore, an expected computational cost necessary to find a sufficiently good, suboptimal solution can be estimated at most. Additionally, due to typical weaknesses of these algorithms, like, for example, the tendency to get stuck in local optima, the expected cost cannot be estimated in general, but just for specific optimization environments. In spite of this, conclusions

K. Trojanowski · T. Kulpa (✉)
Faculty of Mathematics and Natural Sciences, School of Exact Sciences,
Cardinal Stefan Wyszyński University in Warsaw,
Wóycickiego 1/3, 01-938 Warsaw, Poland
e-mail: tomasz.kulpa@uksw.edu.pl

K. Trojanowski
e-mail: k.trojanowski@uksw.edu.pl

© Springer Nature Switzerland AG 2019
J. J. Merelo et al. (eds.), *Computational Intelligence*,
Studies in Computational Intelligence 792,
https://doi.org/10.1007/978-3-319-99283-9_4

from such an analysis can be a source of improvements in real-world applications and thus convergence properties of these methods remain a subject of undiminished interest.

A stochastic population-based optimization approach, precisely, a particle swarm optimization (PSO) is a subject of the presented research. In [1], authors indicate properties affecting the computational cost of finding suboptimal solutions, like particle stability, patterns of particle movements, or a local convergence of a particle and of a swarm. We are interested in the estimation of an expected runtime of the particle, precisely, the number of perturbations and respective fitness function calls necessary for the particle to obtain its stable state. For the stochastic model of the particle movement, new definitions of particle convergence based on the convergence of its expected location and expected variance of the location are proposed and estimations of the number of steps to obtain the stability state are presented.

The chapter consists of eight sections. Section 2 presents a brief review of selected areas of PSO theoretical analysis concerning (1) stability and region of stable particle parameter configurations and (2) runtime analysis, particularly, estimation of time necessary to hit a satisfying solution. Section 3 reminds definition of the upper bound of particle convergence time (*pctb*) in the deterministic model and describes an updated version of this definition. In Sect. 4, a stochastic model of the particle movement is presented. Section 5 introduces two definitions: particle convergence expected time (*pcet*) and particle weak convergence expected time (*pwcet*). Section 6 focuses on the convergence of particle location variance and introduces next two definitions: the particle location variance convergence time $pvct(\delta)$ and its weak version. Results of experimental evaluations are presented in Sect. 7. Section 8 summarizes the presented research.

2 Related Work

The PSO model with inertia weight implements following velocity and position equations:

$$\begin{cases} \mathbf{v}_{t+1} = w \cdot \mathbf{v}_t + \varphi_{t,1} \otimes (\mathbf{y}_t - \mathbf{x}_t) + \varphi_{t,2} \otimes (\mathbf{y}_t^* - \mathbf{x}_t), \\ \mathbf{x}_{t+1} = \mathbf{x}_t + \mathbf{v}_{t+1} \end{cases} \tag{1}$$

where $\mathbf{v}_t$ is a particle's velocity, $\mathbf{x}_t$ — particle's location, $\mathbf{y}_t$ — the best location the particle has found so far, $\mathbf{y}_t^*$ — the best location found by particles in its neighborhood, w – inertia coefficient, $\varphi_{t,1}$ and $\varphi_{t,2}$ control influence of the attractors on the velocity, $\varphi_{t,1} = R_{t,1}c_1$, $\varphi_{t,2} = R_{t,2}c_2$, and c_1,c_2 represent acceleration coefficients, $R_{t,1}$, $R_{t,2}$ are two vectors of random values uniformly generated in range [0, 1] and $\otimes$ denotes pointwise vector product. Values of coefficients w, c_1 and c_2 define convergence properties of the particle.

2.1 Stability and Stable Regions

In the literature, a theoretical analysis of PSO is conducted with the following four assumptions [2]: (1) deterministic assumption, where $\varphi_1 = \varphi_{t,1}$ and $\varphi_2 = \varphi_{t,2}$, for all t, (2) stagnation assumption, where $\mathbf{y}_t = \mathbf{y}$ and $\mathbf{y}_t^* = \mathbf{y}^*$, for all t sufficiently large, (3) weak chaotic assumption, where both $\mathbf{y}_t$ and $\mathbf{y}_t^*$ will occupy an arbitrarily large but finite number of unique positions, and (4) weak stagnation assumption, where the global attractor of the particle has obtained the best objective function evaluation and remains constant for all t sufficiently large.

Under the deterministic assumption the following region of particle convergence was derived [3, 4]:

$$\begin{cases} 0 < \varphi_1 + \varphi_2 < 2(1+w), \\ 0 < w < 1, \quad \varphi_1 > 0 \wedge \varphi_2 > 0 \end{cases} \tag{2}$$

and the stability is defined as $\lim_{t\to\infty} \mathbf{x}_t = \mathbf{y}$.

To deal with randomness of $\varphi_{t,1}$ and $\varphi_{t,2}$ they are replaced with their expectations $c_1/2$ and $c_2/2$, respectively. In this case stability is defined as $\lim_{t\to\infty} E|\mathbf{x}_t| = \mathbf{y}$ [5] and is called the order-1 stability. The region defined with in Ineq. (2) satisfies this stability, thus, it is also called the order-1 stable region. In later publications (e.g. [6–8]) the region is extended to $|w| < 1$ and $0 < \varphi_1 + \varphi_2 < 2(1+w)$.

Unfortunately, the order-1 stability is not enough to ensure convergence, simply the particle may oscillate or even diverge while the expectation converges to a point. Therefore, the convergence of the variance (or standard deviation) must occur as well, which is called the order-2 stability condition [5, 9]. In [9] the stability is defined as $\lim_{t\to\infty} E[\mathbf{x}_t - \mathbf{y}]^2 = 0$ where $\mathbf{y} = \lim_{t\to\infty} E[\mathbf{x}_t]$. In [5] the stability is defined as $\lim_{t\to\infty} E[\mathbf{x}_t^2] = \beta_0$ and $\lim_{t\to\infty} E[\mathbf{x}_t \mathbf{x}_{t-1}] = \beta_1$ where β_0 and β_1 are constant. Eventually, both authors obtained the same set of inequalities which define the so called order-2 stable region:

$$\varphi < \frac{12(1-w^2)}{7-5w} \text{ where } \varphi_1 = \varphi_2 = \varphi. \tag{3}$$

2.2 Runtime Analysis

For applications of PSO in the real-world problems, it is important to estimate how much time a swarm or a particle needs to reach the close vicinity of the optimum. Analysis of this problem appeared in [10] and in [11] authors introduced a formal definition of the first hitting time (FHT) and expected FHT (EFHT). Both concepts refer to an entire swarm, precisely, FHT represents the number the evaluation function f_{eval} calls until the swarm contains for the very first time a particle $\mathbf{x}$ for which $|f_{eval}(\mathbf{x}) - f_{eval}(\mathbf{y}^*)| < \delta$.

Another approach can be found in [12, 13], where distances between subsequent locations of particles are a subject of observation. Authors propose a concept of particle convergence time (pct) as a measure of the speed at which the equilibrium state is reached. In this case, the "equilibrium state" is the state when the distance between current and the next location of the particle is never greater than the given threshold value δ.

Definition 1 (*The Particle Convergence Time*) Let δ be a given positive number and $S(\delta)$ be a set of natural numbers such that:

$$s \in S(\delta) \iff ||\mathbf{x}_{t+1} - \mathbf{x}_t|| < \delta \text{ for all } t \geq s. \tag{4}$$

The particle convergence time ($pct(\delta)$) is the minimal number in the set $S(\delta)$, that is

$$pct(\delta) = \min\{s \in S(\delta)\}. \tag{5}$$

The pct values were studied for the deterministic model of a particle. It was also assumed, that the global attractor remains unchanged (the stagnation assumption), that is, the value of global attractor is never worse than the value of any location visited during the convergence process. As far as this condition is satisfied, the shape of evaluation function f_{eval} is negligible. Under the deterministic and stagnation assumptions, and also the best particle stagnation assumption (that is, $\mathbf{y}_t = \mathbf{y}_t^* = \mathbf{y}$), the explicit version of an upper bound formula of pct called $pctb(\delta)$ is given [12, 13].

3   The Updated Upper Bound of pct in the Deterministic Model

The explicit formula for $pctb(\delta)$ is proposed as follows:

$$pctb(\delta) = \max \left(\frac{\ln \delta - \ln(2|k_2||\lambda_1 - 1|)}{\ln |\lambda_1|} , \frac{\ln \delta - \ln(2|k_3||\lambda_2 - 1|)}{\ln |\lambda_2|} \right) \tag{6}$$

for real value of γ and

$$pctb(\delta) = \frac{\ln \delta - \ln(|\lambda - 1|(|k_2| + |k_3|))}{\ln |\lambda|} \tag{7}$$

for imaginary value of γ, where

$$k_1 = y, \tag{8}$$

$$k_2 = \frac{\lambda_2(x_0 - x_1) - x_1 + x_2}{\gamma(\lambda_1 - 1)}, \tag{9}$$

$$k_3 = \frac{\lambda_1(x_1 - x_0) + x_1 - x_2}{\gamma(\lambda_2 - 1)}, \tag{10}$$

$$x_2 = (1 + w - \phi)x_1 - wx_0 + \phi y, \tag{11}$$

$$\lambda_1 = \frac{1 + w - \phi + \gamma}{2}, \tag{12}$$

$$\lambda_2 = \frac{1 + w - \phi - \gamma}{2}, \tag{13}$$

$$\gamma = \sqrt{(1 + w - \phi)^2 - 4w}. \tag{14}$$

For any fixed values of $[x_0, x_1, w, \phi]$ the function $pctb(\delta)$ is monotonically decreasing. Particularly, for big values of δ one can observe negative values of $pctb$. Example figures depicting this situation are given in Fig. 1.

This means that it is necessary to modify the proposed measure of $pctb$. In [12, 13] it was shown that if the following inequalities are satisfied

$$\begin{cases} t > \dfrac{\ln \delta - \ln(2|k_2||\lambda_1 - 1|)}{\ln |\lambda_1|}, \\ t > \dfrac{\ln \delta - \ln(2|k_3||\lambda_2 - 1|)}{\ln |\lambda_2|}, \end{cases} \tag{15}$$

than

$$|x_{t+1} - x_t| < \delta. \tag{16}$$

It turns out that expressions on the right-hand side of Ineq. (15) may have negative values. Obviously, in this case, the condition (16) is fulfilled for all t which are

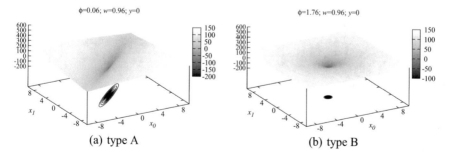

(a) type A   (b) type B

Fig. 1 Graphs of $pctb(x_0, x_1)$ for selected configurations $[\phi, w]$ and $\delta = 1$. Isolines are printed just for negative values of $pctb(x_0, x_1)$

positive integer numbers. Therefore, when both right-hand sides of Ineq. (15) have negative values, the value of *pct* equals 1, and thus, it seems reasonable to set also the respective value of *pctb* to 1. Hence, a modification in the formula of *pctb* which would represent this property is needed.

Estimation of *pctb* should be extended as follows

$$pctub(\delta) = \max \left(\frac{\ln \delta - \ln(2|k_2||\lambda_1 - 1|)}{\ln |\lambda_1|} , \frac{\ln \delta - \ln(2|k_3||\lambda_2 - 1|)}{\ln |\lambda_2|}, 1 \right) \quad (17)$$

for real value of γ and

$$pctub(\delta) = \max \left(\frac{\ln \delta - \ln(|\lambda - 1|(|k_2| + |k_3|))}{\ln |\lambda|}, 1 \right) \quad (18)$$

for imaginary value of γ.

Example graphs of the updated upper bound of *pct*, namely *pctub*, are given in Figs. 2 and 3. A graph of *pctub* given in Fig. 3 was generated just for example values of x_0 and x_1. However, it is necessary to note here, that graphs for any other values of x_0 and x_1 have almost identical shape. In this figure, values of *pctub* for the configurations (ϕ, w) from outside the convergence region has constant values of 3000.

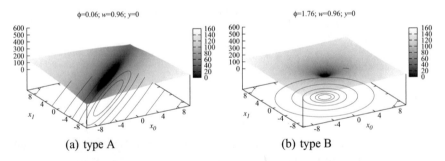

(a) type A   (b) type B

Fig. 2 Graphs of $pctub(x_0, x_1)$ for two example configurations $[\phi, w]$ and $\delta = 1$

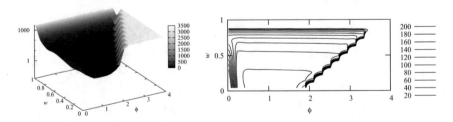

Fig. 3 Graphs of $pctub(\phi, w)$ for example configuration $x_0 = 1$, $x_1 = -8.1$, $\delta = 0.0001$; 3D shape with logarithmic scale for $pctub(\phi, w)$ (left graph), and isolines from 0 to 200 with step 20 (right graph)

4 The Stochastic Model

Under the best particle stagnation assumption the update equation of the particle location in one-dimensional search space can be reformulated as follows:

$$x_{t+1} = (1 + w - \phi_t)x_t - wx_{t-1} + \phi_t y, \tag{19}$$

where w is a constant parameter of inertia and ϕ_t is the sum of two independent random variates, $\phi_t = \varphi_{t,1} + \varphi_{t,2}$, $\varphi_{t,i} \sim U(0, c_i)$, $i = 1, 2$. It is also assumed that $\phi_t, t = 1, 2, 3 \dots$ are independent and identically distributed.

Thus, in the further evaluations $E[\phi_t]$ and $E[\phi_t^2]$ equal

$$E[\phi_t] = E[\phi_{t,1}] + E[\phi_{t,2}] = \frac{c_1 + c_2}{2},$$
$$E[\phi_t^2] = \text{Var}[\phi_t] + (E[\phi])^2 =$$
$$\frac{c_1^2}{12} + \frac{c_2^2}{12} + \left(\frac{c_1 + c_2}{2}\right)^2.$$

Set $e_t = E[x_t]$, $m_t = E[x_t^2]$, $h_t = E[x_t x_{t-1}]$, $f = E[\phi_t]$ and $g = E[\phi_t^2]$.

The proposed model is a simplified version of the model presented in [5, 14, 15], particularly, we apply the same analysis of dynamics of first and second moments of the PSO sampling distribution.

We apply the expectation operator to both sides of Eq. (19). Because of the statistical independence between ϕ_t and x_t we obtain

$$e_{t+1} = (1 + w - f)e_t - we_{t-1} + fy. \tag{20}$$

Equation (20) gives us the same model as the model described by Eq. (19), however, instead of the acceleration coefficient ϕ_t and the particle location x_t we have its expected value f and the particle expected location e_t, respectively. We can say that the update of the expected position of a particle follows in the same way as the particle trajectory in the deterministic model described by Eq. (19).

We raise both sides of Eq. (19) to the second power and obtain

$$x_{t+1}^2 = (1 + w - \phi_t)^2 x_t^2 + w^2 x_{t-1}^2 + \phi_t^2 y^2$$
$$- 2(1 + w - \phi_t)wx_t x_{t-1} - 2wy\phi_t x_{t-1} \tag{21}$$
$$+ 2y\phi_t(1 + w - \phi_t)x_t.$$

Applying the expectation operator to both sides of Eq. (21) and again because of the statistical independence between ϕ_t, x_t and x_{t-1} we obtain

$$m_{t+1} = m_t((1+w)^2 - 2(1+w)f + g)$$
$$+ m_{t-1}w^2 - h_t 2w(1+w-f)$$
$$+ e_t 2y(f(1+w) - g) - e_{t-1}2wyf + y^2 g. \tag{22}$$

Multiplying both sides of Eq. (19) by x_t we get

$$x_{t+1}x_t = (1 + w - \phi_t)x_t^2 - wx_t x_{t-1} + \phi_t yx_t. \tag{23}$$

Again, we apply the expectation operator to (23) and obtain

$$h_{t+1} = (1 + w - f)m_t - wh_t + fye_t. \tag{24}$$

Now, a vector $\mathbf{z}_t = (e_t, e_{t-1}, m_t, m_{t-1}, h_t)^T$ can be introduced. Equations (20), (22), and (24) can be rewritten as a matrix equation

$$\mathbf{z}_{t+1} = M_t \mathbf{z}_t + \mathbf{b}, \tag{25}$$

where

$$M_t = \begin{bmatrix} m_{1,1} & -w & 0 & 0 & 0 \\ 1 & 0 & 0 & 0 & 0 \\ m_{3,1} & m_{3,2} & m_{3,3} & w^2 & m_{3,5} \\ 0 & 0 & 1 & 0 & 0 \\ fy & 0 & m_{5,3} & 0 & -w \end{bmatrix} \tag{26}$$

where the matrix components are

$m_{1,1} = 1 + w - f,$
$m_{3,1} = 2y(f(1+w) - g),$
$m_{3,2} = -2wyf,$
$m_{3,3} = (1+w)^2 - 2(1+w)f + g,$
$m_{3,5} = -2w(1+w-f),$
$m_{5,3} = 1 + w - f.$

and

$$\mathbf{b} = (fy, 0, y^2 g, 0, 0)^T. \tag{27}$$

The particle is order-2 stable if e_t, m_t, and h_t converge to stable fixed points. This happens when all absolute values of eigenvalues of M are less than 1.

In that case, there exist a fixed point of the system described by equation

$$\mathbf{z}^* = (I - M)^{-1}\mathbf{b}. \tag{28}$$

When the system is order-2 stable, by the change of variables $\mathbf{u}_t = \mathbf{z}_t - \mathbf{z}^*$, we can rewrite Eq. (25)

$$\mathbf{u}_{t+1} = M\mathbf{u}_t, \tag{29}$$

which can be integrated to obtain the explicit formula

$$\mathbf{u}_t = M^t \mathbf{u}_0. \tag{30}$$

The order-2 analysis of the system described by Eq. (30) is not easy because of complicated formulas for eigenvalues of M. However, the order-1 analysis can be done, because two of them are known as

$$\lambda_1 = \frac{1 + w - f + \gamma}{2},$$
$$\lambda_2 = \frac{1 + w - f - \gamma}{2}, \tag{31}$$

where

$$\gamma = \sqrt{(1 + w - f)^2 - 4w}. \tag{32}$$

For fixed initial values of e_0 and e_1, the explicit formula for e_t, first time obtained by [4], is given by equation

$$e_t = k_1 + k_2 \lambda_1^t + k_3 \lambda_2^t, \tag{33}$$

where

$$k_1 = y,$$
$$k_2 = \frac{\lambda_2(e_0 - e_1) - e_1 + e_2}{\gamma(\lambda_1 - 1)},$$
$$k_3 = \frac{\lambda_1(e_1 - e_0) + e_1 - e_2}{\gamma(\lambda_2 - 1)},$$
$$e_2 = (1 + w - f)e_1 - we_0 + fy. \tag{34}$$

5 Particle Convergence Expected Time

Due to the analogy between the deterministic model based on the update equation of the particle location (19) and the studied order-1 stochastic model of PSO described by Eq. (20), we can define a measure of particle convergence expected time (*pcet*) respectively to the idea given in Definition 1

Definition 2 (*The Particle Convergence Expected Time*) Let δ be a given positive number and $S(\delta)$ be a set of natural numbers such that:

$$s \in S(\delta) \iff |e_{t+1} - e_t| < \delta \text{ for all } t \geq s. \tag{35}$$

The particle convergence expected time ($pcet(\delta)$) is the minimal number in the set $S(\delta)$, that is

$$pcet(\delta) = \min\{s \in S(\delta)\}. \tag{36}$$

Briefly, the particle convergence expected time *pcet* is the minimal number of steps necessary for the expected particle location to obtain its stable state as defined above.

The explicit formula for solutions of the recurrence Eq. (19) is given in [4]. This formula is used in [12] to find an upper bound formula of *pct*. Because of the analogy between the models described by Eqs. (19) and (20) we obtain the following upper bound for *pcet*, namely *pcetub*

$$pcetub(\delta) = \max \left(\frac{\ln \delta - \ln(2|k_2||\lambda_1 - 1|)}{\ln |\lambda_1|} , \frac{\ln \delta - \ln(2|k_3||\lambda_2 - 1|)}{\ln |\lambda_2|}, 1 \right) \quad (37)$$

for real value of γ given by (32) and

$$pcetub(\delta) = \max \left(\frac{\ln \delta - \ln(|\lambda_1 - 1|(|k_2| + |k_3|))}{\ln |\lambda_1|}, 1 \right) \quad (38)$$

for imaginary value of γ, where λ_1 and λ_2 are given by Eq. (31) and k_1, k_2 and k_3 are given by Eq. (34).

Obviously, characteristics of $pcetub(\delta)$ depicted in Fig. 4 (generated for $\delta = 0.0001$) looks the same as the characteristics of *pctub* (see [12] for comparisons)

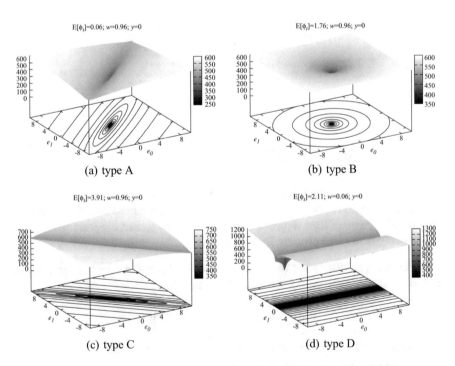

Fig. 4 Graphs of $pcetub(e_0, e_1)$ for selected configurations ($E[\phi_t]$, w) and $\delta = 0.0001$

and have the same distinctive shape of a funnel. Thus, as in the case of *pctub*, they can also be classified into four main types.

Empirical evaluation of *pcet* is difficult, so, we introduce the less restrictive measure, that is, a particle weak convergence time.

Definition 3 (*The Particle Weak Convergence Expected Time*) Let δ be a given positive number. The particle weak convergence expected time $pwcet(\delta)$ is the minimal number of steps necessary to get the expected value of difference between subsequent particle locations lower than δ, that is

$$pwcet(\delta) = min\{t : |e_t - e_{t+1}| < \delta\}. \tag{39}$$

It is obvious that $pwcet(\delta) \leq pcet(\delta)$ and equality generally does not hold. Empirical characteristics of *pwcet* are depicted in Figs. 5 and 6. The characteristics were obtained with Algorithm 1.

Algorithm 1. Particle weak convergence expected time evaluation procedure.

1: Initialize: $T_{max} = $ 1e+5, two successive expected locations e_0 and e_1, and an attractor of a particle, for example, $y = 0$.
2: $s_1 = e_1 - e_0$
3: $f = (c_1 + c_2)/2$
4: $t = 1$
5: **repeat**
6:     $e_{t+1} = (1 + w - f)e_t - we_{t-1} + fy$
7:     $s_{t+1} = e_{t+1} - e_t$
8:     $t = t + 1$
9: **until** $(s_t > \delta) \wedge (s_t < $ 1e+10$) \wedge (t < T_{max})$
10: **if** $s_t < $ 1e+10 **then**
11:     **return** t
12: **else**
13:     **return** T_{max}
14: **end if**

Figure 5 depicts the values of *pwcet* generated for $\delta = 0.0001$ as a function of initial location and velocity represented by expected locations e_0 and e_1 where $E[\phi_t]$ and w are fixed. A grid of pairs $[e_0, e_1]$ consists of 40000 points (200 × 200) varying from -10 to 10 for both e_0 and e_1.

Figure 6 shows the values of *pwcet* also for $\delta = 0.0001$ obtained for a grid of configurations $(E[\phi_t], w)$ starting from $[E[\phi_t] = 0.0, w = -1.0]$ and changing with step 0.02 for w and step 0.04 for $E[\phi_t]$ (which gave 200 × 100 points). The configurations generating *pwcet* > 100000 have assigned a constant value of 100000. It is also assumed that $c_1 = c_2 = \phi_{max}/2$.

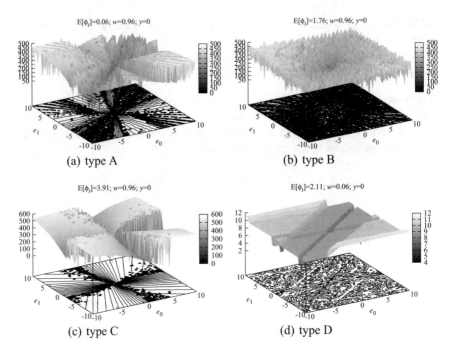

$E[\phi_t]$=0.06; w=0.96; y=0

(a) type A

$E[\phi_t]$=1.76; w=0.96; y=0

(b) type B

$E[\phi_t]$=3.91; w=0.96; y=0

(c) type C

$E[\phi_t]$=2.11; w=0.06; y=0

(d) type D

Fig. 5 Graphs of recorded values of $pwcet(e_0, e_1)$ for selected configurations $(E[\phi_t], w)$ and $\delta = 0.0001$ extracted from [16]

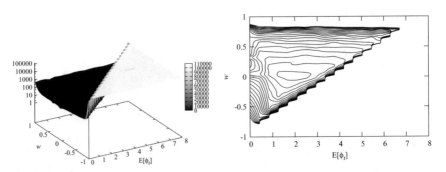

Fig. 6 Recorded convergence times of the particle location $pwcet(E[\phi_t], w)$ for example starting conditions: $e_0 = -9$ and $e_1 = -5$; 3D shape with logarithmic scale for $pwcet(E[\phi_t], w)$ (left graph), and isolines from 0 to 100 with step 5 (right graph) extracted from [16]

6 Convergence of Variance of Particle Location Distribution

Convergence of the expected value of the particle location still does not guarantee the convergence of the particle position. This is the case, for example, where the particle oscillates symmetrically and the oscillations do not fade. In [5] author studied the convergence of the variance and standard deviation of the particle location in the model with the best particle stagnation assumption described by Eq. (19). Obtained region of configurations (Ineq. (3)) guarantees the order-2 stability of the system, that is, the variance of the particle location converges to zero.

It is interesting to show how fast the variance of a particle location fades. Formally, we are interested in the evaluation of the particle location variance convergence time. Below, d_t denotes variance of particle location in time t, that is

$$d_t = Var[x_t] = m_t - e_t^2. \tag{40}$$

Definition 4 (*The Particle Location Variance Convergence Time*) Let δ be a given positive number. The particle location variance convergence time $pvct(\delta)$ is the minimal number of steps necessary to get the variance of particle location lower than δ for all subsequent time steps, that is

$$pvct(\delta) = min\{t \,:\, d_s < \delta \; for \; all \; s \geq t\}. \tag{41}$$

Empirical evaluation of $pvct$ is difficult, so, we introduce the less restrictive measure, that is, a particle location variance weak convergence time.

Definition 5 (*The Particle Location Variance Weak Convergence Time*) Let δ be a given positive number. The particle location variance weak convergence time $pvwct(\delta)$ is the minimal number of steps necessary to get the variance of particle location lower than δ, that is

$$pvwct(\delta) = min\{t \,:\, d_t < \delta\}. \tag{42}$$

As in the case of $pwcet(\delta)$ and $pwct(\delta)$ it is also obvious that $pvwct(\delta) \leq pvct(\delta)$ and equality generally does not hold.

When $pvwct(\delta)$ has to be calculated according to Definition 5, it is important to select appropriately initial values of the algorithm parameters: h_1 and m_1. To do this, lets first note that Eq. (1) can be converted to the form:

$$\begin{cases} v_{t+1} = w \cdot v_t + \varphi_t(y - x_t), \\ x_{t+1} = x_t + v_{t+1}. \end{cases} \tag{43}$$

When we substitute zero for t in Eq. (43) we obtain Eq. (44):

$$x_1 = x_0 + w \cdot v_0 + \varphi_0(y - x_0). \tag{44}$$

Let us assume, that x_0 and v_0 are independent random variables. Applying the expectation operator to both sides of Eq. (44) we get

$$e_1 = e_0(1 - f) + w \cdot s_0 + fy, \tag{45}$$

where $s_0 = Ev_0$. From Eq. (45) we obtain

$$s_0 = \frac{e_1 - e_0(1 - f) - fy}{w}. \tag{46}$$

Multiplying both sides of Eq. (44) by x_0 we get

$$x_1 x_0 = x_0^2(1 - \varphi_0) + w x_0 v_0 + x_0 \varphi_0 y. \tag{47}$$

Applying expectation operator to both sides of the Eq. (47) we obtain

$$h_1 = m_0(1 - f) + w e_0 s_0 + e_0 fy, \tag{48}$$

and substituting expression from Eq. (46) for s_0

$$h_1 = m_0(1 - f) + e_0(e_1 - e_0(1 - f) - fy) + e_0 fy. \tag{49}$$

Eventually, above formula can be simplified to the form

$$h_1 = (m_0 - e_0^2)(1 - f) + e_0 e_1, \tag{50}$$

or equivalent

$$h_1 = d_0(1 - f) + e_0 e_1. \tag{51}$$

Next, we raise both sides of Eq. (44) to the second power and obtain

$$\begin{aligned} x_1^2 = {} & x_0^2(1 - \varphi_0)^2 + w^2 v_0^2 + \varphi_0^2 y^2 \\ & + 2w x_0(1 - \varphi_0)v_0 + 2x_0(1 - \varphi_0)v_0 \\ & + 2x_0(1 - \varphi_0)\varphi_0 y + 2w v_0 \varphi_0 y. \end{aligned} \tag{52}$$

Applying the expectation operator to both sides of Eq. (52) and because of the statistical independence of x_0, φ_0 and v_0 we get

$$m_1 = m_0(1 - 2f + g) + w^2 s_2 + gy^2 + 2w e_0(1 - f)s_0 + 2e_0(f - g)y + 2w s_0 fy, \tag{53}$$

where $s_2 = Ev_0^2$. Expression from Eq. (46) can be substituted for s_0 in Eq. (53). This way we obtain

$$m_1 = m_0(1 - 2f + g) + w^2 s_2 + gy^2 + 2(e_1 - e_0(1 - f) - fy)(e_0(1 - f) + fy). \tag{54}$$

Let $d_0 = Var[x_0]$ and $l_o = Var[v_0]$ are given. Then we can calculate

$$m_0 = e_0^2 + d_0$$

and

$$s_2 = s_0^2 + l_0,$$

what can be written in view of Eq. (46) as

$$s_2 = \frac{(e_1 - e_0(1 - f) - fy)^2}{w^2} + l_0. \tag{55}$$

Expression from Eq. (55) can be substituted for s_2 in Eq. (54). This way one can obtain the final version of equation for m_1:

$$m_1 = m_0(1 - 2f + g) + w^2 l_0 + gy^2 + e_1^2 - (e_0(1 - f) + fy)^2. \tag{56}$$

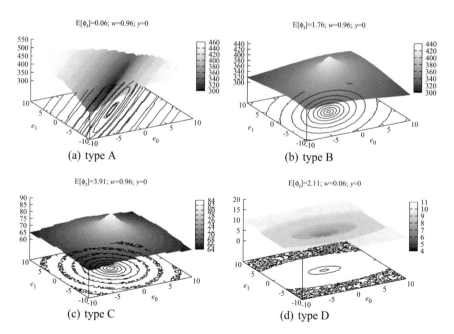

Fig. 7 Graphs of recorded values of the particle location variance $pvwct(E[\phi_t], w)$ for selected configurations $(E[\phi_t], w)$ and $\delta = 0.0001$ extracted from [16]

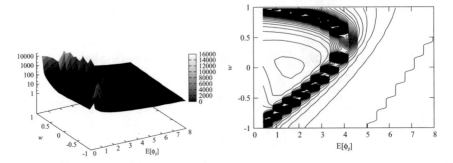

Fig. 8 Recorded convergence times of the particle location variance $pvwct(E[\phi_t], w)$ for example starting conditions: $e_0 = -9$ and $e_1 = -5$; 3D shape with logarithmic scale for $pvwct(E[\phi_t], w)$ (left graph), and isolines from 0 to 20000 with step 10 (right graph) extracted from [16]

Empirical characteristics of the particle location variance weak convergence time ($pvwct$) are given in Figs. 7 and 8.

Figure 7 presents the values of $pvwct$ as a function of e_0 and e_1 where $E[\phi_t]$ and w are fixed. The grid of pairs $[e_0, e_1]$ consists of 40000 points (200×200) varying from -10 to 10 for both e_0 and e_1.

As in the case of empirical characteristics of $pwcet$, Fig. 8 shows the values of $pvwct$ also obtained for a grid of configurations ($E[\phi_t]$, w) starting from [$E[\phi_t] = 0.0$, $w = -1.0$] and changing with step 0.02 but in both directions (which also gave 200×100 points). It is assumed that $c_1 = c_2 = \phi_{max}/2$ and a constant value of 100000 is assigned to the configurations generating $pvwct > 100000$.

The characteristics depicted in Figs. 7 and 8 were obtained with Algorithm 2 for selected values of variance of initial location $d_0 = 0$ and velocity $l_0 = 1$.

7 Experimental Evaluations

In search of dependence between obtained graphs and efficiency of respective configurations of PSO with inertia weight, a set of experimental tests was performed. As an empirical equivalent of particle convergence time, we selected a measure of expected running time (ERT) [17] which refers to a swarm and not a single particle. ERT represents the expected number of times the evaluation function f_{eval} is called until the swarm contains for the very first time a particle $\mathbf{x}$ for which $|f_{eval}(\mathbf{x}) - f_{eval}(\mathbf{x}_{opt})| < \delta$.

In the experimental part a tested swarm always consists of 32 particles with *global best* communication topology. To ensure the presence of randomness in the particle movement on the one hand but on the other hand also that $\varphi_{t,1} + \varphi_{t,2} = \phi_{max}$, the attractor coefficients have been evaluated in every step t as follows: $\varphi_{t,1} = R_t \phi_{max}$ and $\varphi_{t,2} = \mathbf{1}\phi_{max} - \varphi_{t,1}$.

The algorithm optimized the Ackley function in the search space with the number of dimensions varying from 1 to 5. ERT was evaluated for a grid of particle configurations (ϕ_{max}, w) starting from $[\phi_{max} = 0.0, w = 0.0]$ and changing with step 0.01 for ϕ_{max} and w to the final configuration $[\phi_{max} = 4, w = 1]$. For every configuration of a particle, a series of 100 experiments was carried out where values of initial particle coordinates were selected randomly form the range $[-32.768, 32.768]$ every time. For each series a mean value of ERT is calculated. The stopping condition δ was set to 0.0001.

Algorithm 2. Particle location variance weak convergence time evaluation procedure.

1: Initialize: $T_{max} = 1e+5$, two successive expected locations e_0 and e_1, variance of initial location and velocity, for example, $d_0 = 0$ and $l_0 = 1$ respectively, and an attractor of a particle, for example, $y = 0$.
2: $f = (c_1 + c_2)/2$;
3: $g = (c_1)^2/12 + (c_2)^2/12 + ((c_1 + c_2)/2)^2$;
4: $m_0 = e_0^2 + d_0$.
5: $m_1 = m_0(1 - 2f + g) + w^2 l_0 + g y^2 + e_1^2 - (e_0(1 - f) + fy)^2$.
6: $h_1 = d_0(1 - f) + e_0 e_1$.
7: $d_1 = m_1 - e_1^2$.
8: $t = 1$
9: **repeat**
10:     $h_{t+1} = (1 + w - f)m_t - wh_t + fye_t$
11:     $e_{t+1} = (1 + w - f)e_t - we_{t-1} + fy$
12:     $m_{t+1} = m_t((1 + w)^2 - 2(1 + w)f + g) + m_{t-1}w^2 - 2h_t w(1 + w - f) + 2e_t y(f(1 + w) - g) - 2e_{t-1}wyf + y^2 g$
13:     $d_{t+1} = m_{t+1} - e_{t+1}^2$
14:     $t = t + 1$
15: **until** $(d_t > \delta) \wedge (d_t < 1e+10) \wedge (t < T_{max})$
16: **if** $d_t < 1e+10$ **then**
17:     **return** t
18: **else**
19:     **return** T_{max}
20: **end if**

Graphs of recorded mean values of ERT are depicted in Fig. 9. One can see similarities between recorded convergence times of the particle location $pwcet$ depicted in Fig. 6 and ERT in Fig. 9 as well as the upper bound of pct, that is, $pctub$ in Fig. 3. These similarities are getting smaller as the number of dimensions grows. The higher the number of dimensions, the larger a triangular region representing configurations of inefficient swarms with high values of ERT located in the bottom left corner of the domain. This adverse region pushes the region of low values of ERT toward the right top part of the domain. One can say, that when the number of dimensions grows, heavier particles and higher attraction forces are needed to cope with growing complexity of a search space.

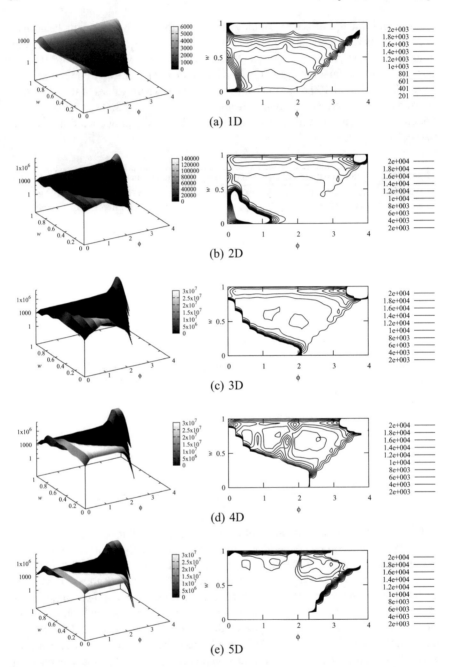

Fig. 9 Graphs of recorded values of the expected running time (ERT) with precision $\delta = 0.0001$ for the domain of configurations (ϕ, w)

8 Conclusions

In PSO, stability properties of a particle contribute towards improving its effectiveness of searching. In this paper, we study the particle stability in two models of PSO: deterministic and stochastic. In the analysis, just a single particle is considered since it is assumed in both cases that particle attractors remain unchanged over time, so no communication between particles occurs. In the case of the deterministic model, for the earlier defined measure of pct we proposed an updated formula of the upper bound of pct, that is, $pctub$. For the stochastic model, new measures of stability based on the order-1 and order-2 convergence analysis are proposed. In the stochastic model, the particle is regarded as stable when its expected location and its expected location variance are convergent. Thus, the first two proposed measures are: (1) particle convergence expected time $pcet(\delta)$ which represents the smallest number of steps necessary for the expected particle location to converge, and (2) the particle location variance convergence time $pvct(\delta)$ which represents the smallest number of steps necessary to get the variance of particle location lower than δ for all subsequent steps.

The proposed two measures are hard to apply in simulations, therefore weak versions of $pcet(\delta)$ and $pvct(\delta)$, that is, $pwcet(\delta)$ and $pvwct(\delta)$ are also proposed and empirical characteristics of $pwcet(\delta)$ and $pvwct(\delta)$ are presented.

Additionally, in a series of experiments with a homogeneous swarm consisting of 32 particles, an expected running time (ERT) was evaluated and compared with respective graphs for $pwcet$ and $pctub$. A similarity can be observed for graphs of ERT obtained for one-dimensional search space and the remaining two measures.

Acknowledgements Authors would like to thank Krzysztof Jura from Cardinal Stefan Wyszyński University in Warsaw, Poland, for his assistance in computer simulations.

References

1. Bonyadi, M.R., Michalewicz, Z.: Particle swarm optimization for single objective continuous space problems: a review. Evol. Comput. **25**, 1–54 (2017)
2. Cleghorn, C.W., Engelbrecht, A.P.: Particle swarm variants: standardized convergence analysis. Swarm Intell. **9**, 177–203 (2015)
3. Trelea, I.C.: The particle swarm optimization algorithm: convergence analysis and parameter selection. Inf. Process. Lett. **85**, 317–325 (2003)
4. van den Bergh, F., Engelbrecht, A.P.: A study of particle swarm optimization particle trajectories. Inf. Sci. **176**, 937–971 (2006)
5. Poli, R.: Mean and variance of the sampling distribution of particle swarm optimizers during stagnation. IEEE Trans. Evol. Comput. **13**, 712–721 (2009)
6. Cleghorn, C.W., Engelbrecht, A.P.: A generalized theoretical deterministic particle swarm model. Swarm Intell. **8**, 35–59 (2014)
7. Bonyadi, M.R., Michalewicz, Z.: Analysis of stability, local convergence, and transformation sensitivity of a variant of particle swarm optimization algorithm. IEEE Trans. Evol. Comput. **20**, 370–385 (2016)

8. Liu, Q.: Order-2 stability analysis of particle swarm optimization. Evol. Comput. **23**, 187–216 (2015)
9. Jiang, M., Luo, Y.P., Yang, S.Y.: Stochastic convergence analysis and parameter selection of the standard particle swarm optimization algorithm. Inf. Process. Lett. **102**, 8–16 (2007)
10. Witt, C.: Why standard particle swarm optimisers elude a theoretical runtime analysis. In: Proceedings of the 10th ACM SIGEVO Workshop on Foundations of Genetic Algorithms, FOGA '09, pp. 13–20. ACM (2009)
11. Lehre, P.K., Witt, C.: Finite First Hitting Time Versus Stochastic Convergence in Particle Swarm Optimisation, pp. 1–20. Springer, New York (2013)
12. Trojanowski, K., Kulpa, T.: Particle convergence time in the PSO model with inertia weight. In: Proceedings of the 7th International Joint Conference on Computational Intelligence (IJCCI 2015) - Volume 1: ECTA, SCITEPRESS, pp. 122–130 (2015)
13. Trojanowski, K., Kulpa, T.: Particle convergence time in the deterministic model of PSO. In: Computational Intelligence. Studies in Computational Intelligence, vol. 669, pp. 175–194. Springer, Berlin (2017)
14. Poli, R., Broomhead, D.: Exact analysis of the sampling distribution for the canonical particle swarm optimiser and its convergence during stagnation. In: GECCO'07: Proceedings of the 9th Annual Conference on Genetic and Evolutionary Computation, vol. 1, pp. 134–141. ACM Press (2007)
15. Poli, R.: Dynamics and stability of the sampling distribution of particle swarm optimisers via moment analysis. J. Artif. Evol. Appl. **2008** (2008)
16. Trojanowski, K., Kulpa, T.: Particle convergence expected time in the PSO model with inertia weight. In: Proceedings of the 8th International Joint Conference on Computational Intelligence (IJCCI 2016) - Volume 1: ECTA, SCITEPRESS, pp. 69–77 (2016)
17. Price, K.V.: Differential evolution vs. the functions of the 2nd ICEO. In: IEEE International Conference on Evolutionary Computation, 1997, pp. 153–157. IEEE Publishing (1997)

Evolution of Mobile Strategies in Social Dilemma Games: An Analysis of Cooperative Cluster Formation

Maud D. Gibbons, Colm O'Riordan and Josephine Griffith

Abstract This paper analyses the formation of cooperative clusters toward the emergence of cooperative clusters in evolutionary spatial game theory. In the model considered, agents inhabit a toroidal lattice grid, in which they participate in a social dilemma games, and have the ability to move in response to environmental stimuli. In particular, using the classical 2-player prisoner's dilemma and a generalised N-player prisoner's dilemma, we compare and contrast the evolved movement strategies, and the cooperative clusters formed therein. Additionally, we explore the effect of varying agent density on the evolution of cooperation, cluster formation, and the movement strategies that are evolved for both cooperative and non-cooperative strategies.

1 Introduction

Questions relating to cooperation and its emergence have been studied in a range of domains including economics, psychology, theoretical biology, and computer science. Researchers have explored the conditions necessary for cooperation to emerge among groups or societies of self-interested agents. Social dilemma games, such as the Prisoner's Dilemma [1], have been adopted as a succinct representation of the conflict between individually selfish behaviours and collectively rational behaviours. Evolutionary game theory has been studied since the 1980s when ideas from evolutionary theory were incorporated into game theory [2].

A variety of social dilemmas have been studied with the majority of attention afforded to the 2-player prisoner's dilemma. Many variations of this game exist, which allow researchers to explore questions regarding cooperation in the presence

M. D. Gibbons (✉) · C. O'Riordan · J. Griffith
National University of Ireland, Galway, Ireland
e-mail: m.gibbons11@nuigalway.ie

C. O'Riordan
e-mail: colm.oriordan@nuigalway.ie

J. Griffith
e-mail: josephine.griffith@nuigalway.ie

© Springer Nature Switzerland AG 2019
J. J. Merelo et al. (eds.), *Computational Intelligence*,
Studies in Computational Intelligence 792,
https://doi.org/10.1007/978-3-319-99283-9_5

of noise, trust, spatial mechanisms and other extensions. One interesting extension that has been explored in the literature is that of N-player social dilemmas [3] where N agents participate simultaneously in the interaction. Each agent can cooperate or defect, and receives a reward based on the number of cooperators present. Additionally, cooperators incur a cost to interact while defectors do not.

In this work, we consider populations of agents participating in both the 2-player and N-player versions of the prisoner's dilemma, and the clusters of cooperators formed therein. We adopt a spatial model where agents' interactions are defined by some topological constraints. Much recent work has focused on the effect of such constraints [4–6]. We use a toroidal lattice where agents may interact with their immediate eight neighbours, if any. We further imbue the agents with the ability to move based on environmental stimuli. The role of mobility in the evolution of cooperation has grown in importance and recognition in recent decades with several researchers demonstrating its use in the promotion of cooperation in artificial life simulations [7, 8]. We adopt an evolutionary framework where successive populations are evolved; the strategy for interacting in the games and the mobility strategy are both subject to evolution. Finally, we define a set of metrics to both qualitatively and quantitatively evaluate the formation of cooperative clusters.

The N-player prisoner's dilemma has not been widely studied in evolutionary models where agents are spatially situated with the inclusion of mobility. In previous work [9], we investigated the significant differences prevalent between the 2-player and N-player dilemmas in this context. In this work, we wish to further explore these differences in the context of cooperative cluster formation, and in addition, examine the effect of varying the density of the agents in the environment. Finally we wish to analyse the movement strategies evolved in these conditions.

In this paper, we show through simulation that there is in fact a substantial difference between the 2-player and the N-player scenarios in terms of the likelihood of cooperation emerging for varying density levels. We demonstrate that for a range of density levels, cooperation emerges in the N-player case. Finally, we analyse the formation and proliferation of cooperative clusters in simulations where cooperation emerges.

The paper outline is as follows: the next section discusses some related work in the field, Sect. 3 outlines our model and approach, and Sect. 4 presents and discusses our results. Finally conclusions and some potential future directions are presented.

2   Related Work

In this section we review some of the relevant research in the literature; we introduce some concepts pertaining to social dilemmas and discuss some work on spatial and evolutionary game theory and the role of mobility.

2.1 Social Dilemma Games

Social dilemma games (most famously the prisoner's dilemma and its variants) have been studied in a wide range of domains due to their usefulness in capturing the conflict between individual and collectively rational behaviours. The prisoner's dilemma in the classical game is described as follows: two players make a choice simultaneously to either cooperate or defect. Mutual cooperation yields a reward R for both participants. However, unilateral defection results in a greater payoff, T, for the defector and a worse payoff, S, for the cooperator (the sucker's payoff). If both defect, both receive P as a payoff such that: $T > R > P > S$.

It has been argued that the N-player variant captures a wider set of dilemmas (e.g. donating to charity organisations, environmental issues etc.). In the N-player dilemma game there are N participants, and again, each player is confronted with a choice: to either cooperate or defect. In one formalism of the game [10], all players receive a benefit based on the number of cooperators present. Cooperators have to pay a cost. No such cost is borne by defecting players. For instance, let B represent some fixed benefit, N the number of players, c the cost and i the number of cooperators. Participants receive $(B \times i)/N$. Cooperators must pay c and thus receive a net reward of $((B \times i)/N) - c$. This, or similar, formulas have been adopted in several other works [3, 11, 12].

We represent the payoff obtained by a strategy which defects given i cooperators as $D(i)$ and the payoff obtained by a cooperative strategy given i cooperators as $C(i)$. Defection represents a dominant strategy, that is, for any individual, moving from cooperation to defection is beneficial for that player in that they still receive a benefit without the cost:

$$D(i) > C(i) \quad 0 < i \leq N - 1 \tag{1}$$

However, if all participants adopted this dominant strategy, the resulting scenario would be a sub-optimal, and from a group point of view, irrational outcome:

$$C(N) > D(0) \tag{2}$$

If any player changes from defection to cooperation, the society performs better:

$$(i + 1)C(i + 1) + (N - i - 1)D(i + 1) > (i)C(i) + (N - i)D(i)$$

In multi-person games, the problem of avoiding exploitation, or free riders, is more difficult, and cooperation may be harder to achieve. In 2-player games, reciprocity has been explored as a means to engender cooperation [13]. However, in N-person games reciprocity may be less advantageous. In order for an agent to punish a defector by defecting in retaliation, the agent must also punish all those that did cooperate.

2.2   Evolutionary N-Player Games

There have been several other notable approaches to exploring the N-player prisoner's dilemma using the tools and approaches in evolutionary game theory. Yao and
Darwen [3] explore the effect of group size in the evolution of cooperation. Strategies
are represented using a generalised form of the representation employed by Axelrod
and Dion [14]. In their experiments, it is shown that cooperation can be evolved in
groups but that it becomes more difficult with increasing group size.

The effects of spatial influences on the evolution of cooperation among strategies participating in the N-players prisoner's dilemma is explored by Suzuki and
Arita [12]. The two spatial factors under investigation are on the *scale of interaction*
(determines which neighbours to play with) and *scale of influence* (specifies which
neighbouring candidates to choose for offspring). Results for simulations involving
a *tit-for-tat* like strategy showed that cooperation becomes most wide-spread for a
modest value of scale of interaction and that, as the cost of cooperation increases,
the levels of cooperation decrease and a higher value of the scale of interaction is
found. Results also indicate that higher cooperation levels are achieved for higher
values of the *scale of influence*.

2.3   Mobility

Traditional spatial models promote the evolution of cooperation by constraining agent
interactions to a particular static topology. Previous work has investigated structures
such as lattices [15], small-world graphs [16], and scale-free graphs [17]. However,
the inclusion of movement creates a more realistic model by allowing agents to
respond to their current neighbourhood by moving within their environment.

Mobility is a form of network reciprocity [13], which has gone from being perceived as a hindrance to the emergence of cooperation to a key concept in its promotion. While unrestrained movement can, and does, lead to the 'free-rider' effect
[18], allowing highly mobile defectors to go unpunished, using simple strategy rules
[7, 19] or using mobility rates [8, 20] significantly curb the free-rider phenomenon
allowing self-preserving cooperator clusters to form, and cooperation to proliferate.

Several mechanisms for the emergence of cooperation exist, but all essentially
express a need for cooperators to either avoid interactions with defectors or increase
and sustain interactions with other cooperators. Research in this domain is largely
divided into two categories based on authors' definition of mobility; all movement
should be random [8, 20–22], or should be purposeful or strategically driven, but may
indeed contain random elements [7, 23–27]. Random mobility can be used to describe
the minimal conditions for the evolution of cooperation. Alternatively, contingent
mobility has the capacity to be proactive. This is where individuals deliberately
seek better neighbourhoods, rather than simply reacting to stimuli and randomly
relocating.

The majority of the contingent mobility strategies in the literature are hand crafted or guided by heuristics. However, there has been some research [28–30] using evolutionary models to evolve movement strategies that are conducive to the emergence of cooperation. Ichinose et al. [19] also use an evolutionary model and investigates the coevolution of migration and cooperation. Agents play an N-player Prisoner's Dilemma game after which they move locally according to an evolved probability vector. All agents are evolved to collectively follow or chase cooperators. The authors highlight the importance of flexibility in the direction of migration for the evolution of cooperation.

Chiong et al. [31] describe a random mobility model where a population of agents interact in an N-player Prisoner's Dilemma set in a fully occupied regular lattice. Pairs of agents move by exchanging grid positions. Mobility in this environment is a probability function based on the time an agent has spent in a location, and the relative fitness of the agent at the destination. The agents have a limited memory of past interactions, and past cooperator and defector levels. Cooperation is shown to be promoted under a limited small set of parameters including the cost to benefit ratio of cooperation and the movement radius.

Most recently, Suarez et al. [32] present a contingent mobility model, using the N-Player game, in which agents move toward locations with higher potential payoff. While cooperation does emerge, the authors do not elaborate on the specific effects of mobility, focusing more on the impact of the neighbourhood size.

3   Methodology

3.1   Environment and Agent Representation

The population of agents A inhabits a toroidal shaped diluted lattice with $L \times L$ cells, each of which can be occupied by up to one agent. The interaction and movement radii of agents is determined using the Moore neighbourhood of radius one. This comprises the eight cells surrounding an individual in a cell on the lattice. The agents can only perceive and play with those within this limited radius.

Each agent is represented by a genotype, which determines their strategy to interact with other agents and to move in the environment. The first section of the gene describes their strategy for playing the game: that is to cooperate or defect and the remaining sections determine how an agent will move. The remainder of the genotype encodes actions for a range of scenarios that may arise within the environment, including: encountering a cooperator, encountering a defector, or encountering both at once. If an agent meets a cooperator, they have a set of potential actions. These actions are as follows: remain where they are, move randomly, follow the cooperator or flee from it. Similarly these potential actions are mirrored when an agent meets a defector. The final section is used to determine actions when an agent meets both a defector and a cooperator. The actions are: flee from both cooperator and defector;

follow both cooperator and defector; follow the cooperator and flee from the defector and the converse action (flee from the cooperator and follow the defector). During a simulation run, each potential action of an agent is determined by its genotype.

At each time step, agents participate in a single round of the Prisoner's Dilemma with each of their neighbours, if any. The strategy with which agents play is fixed; either always cooperate or always defect. We choose to implement pure strategies in order to reduce the strategy space allowing us to more clearly examine the effect of mobility in these experiments. Agents are aware of the actions taken by their neighbours in a single round, but these memories do not persist. Following this interaction phase, agents have the opportunity to take one step into an adjacent free cell according to their movement strategy. Movement will not occur if there is no adjacent free space, or if their strategy dictates that they remain in their current location. Isolated agents will take one step in a random direction.

3.2   Evolutionary Dynamics

The movement strategies adopted by the population are explored by using an ALife inspired evolutionary model. In a single generation, agents accumulate their payoffs received from playing the Prisoner's Dilemma with their neighbours. This is used as a measure of fitness, and at the end of each generation, the agents are ranked according to this score. The bottom 20% are replaced with copies of the top 20%. This replacement strategy was chosen as it has been previously shown to produce a fair sampling of the population's fitness while still allowing for convergence in a reasonable time frame. No other genetic operators are utilized. These offspring are randomly placed on the grid, and the other agents remain in the same place, thus maintaining any spatial clustering between generations. Following reproduction, the fitness score of the whole population is reset and a new generation begins.

3.3   Interaction Model

In keeping with previous work, we adopt a well known formalism for the N-player prisoner's dilemma. Letting B be a constant representing social benefit, c be the cost of cooperation and i the number of cooperators from a group of N agents, the following payoffs are used:

$$C(i) = \frac{B \times i}{N} - c$$

$$D(i) = \frac{B \times i}{N}$$

Table 1 Prisoner's dilemma game matrix

	C	D
C	2,2	$-\frac{1}{2}, 2\frac{1}{2}$
D	$2\frac{1}{2}, -\frac{1}{2}$	0,0

The following constraints hold: $B > c$ and both B and c are positive values.

Considering the N-player dilemma, when N = 2 and attempting to align with the classical interpretation of the 2-player prisoner's dilemma, we also require that $B < 2c$. Values chosen in this research that are in keeping with previous studies in the field are: $B = 5, c = 3$. For example, in mapping this back to the two player games, we use the payoff matrix as described in Table 1.

In our simulations, we contrast scenarios with 2-player interactions and N-player interactions. In the N-player case, an agent particpates in the dilemma with all of its immediate neighbours; the number of such neighbours determines the number of participants. In the 2-player case, an agent participates in individual 2-player games with each of its immediate neighbours.

3.4 Cluster Analysis

The aim of this work is to definitively determine that the clustering of cooperators is the primary cause for the proliferation of cooperation throughout a population. It is important to establish a metric for clusters as the concept is oftentimes ill-defined. Much previous research asserts the existence and formation of clusters without explicit measures of the number and type of clusters formed. We define a cooperative cluster as a set, with cardinality greater than one, of spatially contiguous cooperative agents. We also define three cluster metrics which we use to compare cluster formation across experiments. These are: (1) the number of clusters in the population, (2) the average size of clusters, and (3) the mean average neighbourhood size of individuals in each cluster. This value is obtained by first counting the number of neighbours of each agent in a cluster, calculating the average neighbourhood size for that cluster, and then calculating the average across all clusters in the population. This metric shall be referred to as cluster quality from this point onward.

4 Simulation Results

4.1 Experimental Setup

In these experiments, we run two sets of similar simulations, one with *2-Player* interactions the other with *N-Player* interactions, comparing the respective outcomes. It is generally accepted that when comparing the two interaction models inducing cooperation in the N-Player games is considerably harder.

Table 2 2-Player versus
N-Player: % cooperator wins.
extracted from [9]

	Avg	Std dev
2-player	33.2%	4.2%
N-player	25.8%	4.7%

The population of $A = 100$ agents is placed randomly on the $L \times L$ torus with $L = 30$, the strategies (whether to cooperate or to defect) are assigned in equal proportion, and the movement strategies are assigned randomly. A single simulation lasts 1,250 time-steps, in which the agents take 25 steps in each of 50 generations. The distribution of spatial strategies, level of cooperation, time taken for the simulation to converge on cooperation (or defection), and the total number of interactions will all be recorded. Each simulation will be run over a 1000 times to generate statistically valid results.

In order to perform the cluster analysis, a snapshot of the population calculating the cluster analysis metrics, number of clusters, average cluster size and quality, are recorded every 5 steps.

4.2   2-Player Versus N-Player

Cooperative Outcomes. On average in these environmental settings, the 2-Player interaction model is more effective at inducing the spread of cooperation in a larger percentage of simulations. Table 2 shows that in roughly one third of evolutionary simulations using 2-player interaction, cooperation emerges as the outcome, whereas when agents participate in an N-player interaction, cooperation emerges in roughly one quarter of the simulations. On average, simulations using the 2-Player interaction model tend to converge more quickly, and with less variance. The simulations resulting in the emergence of defectors exhibit a faster convergence and less variability in convergence speed regardless of the interaction model.

These results are in keeping with the general consensus that evolving cooperation in the N-player prisoner's dilemma can be more difficult. This previous research did not allow movement of agents, but still captured the difficulty with N-player dilemmas where an agent can exploit multiple participants and achieve a considerable gain in payoff per interaction.

Evolved Movement Strategies. Tables 3, 4 and 5 show the movement behaviours that are evolved for 2-player and N-player situations respectively in those runs when cooperation emerges. One hundred simulation runs resulting in cooperative outcomes are considered.

Upon seeing a cooperator in their neighbourhood, agents evolve to either stay where they are or to follow the cooperator; this occurs in both 2-player and N-player scenarios. When a defector is encountered, agents have evolved to flee or adopt a random movement in 75% of cases in the 2-player game and 97% of cases in the N-player game. For the scenarios where agents see both cooperators and defectors

Table 3 On seeing cooperator : % genes evolved. extracted from [9]

	2-player	N-player
Random	0%	0%
Follow	15%	27%
Flee	0%	0%
Stay	**85%**	**73%**

Table 4 On seeing defector : % genes evolved. extracted from [9]

	2-player	N-player
Random	34%	22%
Follow	16%	2%
Flee	**41%**	**75%**
Stay	9%	1%

Table 5 On seeing cooperator & defector : % genes evolved. extracted from [9]

	2-player	N-player
FollowCFollowD	27%	3%
FollowCFleeD	**44%**	**52%**
FleeCFollowD	9%	11%
FleeCFleeD	20%	34%

we see similar behaviours being evolved. Movement behaviours that promote cooperation and avoid exploitation are selected. We can see that cooperators who interact using the N-Player interaction model have a greater evolutionary incentive to be adverse to defectors.

In all cases agents learn movement behaviours that allow them to continue cooperative interactions and, to a lesser extent, to avoid interactions with defectors. Behaviours that continue defector interactions die off, although at a slower rate. Following cooperators is selected more quickly than fleeing from defectors.

It is important to note that the selective pressure to avoid defectors is removed when the defectors are replaced in the population with cooperators and hence we do not see convergence to 100% for the genes that promote avoiding defector interactions. Adopting a random movement can also often have the same effect as fleeing from or indeed following an individual.

The population did not always evolve a single strategy; random fluctuation and lack of relevant stimuli resulted in simulations in which agents converged on several strategies that were genotypically different, but phenotypically similar.

In non-cooperative runs, defectors learned to (1) follow cooperators, (2) flee from defectors, and to (3) follow both cooperators and defectors.

Cluster Analysis. We can observe in Fig. 1a that the number of clusters, in both 2-player and N-player simulations, initially increases rapidly, then reaches a plateau, and eventually decreases. At the beginning of the simulation, the population is randomly dispersed with many small clusters forming as cooperators move and interact

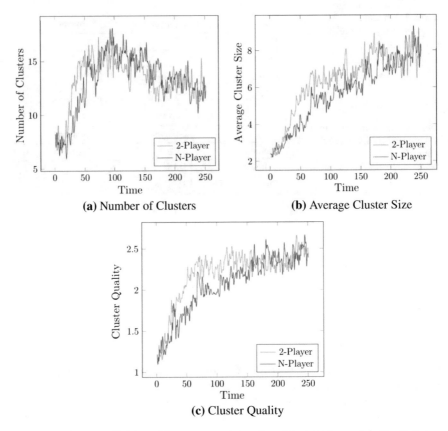

Fig. 1 2-Player versus N-Player: The mean average number of clusters, size, and quality of clusters from a number of simulations resulting in cooperative victories

according to their strategy. As the defectors start to reduce in numbers, the cooperator clusters have the space and freedom to merge and grow larger, thus reducing the overall cluster count. This is backed up by Fig. 1b where we see that the average cluster size, in both cases, increases over time. The cluster quality quantifies the potential fitness of one cluster over another. Figure 1c demonstrates that in both 2-player and N-player simulations cooperator clusters are improving over time, facilitating the further spread of cooperation.

In Fig. 1 we see populations using the 2-player interaction model form a greater number, larger, and high quality cooperator clusters more quickly than those using the N-player version. We see variability in theses graphs because clusters can both be divided and broken up by defectors, and the evolutionary process.

Fig. 2 2-Player versus N-Player: The percentage of simulations resulting in cooperative victories as we vary the grid density starting from random initialization. Extracted from [9]

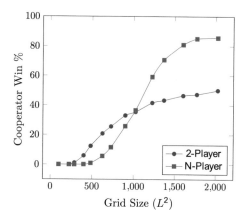

4.3 Variation in Density

In the previous experiments, the percentage of cooperative outcomes and the evolution of movement strategies was a function of the agent interactions. The ratio of cooperative interactions to other types of interactions influences the evolutionary trajectories.

In this experiment we aim to investigate the impact of the density of agents in the environment. We define the density as $D = A/L^2$ where A is the size of the population, and L is the length of the lattice grid. Density is a function of the population size and the size of the grid. We keep the population size constant and vary the size of the grid as a means to vary the density.

The movement strategies of agents are randomly initialized, the strategies for game interactions are assigned in equal proportions and both the movement and interaction strategies are subject to evolution. In one set of simulations, the population interacts using the 2-player interaction model, and the other uses the N-player model.

As shown in Fig. 2, at the highest density level, there is not enough space within the grid for agents to move freely and so defection dominates in the vast majority of simulations. These conditions echo the traditional spatial models with an agent located in every cell where no movement is possible. These findings mirror those results with defection spreading and dominating the population.

As the density is reduced, we see that the evolutionary runs using the 2-Player interaction model are more readily able to induce higher levels of cooperation. However, using the 2-Player interaction model, random initialization in low densities can only achieve cooperation in just above 50% of simulations. With these same settings the N-Player interaction model can induce cooperation in a far greater percentage (80%) of runs.

For a grid size of 32×32 (1024 cells), the N-player interactions overtake the 2-player interaction model in their ability to induce cooperation. This result demonstrates that despite the difficulty of inducing cooperation, cooperation emerges in

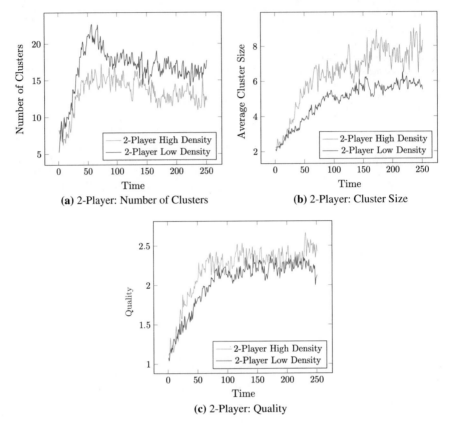

Fig. 3 N-Player High Density versus Low Density: The mean average number of clusters, size, and quality of clusters from a number of simulations resulting in cooperative victories

N-player games, the addition of movement capabilities can support the emergence of cooperation in these conditions.

Cluster Analysis. In low densities, for both 2-player (Fig. 3) and N-player (Fig. 4), cooperators, on average, form greater numbers of smaller clusters. In the lower densities, the population is more dispersed throughout the environment; thus cooperators, on average, meet fewer agents and do not form larger clusters. However, for both density levels, in terms of quality, similar clusters are formed.

4.4 Seeding the Evolved Strategies

In our final experiment, the evolved movement strategies for both cooperators and defectors are seeded in the population and we repeat the density experiment. In the previous experiment both movement strategies were randomly assigned and it took

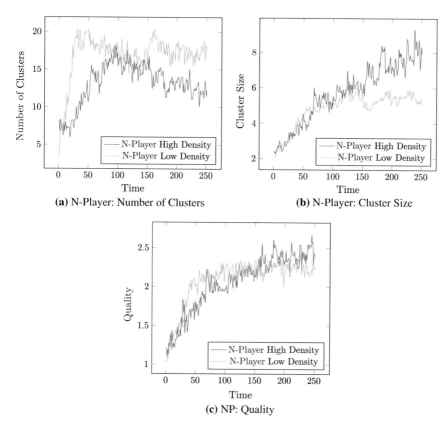

(a) N-Player: Number of Clusters

(b) N-Player: Cluster Size

(c) NP: Quality

Fig. 4 N-Player High Density versus Low Density: The mean average number of clusters, size, and quality of clusters from a number of simulations resulting in cooperative victories

several generations for movement strategies to emerge. A number of these strategies. were identified as being favorable to the emergence of cooperation. The aim of this experiment is to explore the effect of these *good* strategies when they are present in the first generation. If these strategies help cooperators to follow each other and form cooperative clusters, then higher levels of cooperation are expected across the various density levels.

Results show in both sets of simulations that the evolved cooperator movement strategies are able to induce cooperation for a much wider range of densities, as illustrated in Fig. 5. There is a far greater level of cooperation than that which was achieved by either interaction model in the experiment with random initialization. For the N-player interaction model, once the grid size reaches 1024 (density roughly equal to 10%), cooperation is achieved 100% of the time. For the 2-player interaction model, this level of cooperation is also maintained for higher density levels. The agents using the N-player model are more hindered by the exploitative nature of defectors, who are also using a previously evolved movement strategy.

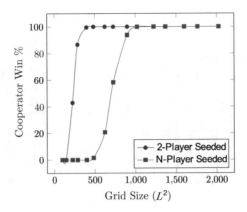

Fig. 5 2-Player versus N-Player: The percentage of cooperative victories, as we vary the grid density, seeding the most prevalent evolved strategies for cooperators and defectors. Extracted from [9]

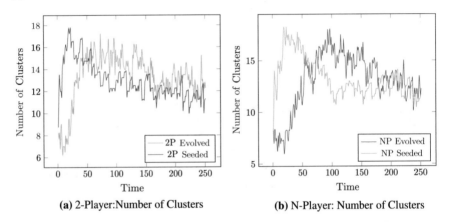

(a) 2-Player:Number of Clusters

(b) N-Player: Number of Clusters

Fig. 6 Evolved versus Seeded: The mean average number of clusters from a number of simulations, using both 2-Player and N-Player interaction models, resulting in cooperative victories

Cluster Analysis. In both the 2-player and N-player scenarios (Fig. 6), the seeded strategies are able to generate a significantly greater number of clusters in the early generations than in the previous experiments where the strategies were unseeded. This difference is due to the fact that in the previous experiments the evolved population hadn't yet learned the optimum movement strategy and hence quick clustering of cooperators was hampered. However, other than the time to reach a level of clustering, there are no significant differences in the level of quality of the clusters.

5 Discussion

Traditionally, it has been difficult to induce cooperation using the N-player Prisoner's Dilemma. However, in our model we observe high levels of cooperation in a range of settings. The incorporation of a contingent mobility allows cooperators to cluster together, and avoid repeated defector interactions. In forming these clusters, these agents can increase their number of mutually cooperative interactions, thereby boosting their score. However, these cooperative clusters can be exploited by defectors unless they employ strategies that can avoid repeated exploitative encounters. We observe high levels of cooperation coupled with evolved movement strategies that encourage the formation of these larger self-preserving clusters free from the influence of defectors.

As expected, the 2-Player interaction model was more successful at inducing cooperation in the higher grid densities when we evolved from random strategies. This is due to the fact that while the chances of encountering a defector are higher, they have less of an exploitative impact on individuals or clusters of cooperators. Surprisingly, the N-player interaction model was significantly more successful at inducing cooperation when the grid density was very low. We attribute this success to the reduced chances of encountering a defector, and increased gains made by mutually cooperative interactions in clusters. Additionally, single defectors benefit by being in the neighbourhood of cooperators but this benefit is reduced in the presence of other defectors.

6 Conclusion

In this paper, we have shown that simple mobile strategies, which use both the 2-Player and N-Player interaction models, are extremely adept at spreading cooperation throughout agent populations by forming cooperator clusters.

Clusters form, and proliferate, when cooperators learn movement strategies that allow them to maintain beneficial cooperator interactions and avoid repeated, exploitative interactions with defectors, which leads to the evolution of cooperation. Over the course of the simulations that result in widespread cooperation, we observe, for all cluster metrics used (the number of clusters, the cluster size, and the cluster quality) increasing evidence of clustering of strategies. Furthermore, when a population is seeded with the movement strategies that promote cluster formation, in a suitable environment (medium to low agent density), it is possible to guarantee the evolution of cooperation. This is achieved using only local information and without the need for complex computation or costly memories.

Through experimentation, we show that the presence of contingent mobility strategies induces cooperation in the N-player Prisoner's Dilemma. Despite the inherent disadvantages of this interaction model, we demonstrate that it is possible to generate very high levels of cooperation from both seeded and, surprisingly, randomly initialised populations. It is clear that density plays a significant role, particularly in the N-Player model, as it possesses the greatest influence over cluster formation.

Future work will involve a more thorough investigation of the nature of the cooperative clusters that form throughout the evolutionary runs. We wish to explore a larger set of N-player social dilemmas and explore more expressive spatial topologies. We also intend to determine the conditions under which cooperative clusters fail to form, and investigate means to encourage their formation.

Acknowledgements This work is funded by the Hardiman Research Scholarship, NUI Galway.

References

1. Axelrod, R.M.: The Evolution of Cooperation. Basic Books (1984)
2. Maynard Smith, J.: Evolution and the Theory of Games. Cambridge University Press, New York (1982)
3. Yao, X., Darwen, P.J.: An experimental study of N-person iterated prisoners dilemma games. Informatica **18**, 435–450 (1994)
4. Szolnoki, A., Perc, M., Szabó, G., Stark, H.U.: Impact of aging on the evolution of cooperation in the spatial prisoners dilemma game. Phys. Rev. E **80**, 021901 (2009)
5. Ohtsuki, H., Hauert, C., Lieberman, E., Nowak, M.A.: A simple rule for the evolution of cooperation on graphs and social networks. Nature **441**, 502–505 (2006)
6. Lieberman, E., Hauert, C., Nowak, M.A.: Evolutionary dynamics on graphs. Nature **433**, 312–316 (2005)
7. Aktipis, C.A.: Know when to walk away: contingent movement and the evolution of cooperation. J. Theor. Biol. **231**, 249–260 (2004)
8. Vainstein, M.H., Silva, A.T.C., Arenzon, J.J.: Does mobility decrease cooperation? J. Theor. Biol. **244**, 722–728 (2007)
9. Gibbons, M.D., O'Riordan, C., Griffith, J.: Evolution of cooperation in n-player social dilemmas: the importance of being mobile. In: Proceedings of the 8th International Joint Conference on Computational Intelligence - Volume 1: ECTA, (IJCCI 2016), INSTICC, pp. 78–85. ScitePress (2016)
10. Boyd, R., Richerson, P.J.: The evolution of reciprocity in sizable groups. J. Theor. Biol. **132**, 337–356 (1988)
11. O'Riordan, C., Sorensen, H.: Stable cooperation in the N-player prisoners dilemma: the importance of community structure. In: Adaptive Agents and Multi-Agent Systems III. Adaptation and Multi-Agent Learning, pp. 157–168. Springer (2008)
12. Suzuki, R., Arita, T.: Evolutionary analysis on spatial locality in n-person iterated prisoner's dilemma. Int. J. Comput. Int. Appl. **3**, 177–188 (2003)
13. Nowak, M.A.: Five rules for the evolution of cooperation. Science **314**, 1560–3 (2006)
14. Axelrod, R., Dion, D.: The further evolution of cooperation. Science **242**, 1385–1390 (1988)
15. Nowak, M.A., May, R.M.: Evolutionary games and spatial chaos. Nature **359**, 826–829 (1992)
16. Santos, F., Rodrigues, J., Pacheco, J.: Graph topology plays a determinant role in the evolution of cooperation. Proc. R. Soc. B: Biol. Sci. **273**, 51–55 (2006)
17. Poncela, J., Gómez-Gardeñes, J., Floría, L.M., Moreno, Y., Sánchez, A.: Cooperative scale-free networks despite the presence of defector hubs. EPL (Europhys. Lett.) **88**, 38003 (2009)
18. Enquist, M., Leimar, O.: The evolution of cooperation in mobile organisms. Anim. Behav. **45**, 747–757 (1993)
19. Ichinose, G., Saito, M., Suzuki, S.: Collective chasing behavior between cooperators and defectors in the spatial prisoner's dilemma. PLoS ONE **8**, 28–31 (2013)

20. Meloni, S., Buscarino, A., Fortuna, L., Frasca, M., Gómez-Gardeñes, J., Latora, V., Moreno, Y.: Effects of mobility in a population of prisoner's dilemma players. Phys. Rev. E - Stat. Nonlinear, Soft Matter Phys. **79**, 3–6 (2009)
21. Sicardi, E.A., Fort, H., Vainstein, M.H., Arenzon, J.J.: Random mobility and spatial structure often enhance cooperation. J. Theor. Biol. **256**, 240–246 (2009)
22. Antonioni, A., Tomassini, M., Buesser, P.: Random diffusion and cooperation in continuous two-dimensional space. J. Theor. Biol. **344**, 40–48 (2014)
23. Helbing, D., Yu, W.: Migration as a mechanism to promote cooperation. Adv. Complex Syst. **11**, 641–652 (2008)
24. Helbing, D., Yu, W.: The outbreak of cooperation among success-driven individuals under noisy conditions. Proc. Natl. Acad. Sci. USA **106**, 3680–3685 (2009)
25. Jiang, L.L., Wang, W.X., Lai, Y.C., Wang, B.H.: Role of adaptive migration in promoting cooperation in spatial games. Phys. Rev. E **81**, 036108 (2010)
26. Yang, H.X., Wu, Z.X., Wang, B.H.: Role of aspiration-induced migration in cooperation. Phys. Rev. E **81**, 065101 (2010)
27. Tomassini, M., Antonioni, A.: Lévy flights and cooperation among mobile individuals. J. Theor. Biol. **364**, 154–161 (2015)
28. Joyce, D., Kennison, J., Densmore, O., Guerin, S., Barr, S., Charles, E., Thompson, N.S.: My way or the highway: a more naturalistic model of altruism tested in an iterative prisoners' dilemma. J. Artif. Soc. Soc. Simul. **9**, 4 (2006)
29. Gibbons, M., O'Riordan, C.: Evolution of coordinated behaviour in artificial life simulations. In: Proceedings of the International Conference on Theory and Practice in Modern Computing. (2014)
30. Gibbons, M.D., O'Riordan, C., Griffith, J.: Follow flee: A contingent mobility strategy for the spatial prisoners dilemma. In: International Conference on Simulation of Adaptive Behavior, pp. 34–45. Springer (2016)
31. Chiong, R., Kirley, M.: Random mobility and the evolution of cooperation in spatial N-player iterated prisoners dilemma games. Phys. A: Stat. Mech. Appl. **391**, 3915–3923 (2012)
32. Suarez, D., Suthaharan, P., Rowell, J., Rychtar, J.: Evolution of cooperation in mobile populations. Spora-A J. Biomath. **1**, 2–7 (2015)

The Impact of Coevolution
and Abstention on the Emergence
of Cooperation

Marcos Cardinot, Colm O'Riordan and Josephine Griffith

Abstract This paper explores the Coevolutionary Optional Prisoner's Dilemma (COPD) game, which is a simple model to coevolve game strategy and link weights of agents playing the Optional Prisoner's Dilemma game, which is also known as the Prisoner's Dilemma with voluntary participation. A number of Monte Carlo simulations are performed to investigate the impacts of the COPD game on the emergence of cooperation. Results show that the coevolutionary rules enable cooperators to survive and even dominate, with the presence of abstainers in the population playing a key role in the protection of cooperators against exploitation from defectors. We observe that in adverse conditions such as when the initial population of abstainers is too scarce/abundant, or when the temptation to defect is very high, cooperation has no chance of emerging. However, when the simple coevolutionary rules are applied, cooperators flourish.

Keywords Optional Prisoner's Dilemma game · Voluntary participation · Evolutionary game theory

1 Introduction

Evolutionary game theory in spatial environments has attracted much interest from researchers who seek to understand cooperative behaviour among rational individuals in complex environments. Many models have considered the scenarios where participants interactions are constrained by particular graph topologies, such as lattices [14, 17], small-world graphs [6, 8], scale-free graphs [19, 21] and, bipartite graphs [10]. It has been shown that the spatial organisation of strategies on these topologies affects the evolution of cooperation [3].

The Prisoner's Dilemma (PD) game remains one of the most studied games in evolutionary game theory as it provides a simple and powerful framework to illustrate

M. Cardinot (✉) · C. O'Riordan · J. Griffith
Department of Information Technology, National University of Ireland, Galway, Ireland
e-mail: marcos.cardinot@nuigalway.ie

© Springer Nature Switzerland AG 2019
J. J. Merelo et al. (eds.), *Computational Intelligence*,
Studies in Computational Intelligence 792,
https://doi.org/10.1007/978-3-319-99283-9_6

the conflicts inherent in the formation of cooperation. In addition, some extensions of the PD game, such as the Optional Prisoner's Dilemma (OPD) game, have been studied in an effort to investigate how levels of cooperation can be increased. In the OPD game, participants are afforded a third option — that of abstaining and not playing and thus obtaining the loner's payoff (L). Incorporating this concept of abstention leads to a three-strategy game where participants can choose to cooperate, defect or abstain from a game interaction.

The vast majority of the spatial models in previous work have used static and unweighted networks. However, in many social scenarios that we wish to model, such as social networks and real biological networks, the number of individuals, their connections and environment are often dynamic. Thus, recent studies have also investigated the effects of evolutionary games played on dynamically weighted networks [2, 12, 18, 20, 22, 23] where it has been shown that the coevolution of both networks and game strategies can play a key role in resolving social dilemmas in a more realistic scenario.

In this paper we explore the Coevolutionary Optional Prisoner's Dilemma (COPD) game, which is a simple coevolutionary spatial model where both the game strategies and the link weights between agents evolve over time. In this model, the interaction between agents is described by an OPD game. Previous research on spatial games has shown that when the temptation to defect is high, defection is the dominant strategy in most cases. However, it is been discussed that the combination of both optional games and coevolutionary rules can help in the emergence of cooperation in a wider range of scenarios [4, 5].

Thus, given the Coevolutionary Optional Prisoner's Dilemma game (i.e., an OPD game in a spatial environment, where links between agents can be evolved), the aims of the work are to understand the effect of varying the parameters T (temptation to defect), L (loner's payoff), Δ and δ for both unbiased and biased environments.

By investigating the effect of these parameters, we aim to:

- Compare the outcomes of the COPD game with other games.
- Explore the impact of the link update rules and its properties.
- Investigate the evolution of cooperation when abstainers are present in the population.
- Investigate how many abstainers would be necessary to guarantee robust cooperation.

Results show that cooperation emerges even in extremely adverse scenarios where the temptation to defect is almost at its maximum. It can be observed that the presence of the abstainers are fundamental in protecting cooperators from invasion. In general, it is shown that, when the coevolutionary rules are used, cooperators do much better, being also able to dominate the whole population in many cases. Moreover, for some settings, we also observe interesting phenomena of cyclic competition between the three strategies, in which abstainers invade defectors, defectors invade cooperators and cooperators invade abstainers.

The paper outline is as follows: Section 2 presents a brief overview of the previous work in both spatial evolutionary game theory with dynamic networks and in the

Optional Prisoner's Dilemma game. Section 3 gives an overview of the methodology employed, outlining the Optional Prisoner's Dilemma payoff matrix, the coevolutionary model used (Monte Carlo simulation), the strategy and link weight update rules, and the parameter values that are varied in order to explore the effect of coevolving both strategies and link weights. Section 4 discusses the benefits of combining the concept of abstention and coevolution. Section 5 further explores the effect of using the COPD game in an unbiased environment. Section 6 investigates the robustness of cooperative behaviour in a biased environment. Finally, Sect. 7 summarizes the main conclusions and outlines future work.

2 Related Work

The use of coevolutionary rules constitute a new trend in evolutionary game theory. These rules were first introduced by Zimmermann et al. [22], who proposed a model in which agents can adapt their neighbourhood during a dynamical evolution of game strategy and graph topology. Their model uses computer simulations to implement two rules: firstly, agents playing the Prisoner's Dilemma game update their strategy (cooperate or defect) by imitating the strategy of an agent in their neighbourhood with a higher payoff; and secondly, the network is updated by allowing defectors to break their connection with other defectors and replace the connection with a connection to a new neighbour selected randomly from the whole network. Results show that such an adaptation of the network is responsible for an increase in cooperation.

In fact, as stated by Perc and Szolnoki [16], the spatial coevolutionary game is a natural upgrade of the traditional spatial evolutionary game initially proposed by Nowak and May [14], who considered static and unweighted networks in which each individual can interact only with its immediate neighbours. In general, it has been shown that coevolving the spatial structure can promote the emergence of cooperation in many scenarios [2, 20], but the understanding of cooperative behaviour is still one of the central issues in evolutionary game theory.

Szolnoki and Perc [18] proposed a study of the impact of coevolutionary rules on the spatial version of three different games, i.e., the Prisoner's Dilemma, the Snow Drift and the Stag Hunt game. They introduce the concept of a teaching activity, which quantifies the ability of each agent to enforce its strategy on the opponent. It means that agents with higher teaching activity are more likely to reproduce than those with a low teaching activity. Differing from previous research [22, 23], they also consider coevolution affecting either only the defectors or only the cooperators. They discuss that, in both cases and irrespective of the applied game, their coevolutionary model is much more beneficial to the cooperators than that of the traditional model.

Huang et al. [12] present a new model for the coevolution of game strategy and link weight. They consider a population of 100×100 agents arranged on a regular lattice network which is evolved through a Monte Carlo simulation. An agent's interaction is described by the classical Prisoner's Dilemma with a normalized payoff matrix. A new parameter, Δ/δ, is defined as the link weight amplitude and is calculated as the ratio of Δ/δ. They found that some values of Δ/δ can provide the best environment for the evolution of cooperation. They also found that their coevolutionary model can promote cooperation efficiently even when the temptation of defection is high.

In addition to investigations of the classical Prisoner's Dilemma on spatial environments, some extensions of this game have also been explored as a means to favour the emergence of cooperative behaviour. For instance, the Optional Prisoner's Dilemma game, which introduces the concept of abstention, has been studied since Batali and Kitcher [1]. In their work, they proposed the opt-out or "loner's" strategy in which agents could choose to abstain from playing the game, as a third option, in order to avoid cooperating with known defectors. There have been a number of recent studies exploring this type of game [9, 11, 13, 15, 21]. Cardinot et al. [3] discuss that, with the introduction of abstainers, it is possible to observe new phenomena and, in a larger range of scenarios, cooperators can be robust to invasion by defectors and can dominate.

Although recent work has discussed the inclusion of optional games with coevolutionary rules [4, 5], this still needs to be investigated in a wider range of scenarios. Therefore, the present work aims to combine both of these trends in evolutionary game theory in order to identify favourable configurations for the emergence of cooperation in adverse scenarios, where, for example, the temptation to defect is very high or when the initial population of abstainers is either very scarce or very abundant.

3   Methodology

This section includes a complete description of the Optional Prisoner's Dilemma game, the spatial environment and the coevolutionary rules for both the strategy and link weights. Finally, we also outline the experimental set-up.

In the classical version of the Prisoner's Dilemma (PD), two agents can choose either cooperation or defection. Hence, there are four payoffs associated with each pairwise interaction between the two agents. In consonance with common practice [5, 12, 14], payoffs are characterized by the reward for mutual cooperation ($R = 1$), punishment for mutual defection ($P = 0$), sucker's payoff ($S = 0$) and temptation to defect ($T = b$, where $1 < b < 2$). Note that this parametrization refers to the weak version of the Prisoner's Dilemma game, where P can be equal to S without destroying the nature of the dilemma. In this way, the constraints $T > R > P \geq S$ maintain the dilemma.

The Optional Prisoner's Dilemma (OPD) game is an extended version of the PD game in which agents can not only cooperate (C) or defect (D) but can also choose to abstain (A) from a game interaction, obtaining the loner's payoff ($L = l$) which is awarded to both players if one or both abstain. As defined in other studies [3, 17], abstainers receive a payoff greater than P and less than R (i.e., $P < L < R$). Thus, considering the normalized payoff matrix adopted, $0 < l < 1$. The payoff matrix and the associated values are illustrated in Table 1.

Table 1 The Optional Prisoner's Dilemma game matrix [5]

	C	D	A
C	R / R	T / S	L / L
D	S / T	P / P	L / L
A	L / L	L / L	L / L

(a) Extended game matrix.

Payoff	Value
Temptation to defect (T)	$]1, 2[$
Reward for mutual cooperation (R)	1
Punishment for mutual defection (P)	0
Sucker's payoff (S)	0
Loner's payoff (L)	$]0, 1[$

(b) Payoff values.

In this work, the following parameters are used: an $N = 102 \times 102$ regular lattice grid with periodic boundary conditions is created and fully populated with agents, which can play with their eight immediate neighbours (Moore neighbourhood). We investigate both unbiased (i.e., initially each agent is designated as C, D or A with equal probability) and biased (i.e., varying the initial percentage of abstainers) environments.

Also, each edge linking agents has the same initial weight $w = 1$, which will adaptively change in accordance with the interaction.

Monte Carlo methods are used to perform the Coevolutionary Optional Prisoner's Dilemma game. In one Monte Carlo (MC) step, each player is selected once on average. This means that one MC step comprises N inner steps where the following calculations and updates occur:

- Select an agent (x) at random from the population.
- Calculate the utility u_{xy} of each interaction of x with its eight neighbours (each neighbour represented as agent y) as follows:

$$u_{xy} = w_{xy} P_{xy}, \tag{1}$$

where w_{xy} is the edge weight between agents x and y, and P_{xy} corresponds to the payoff obtained by agent x on playing the game with agent y.
- Calculate U_x the accumulated utility of x, that is:

$$U_x = \sum_{y \in \Omega_x} u_{xy}, \tag{2}$$

where Ω_x denotes the set of neighbours of the agent x.
- In order to update the link weights, w_{xy}, between agents, compare the values of u_{xy} and the average accumulated utility (i.e., $\overline{U}_x = U_x/8$) as follows:

$$w_{xy} = \begin{cases} w_{xy} + \Delta & \text{if } u_{xy} > \overline{U}_x \\ w_{xy} - \Delta & \text{if } u_{xy} < \overline{U}_x, \\ w_{xy} & otherwise \end{cases} \tag{3}$$

where Δ is a constant such that $0 \leq \Delta \leq \delta$.

- In line with previous research [5, 12, 20], w_{xy} is adjusted to be within the range

$$1 - \delta \leq w_{xy} \leq 1 + \delta, \tag{4}$$

where δ ($0 \leq \delta < 1$) defines the weight heterogeneity. Note that when Δ or δ are equal to 0, the link weight keeps constant ($w = 1$), which results in the traditional scenario where only the strategies evolve.

- In order to update the strategy of x, the accumulated utility U_x is recalculated (based on the new link weights) and compared with the accumulated utility of one randomly selected neighbour (U_y). If $U_y > U_x$, agent x will copy the strategy of agent y with a probability proportional to the utility difference (Eq. 5), otherwise, agent x will keep its strategy for the next step.

$$p(s_x = s_y) = \frac{U_y - U_x}{8(T - P)}, \tag{5}$$

where T is the temptation to defect and P is the punishment for mutual defection.

Simulations are run for 10^5 MC steps and the fraction of cooperation is determined by calculating the average of the final 10^3 MC steps. To alleviate the effect of randomness in the approach, the final results are obtained by averaging 10 independent runs. The following scenarios are investigated:

- The benefits of coevolution and abstention.
- Presence of abstainers in the coevolutionary model.
- Inspecting the coevolutionary environment.
- Investigating the properties of the parameters Δ and δ.
- Varying the number of states.
- Investigating the relationship between Δ/δ, b and l.
- Investigating the robustness of cooperation in a biased environment.

4   The Benefits of Coevolution and Abstention

This section presents some of the main differences between the outcomes obtained by the proposed Coevolutionary Optional Prisoner's Dilemma (COPD) game and other models which do not adopt the concept of coevolution and/or abstention. In the COPD game, we also investigate how a population in an unbiased environment evolves over time.

4.1 Presence of Abstainers in the Coevolutionary Model

In order to provide a means to effectively explore the impact of our coevolutionary model, i.e., the Coevolutionary Prisoner's Dilemma (COPD) game, in the emergence of cooperation, we start by investigating the performance of some of the existing models. Namely, the Coevolutionary Prisoner's Dilemma (CPD) game (i.e., same coevolutionary model as the COPD but without the concept of abstention), the traditional Prisoner's Dilemma (PD) game, and the Optional Prisoner's Dilemma game.

As shown in Fig. 1, it can be observed that for both PD and CPD games, when the defector's payoff is very high (i.e., $b > 1.7$) defectors spread quickly and dominate the environment. On the other hand, when abstainers are present in a static and unweighted network, i.e., playing the OPD game, we end up with abstainers dominating the environment. Undoubtedly, in many scenarios, having a population of abstainers is better than a population of defectors. However, it provides clear evidence that all these three models fail to sustain cooperation. In fact, results show that in this type of adverse environment (i.e., with a high temptation to defect), cooperation has no chance of emerging.

Figure 2 shows a typical phase diagram for both CPD and COPD games for a fixed value of $\delta = 0.8$ and $l = 0.6$ (on the COPD game). It can be observed that if a given environmental setting (i.e, b, Δ and δ) produces a stable population of cooperators in the CPD game, then the presence of abstainers will not change it. In other words, the COPD game does not affect the outcome of scenarios in which cooperation is stable in the absence of abstainers. Thus, the main changes occur in scenarios in which defection dominates or survives ($b > 1.5$).

Surprisingly, as shown in Fig. 2, when considering the Coevolutionary Optional Prisoner's Dilemma (COPD) game for the same environmental settings of Fig. 1 (i.e., $l = 0.6$, $\Delta = 0.72$ and $\Delta = 0.72$), with the temptation of defection almost at its peak (i.e., $b = 1.9$), it is possible to reach high levels of cooperation.

To summarize, despite the fact that the Coevolutionary Prisoner's Dilemma (CPD) game succeeds in the promotion of cooperation in a wide range of scenarios, it is still not able to avoid the invasion by defectors in cases where $b > 1.5$, which does not happen when the abstainers are present (i.e., COPD game).

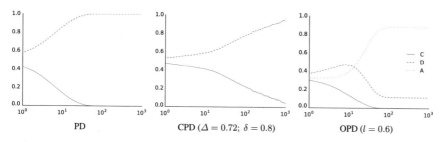

Fig. 1 Comparison of the Prisoner's Dilemma (PD), the Coevolutionary Prisoner's Dilemma (CPD) and the Optional Prisoner's Dilemma (OPD) games. All with the same temptation to defect, $b = 1.9$

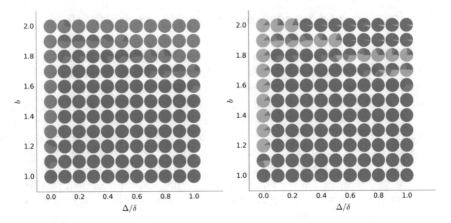

Fig. 2 Typical phase diagram for an initial balanced population playing the Coevolutionary Prisoner's Dilemma game (left) and the Coevolutionary Optional Prisoner's Dilemma game with $l = 0.6$ (right), both with $\delta = 0.8$

4.2 Inspecting the Coevolutionary Environment

In order to further explain the results witnessed in the previous experiments, we investigate how the population evolves over time for the Coevolutionary Optional Prisoner's Dilemma game. Figure 3 features the time course of cooperation for three different values of $\Delta/\delta = \{0.0, 0.2, 1.0\}$, which are some of the critical points when $b = 1.9$, $l = 0.6$ and $\delta = 0.8$. Based on these results, in Fig. 4 we show snapshots for the Monte Carlo steps 0, 45, 1113 and 10^5 for the three scenarios shown in Fig. 3.

We see from Fig. 4 that for the traditional case (i.e., $\Delta/\delta = 0.0$), abstainers spread quickly and reach a stable state in which single defectors are completely isolated by abstainers. In this way, as the payoffs obtained by a defector and an abstainer are the same, neither will ever change their strategy. In fact, even if a single cooperator survives up to this stage, for the same aforementioned reason, its strategy will not change either. In fact, the same behaviour is noticed for any value of $b > 1.2$ and $\Delta/\delta = 0$ (COPD in Fig. 2).

When $\Delta/\delta = 0.2$, it is possible to observe some sort of equilibrium between the three strategies. They reach a state of cyclic competition in which abstainers invade defectors, defectors invade cooperators and cooperators invade abstainers.

This behaviour, of balancing the three possible outcomes, is very common in nature where species with different reproductive strategies remain in equilibrium in the environment. For instance, the same scenario was observed as being responsible for preserving biodiversity in the neighbourhoods of the *Escherichia coli*, which is a bacteria commonly found in the lower intestine of warm-blooded organisms. According to Fisher [7], studies were performed with three natural populations of this bacteria: (i) produces a natural antibiotic but is immune to its effects, (ii) is sensitive to the antibiotic but can grow faster than the third population, which (iii)

Fig. 3 Progress of the
fraction of cooperation ρ_c
during a Monte Carlo
simulation for $b = 1.9$,
$l = 0.6$ and $\delta = 0.8$ [5]

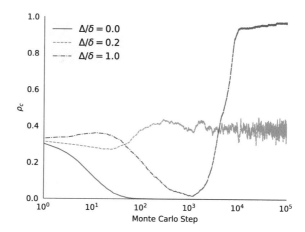

is resistant to the antibiotic. They observed that when these populations are mixed
together, each of them ends up establishing its own territory in the environment.
It happens because the first population kill off any other bacteria sensitive to the
antibiotic, the second population uses their faster growth rate to displace the bacteria
which are resistant to the antibiotic, and the third population could use their immunity
to displace the first population.

Another interesting behaviour is noticed for $\Delta/\delta = 1.0$. In this scenario, defectors
are dominated by abstainers, allowing a few clusters of cooperators to survive. As
a result of the absence of defectors, cooperators invade abstainers and dominate the
environment.

5 Exploring the Coevolutionary Optional Prisoner's Dilemma Game

In this section, we present some of the relevant experimental results of the Monte
Carlo simulations of the Coevolutionary Optional Prisoner's Dilemma game in an
unbiased environment. That is, a well-mixed initial population with a balanced
amount of cooperators, defectors and abstainers.

5.1 Investigating the Properties of Δ and δ

This section aims to investigate the properties of the presented model (Sect. 3) in
regard to the parameters Δ and δ. These parameters play a key role in the evolutionary
dynamics of this model because they define the number of possible link weights that
an agent is allowed to have (i.e., they define the number of states).

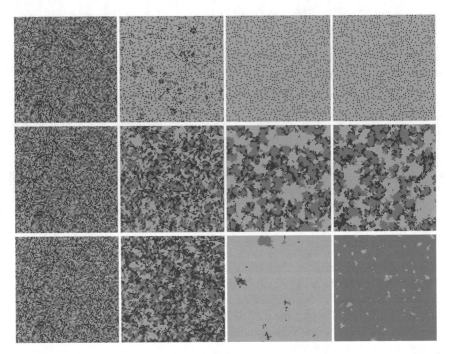

Fig. 4 Snapshots of the distribution of the strategy in the Monte Carlo steps 0, 45, 1113 and 10^5 (from left to right) for Δ/δ equal to 0.0, 0.2 and 1.0 (from top to bottom). In this Figure, cooperators, defectors and abstainers are represented by the colours blue, red and green respectively. All results are obtained for $b = 1.9$, $l = 0.6$ and $\delta = 0.8$ [5]

Despite the fact that the number of states is discrete, the act of counting them is not straightforward. For instance, when counting the number of states between $1 - \delta$ and $1 + \delta$ for $\Delta = 0.2$ and $\delta = 0.3$, we could incorrectly state that there are four possible states for this scenario, i.e., {0.7, 0.9, 1.1, 1.3}. However, considering that the link weights of all edges are initially set to $w = 1$, and due to the other constraints (Eqs. 3 and 4), the number of states is actually seven, i.e., {0.7, 0.8, 0.9, 1.0, 1.1, 1.2, 1.3}.

In order to better understand the relationship between Δ and δ, we plot Δ, δ and Δ/δ as a function of the number of states (numerically counted) for a number of different values of both parameters (Fig. 5). It was observed that given the pairs (Δ_1, δ_1) and (Δ_2, δ_2), if Δ_1/δ_1 is equal to Δ_2/δ_2, then the number of states of both settings is the same.

Figure 5 shows the ratio Δ/δ as a function of the number of states. As we can see, although the function is non-linear and non-monotonic, in general, higher values of Δ/δ have less states.

Fig. 5 The ratio Δ/δ as a function of the number of states. For any combination of Δ and δ, the ration Δ/δ will always have the same number of states

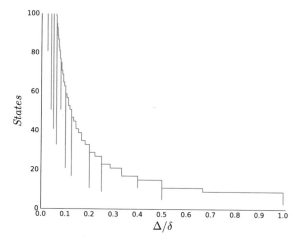

5.2 Varying the Number of States

Figure 6 shows the impact of the coevolutionary model on the emergence of cooperation when the ratio Δ/δ varies for a range of fixed values of the loner's payoff (l), temptation to defect (b) and δ. In this experiment, we observe that when $l = 0.0$, the outcomes of the Coevolutionary Optional Prisoner's Dilemma (COPD) game are very similar to those observed by Huang et al. [12] for the Coevolutionary Prisoner's Dilemma (CPD) game. This result can be explained by the normalized payoff matrix adopted in this work (Table 1). Clearly, when $l = 0.0$, there is no advantage in abstaining from playing the game, thus agents choose the option to cooperate or defect [5].

Results indicate that, in cases where the temptation to defect is very low (e.g, $b \leq 1.34$), the level of cooperation does not seem to be affected by the increment of the loner's payoff, except when the advantage of abstaining is very high (e.g, $l > 0.8$). However, these results highlight that the presence of the abstainers may protect cooperators from invasion. Moreover, the difference between the traditional Optional Prisoner's Dilemma (i.e., $\Delta/\delta = 0.0$) for $l = \{0.0, 0.6\}$ and all other values of Δ/δ is strong evidence that our coevolutionary model is very advantageous to the promotion of cooperative behaviour.

Namely, when $l = 0.6$, in the traditional case with a static and unweighted network ($\Delta/\delta = 0.0$), the cooperators have no chance of surviving; except, of course, when b is very close to the reward for mutual cooperation R, where it is possible to observe scenarios of quasi-stable states of the three strategies or between cooperators and defectors. In fact, in the traditional OPD ($\Delta/\delta = 0.0$), when $l > 0.0$ and $b > 1.2$, abstainers are always the dominant strategy. However, as discussed in previous work [5], when the coevolutionary rules are used, cooperators do much better, being also able to dominate the whole population in many cases.

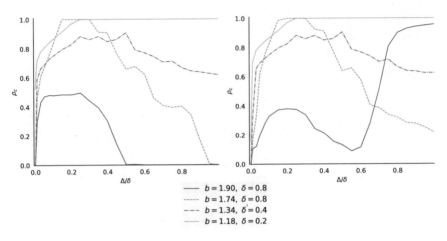

Fig. 6 Relationship between cooperation and the ratio Δ/δ when the loner's payoff (l) is equal to 0.0 (left) and 0.6 (right) [5]

It is noteworthy that the curves in Fig. 6 are usually non-linear and/or non-monotonic because of the properties of the ratio Δ/δ in regard to the number of states of each combination of Δ and δ (Sect. 5.1).

5.3   Investigating the Relationship Between Δ/δ, b and l

To investigate the outcomes in other scenarios, we explore a wider range of settings by varying the values of the temptation to defect (b), the loner's payoff (l) and the ratio Δ/δ for a fixed value of $\delta = 0.8$.

As shown in Fig. 7, cooperation is the dominant strategy in the majority of cases. Note that in the traditional case, with an unweighted and static network, i.e., $\Delta/\delta = 0.0$, abstainers dominate in all scenarios illustrated in this ternary diagram. In addition, it is also possible to observe that certain combinations of l, b and Δ/δ guarantee higher levels of cooperation. In these scenarios, cooperators are protected by abstainers against exploitation from defectors.

Complementing previous findings [5], another observation is that defectors are attacked more efficiently by abstainers as we increase the loner's payoff (l). Simulations reveal that, for any scenario, if the loner's payoff is greater than 0.7 ($l > 0.7$), defectors have no chance of surviving. However, the drawback of increasing the value of l is that it makes it difficult for cooperators to dominate abstainers, which might produce a quasi-stable population of cooperators and abstainers. It is noteworthy that it is a counter-intuitive result from the COPD game, since the loner's payoff is always less than the reward for mutual cooperation (i.e., $L < R$), even for extremely high values of L. This scenario (population of cooperators and abstainers) should always lead cooperators to quickly dominate the environment.

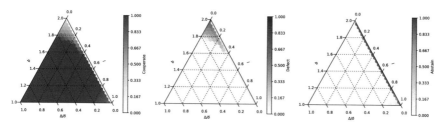

Fig. 7 Ternary diagrams of different values of b, l and Δ/δ for $\delta = 0.8$ [5]

In fact, it is still expected that, in the COPD game, cooperators dominate abstainers, but depending on the value of the loner's payoff, or the amount of abstainers in the population at this stage, it might take several Monte Carlo steps to reach a stable state, which is usually a state of cooperation fully dominating the population.

An interesting behaviour is noticed when $l = [0.45, 0.55]$ and $b > 1.8$. In this scenario, abstainers quickly dominate the population, making a clear division between two states: before this range (defectors hardly die off) and after this range (defectors hardly survive). In this way, a loner's payoff value greater than 0.55 ($l > 0.55$) is usually the best choice to promote cooperation. This result is probably related to the difference between the possible utilities for each type of interaction, which still needs further investigation in future.

Although the combinations shown in Fig. 7 for higher values of b ($b > 1.8$) are just a small subset of an infinite number of possible values, it is clearly shown that a reasonable fraction of cooperators can survive even in an extremely adverse situation where the advantage of defecting is very high. Indeed, our results show that some combinations of high values of l and δ, such as for $\delta = 0.8$ and $l = 0.7$, can further improve the levels of cooperation, allowing for the full dominance of cooperation.

6 Investigating the Robustness of Cooperation in a Biased Environment

The previous experiments revealed that the presence of abstainers together with simple coevolutionary rules (i.e., the COPD game) act as a powerful mechanism to avoid the spread of defectors, which also allows the dominance of cooperation in a wide range of scenarios. However, the distribution of the strategies in the initial population used in all of the previous experiments was uniform. That is, we have explored cases in which the initial population contained a balanced amount of cooperators, defectors and abstainers. Thus, in order to explore the robustness of these outcomes in regard to the initial amount of abstainers in the population, we now aim to investigate how many abstainers would be necessary to guarantee robust cooperation.

Figure 8 features the fraction of each strategy in the population (i.e., cooperators, defectors and abstainers) over time for fixed values of $b = 1.9$, $\Delta = 0.72$ and $\delta = 0.8$.

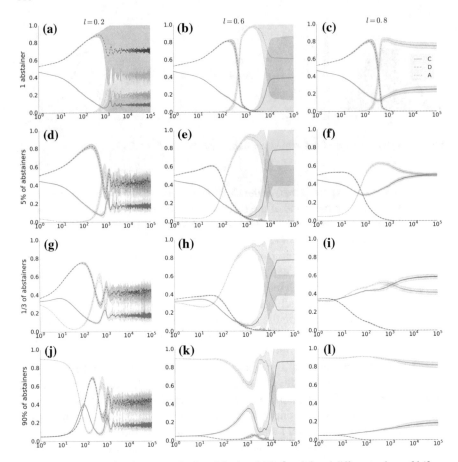

Fig. 8 Time course of each strategy for $b = 1.9$, $\Delta = 0.72$, $\delta = 0.8$ and different values of l (from left to right, $l = \{0.2,\ 0.6,\ 0.8\}$). The same settings are also tested on populations seeded with different amount of abstainers (i.e, from top to bottom: 1 abstainer, 5% of the population, 1/3 of the population, 90% of the population)

In this experiment, several independent simulations were performed, in which the loner's payoff (l) and the number of abstainers in the initial population were varied from 0.0 to 1.0 and from 0.1% to 99.9%, respectively. Other special cases were also analyzed, such as placing only one abstainer into a balanced population of cooperators and defectors, and placing only one defector and one cooperator in a population of abstainers. For the sake of simplicity, we report only the values of $l = \{0.2, 0.6, 0.8\}$ for an initial population with one, 5%, 33% and 90% abstainer(s), which are representative of the outcomes at other values also. Note that, for all these simulations, the initial population of cooperators and defectors remained in balance. For instance, an initial population with 50% of abstainers, will consequently have 25% of cooperators and 25% of defectors.

Experiments reveal that the COPD game is actually extremely robust to radical changes in the initial population of abstainers. It has been shown that if the loner's payoff is greater than 0.55 ($l > 0.55$), then one abstainer might alone be enough to protect cooperators from the invasion of defectors (see Fig. 8a, b and c). However, this outcome is only possible if the single abstainer is in the middle of a big cluster of defectors.

This outcome can happen because the payoff obtained by the abstainers is always greater than the one obtained by pairs of defectors (i.e., $L < P$). Thus, in a cluster of defectors, abstention is always the best choice. However, as this single abstainer reduces the population of defectors, which consequently increases the population of abstainers and cooperators in the population, defection may start to be a good option again due to the increase of cooperators. Therefore, the exploitation of defectors by abstainers must be as fast as possible, otherwise, they might not be able to effectively attack the population of defectors. In this scenario, the loner's payoff is the key parameter to control the speed in which abstainers invade defectors. This explains why a single abstainer is usually not enough to avoid the dominance of defectors when $l < 0.55$.

In this way, as the loner's payoff is the only parameter that directly affects the evolutionary dynamics of the abstainers, intuition might lead one to expect to see a clear and perhaps linear relationship between the loner's payoff and the initial number of abstainers in the population. That is, given the same set of parameters, increasing the initial population of abstainers or the loner's payoff would probably make it easier for abstainers to increase or even dominate the population. Despite the fact that it might be true for high values of the loner's payoff (i.e., $l \geq 0.8$, as observed in Fig. 8), it is not applicable to other scenarios. Actually, as it is also shown in Fig. 8, if the loner's payoff is less than 0.55, changing the initial population of abstainers does not change the outcome at all. When $0.55 \leq l < 0.8$, a huge initial population of abstainers can actually promote cooperation best.

As discussed in Sect. 5.3, populations of cooperators and abstainers tend to converge to cooperation. In this way, the scenario showed in Fig. 8 for $l = 0.8$ will probably end up with cooperators dominating the population, but as the loner's payoff is close to the reward for mutual cooperation, the case in Fig. 8i will converge faster than the one showed in Fig. 8l.

Another very counter-intuitive behaviour occurs in the range $l = [0.45, 0.55]$ (this range may shift a little bit depending on the value of b), where the outcome is usually of abstainers quickly dominating the population (Sect. 5.3). In this scenario, we would expect that changes in the initial population of abstainers would at least change the speed in which the abstainers fixate in the population. That is, a huge initial population of abstainers would probably converge quickly. However, it was observed that the convergence speed is almost the same regardless of the size of the initial population of abstainers.

In summary, results show that an initial population with 5% of abstainers is usually enough to make reasonable changes in the outcome, increasing the chances of cooperators surviving or dominating the population.

7 Conclusions and Future Work

This paper studies the impact of a simple coevolutionary model in which not only the agents' strategies but also the network evolves over time. The model consists of placing agents playing the Optional Prisoner's Dilemma game in a dynamic spatial environment, which in turn, defines the Coevolutionary Optional Prisoner's Dilemma (COPD) game [5].

In summary, based on the results of several Monte Carlo simulations, it was shown that the COPD game allows for the emergence of cooperation in a wider range of scenarios than the Coevolutionary Prisoner's Dilemma (CPD) game [12], i.e., the same coevolutionary model in populations which do not have the option to abstain from playing the game. Results also showed that COPD performs much better than the traditional version of these games, i.e., the Prisoner's Dilemma (PD) and the Optional Prisoner's Dilemma (OPD) games, where only the strategies evolve over time in a static and unweighted network. Moreover, we observed that the COPD game is actually able to reproduce outcomes similar to other games by setting the parameters as follows:

- CPD: $l = 0$.
- OPD: $\Delta = 0$ (or $\delta = 0$).
- PD: $l = 0$ and $\Delta = 0$ (or $\delta = 0$).

Also, it was possible to observe that abstention acts as an important mechanism to avoid the dominance of defectors. For instance, in adverse scenarios such as when the defector's payoff is very high ($b > 1.7$), for both PD and CPD games, defectors spread quickly and dominated the environment. On the other hand, when abstainers were present (COPD game), cooperation was able to survive and even dominate.

Furthermore, simulations showed that defectors die off when the loner's payoff is greater than 0.7 ($l > 0.7$). However, it was observed that increasing the loner's payoff makes it difficult for cooperators to dominate abstainers, which is a counter-intuitive result, since the loner's payoff is always less than the reward for mutual cooperation (i.e., $L < R$), this scenario should always lead cooperators to dominance very quickly. In this scenario, cooperation is still the dominant strategy in most cases, but it might require several Monte Carlo steps to reach a stable state.

Results revealed that the COPD game also allows scenarios of cyclic dominance between the three strategies (i.e., cooperation, defection and abstention), indicating that, for some parameter settings, the COPD game is intransitive. That is, the population remains balanced in such a way that cooperators invade abstainers, abstainers invade defectors and defectors invade cooperators, closing a cycle.

We also explored the robustness of these outcomes in regard to the initial amount of abstainers in the population (biased population). In summary, it was shown that, in some of the scenarios, even one abstainer might alone be enough to protect coopera-tors from the invasion of defectors, which in turn increases the chances of cooperators surviving or dominating the population. We conclude that the combination of both of these trends in evolutionary game theory may shed additional light on gaining an in-depth understanding of the emergence of cooperative behaviour in real-world scenarios.

Future work will consider the exploration of different topologies and the influence of a wider range of scenarios, where, for example, agents could rewire their links, which, in turn, adds another level of complexity to the model. Future work will also involve applying our studies and results to realistic scenarios, such as social networks and real biological networks.

Acknowledgements This work was supported by the National Council for Scientific and Technological Development (CNPq-Brazil). Grant number: 234913/20142.

References

1. Batali, J., Kitcher, P.: Evolution of altruism in optional and compulsory games. J. Theor. Biol. **175**(2), 161–171 (1995)
2. Cao, L., Ohtsuki, H., Wang, B., Aihara, K.: Evolution of cooperation on adaptively weighted networks. J. Theor. Biol. **272**(1), 8–15 (2011)
3. Cardinot, M., Gibbons, M., O'Riordan, C., Griffith, J.: Simulation of an optional strategy in the prisoner's dilemma in spatial and non-spatial environments. From Animals to Animats 14 (SAB 2016), pp. 145–156. Springer International Publishing, Cham (2016)
4. Cardinot, M., Griffith, J., O'Riordan, C.: Cyclic dominance in the spatial coevolutionary optional prisoner's dilemma game. In: Greene, D., Namee, B.M., Ross, R. (eds.) Artificial Intelligence and Cognitive Science 2016. CEUR Workshop Proceedings, vol. 1751, pp. 33–44. Dublin, Ireland (2016)
5. Cardinot, M., O'Riordan, C., Griffith, J.: The optional prisoner's dilemma in a spatial environment: coevolving game strategy and link weights. In: Proceedings of the 8th International Joint Conference on Computational Intelligence (IJCCI 2016), pp. 86–93 (2016)
6. Chen, X., Wang, L.: Promotion of cooperation induced by appropriate payoff aspirations in a small-world networked game. Phys. Rev. E **77**, 017103 (2008)
7. Fisher, L.: Rock, Paper, Scissors: Game Theory in Everyday Life. Basic Books, New York (2008)
8. Fu, F., Liu, L.H., Wang, L.: Evolutionary prisoner's dilemma on heterogeneous Newman-Watts small-world network. Eur. Phys. J. B **56**(4), 367–372 (2007)
9. Ghang, W., Nowak, M.A.: Indirect reciprocity with optional interactions. J. Theor. Biol. **365**, 1–11 (2015)
10. Gómez-Gardeñes, J., Romance, M., Criado, R., Vilone, D., Sánchez, A.: Evolutionary games defined at the network mesoscale: the public goods game. Chaos **21**(1), 016113 (2011)
11. Hauert, C., Traulsen, A., Brandt, H., Nowak, M.A.: Public goods with punishment and abstaining in finite and infinite populations. Biol. Theory **3**(2), 114–122 (2008)
12. Huang, K., Zheng, X., Li, Z., Yang, Y.: Understanding cooperative behavior based on the coevolution of game strategy and link weight. Sci. Rep. **5**, 14783 (2015)
13. Jeong, H.C., Oh, S.Y., Allen, B., Nowak, M.A.: Optional games on cycles and complete graphs. J. Theor. Biol. **356**, 98–112 (2014)
14. Nowak, M.A., May, R.M.: Evolutionary games and spatial chaos. Nature **359**(6398), 826–829 (1992)
15. Olejarz, J., Ghang, W., Nowak, M.A.: Indirect reciprocity with optional interactions and private information. Games **6**(4), 438–457 (2015)
16. Perc, M., Szolnoki, A.: Coevolutionary games - a mini review. Biosystems **99**(2), 109–125 (2010)

17. Szabó, G., Hauert, C.: Evolutionary prisoner's dilemma games with voluntary participation. Phys. Rev. E **66**, 062903 (2002)
18. Szolnoki, A., Perc, M.: Promoting cooperation in social dilemmas via simple coevolutionary rules. Eur. Phys. J. B **67**(3), 337–344 (2009)
19. Szolnoki, A., Perc, M.: Leaders should not be conformists in evolutionary social dilemmas. Sci. Rep. **6**, 23633 (2016)
20. Wang, Z., Szolnoki, A., Perc, M.: Self-organization towards optimally interdependent networks by means of coevolution. New J. Phys. **16**(3), 033041 (2014)
21. Xia, C.Y., Meloni, S., Perc, M., Moreno, Y.: Dynamic instability of cooperation due to diverse activity patterns in evolutionary social dilemmas. EPL **109**(5), 58002 (2015)
22. Zimmermann, M.G., Eguíluz, V.M., San Miguel, M.: Cooperation, Adaptation and the Emergence of Leadership, pp. 73–86. Springer, Berlin (2001)
23. Zimmermann, M.G., Eguíluz, V.M., San Miguel, M.: Coevolution of dynamical states and interactions in dynamic networks. Phys. Rev. E **69**, 065102 (2004)

Advances in the Evolution of Complex Cellular Automata

Michal Bidlo

Abstract In this study we present some advanced experiments dealing with the evolutionary design of multi-state uniform cellular automata. The generic square calculation problem in one-dimensional automata will be treated as one of the case studies. An analysis of the evolutionary experiments will be proposed and properties of the resulting cellular automata will be discussed. It will be demonstrated that various approaches to the square calculations in cellular automata exist, some of which substantially overcome the known solution. The second case study deals with a non-trivial pattern development problem in two-dimensional automata. Some of the results will be presented which indicate that an exact behaviour can be automatically designed even for cellular automata working with more than ten cell states. A discussion for both case studies is included and potential areas of further research are highlighted.

Keywords Evolutionary algorithm · Cellular automaton · Transition function · Conditional rule · Square calculation · Pattern development

1 Introduction

The concept of cellular automata was introduced by von Neumann in [25]. One of the aspects widely studied in his work was the problem of (universal) computational machines and the question about their ability to make copies of themselves (i.e. to self-reproduce). Von Neumann proposed a model with 29 cell states to perform this task. Later Codd proposed another approach and showed that the problem of computation and construction can be performed by means of a simplified model working with 8 states only [7].

M. Bidlo (✉)
Faculty of Information Technology, IT4Innovations Centre of Excellence,
Brno University of Technology, Božetěchova 2, 61266 Brno, Czech Republic
e-mail: bidlom@fit.vutbr.cz
URL: http://www.fit.vutbr.cz/~bidlom

© Springer Nature Switzerland AG 2019
J. J. Merelo et al. (eds.), *Computational Intelligence*,
Studies in Computational Intelligence 792,
https://doi.org/10.1007/978-3-319-99283-9_7

Several other researchers studied cellular automata usually by means of various rigorous techniques. For instance, Sipper studied computational properties of *binary* cellular automata (i.e. those working with 2 cell states only) and proposed a concept of universal computing platform using a two-dimensional (2D) CA with non-uniform transition function (i.e. each cell can, in general, be controlled by a different set of transition rules) [20]. Sipper showed that, by introducing the non-uniform concept to the binary CAs, universal computation can be realised, which was not possible using the Codd's model. In fact, Sipper's work significantly reduced the complexity of the CA in comparison with the models published earlier. Nevertheless even the binary uniform 2D CAs can be computationally universal if 9-cell neighbourhood is considered. Such CA was implemented using the famous rules of the Game of Life [2] (original proof of the concept was published in 1982 and several times revisited – e.g. see [8, 11, 16, 17]).

Although binary CAs may be advantageous due to simple elementary rules and hardware implementations in particular, many operations and real-world problems can effectively be solved by *multi-state* cellular automata (i.e. those working with more than 2 cell states) rather than those using just two states. For example, a technique for the construction of computing systems in a 2D CA was demonstrated in [22] using rules of a simple game called Scintillae working with 6 cell states. Computational universality was also studied with respect to one-dimensional (1D) CA, e.g. in [12, 27].

However, in some cases application specific operations (algorithms) may be more suitable than programming a universal system, allowing to better optimize various aspects of the design (e.g. resources, efficiency, data encoding etc.). For example, Tempesti [24] and Perrier et al. [15] showed that specific arrangements of cell states can encode sequences of instructions (programs) to perform a given operation. Wolfram presented various transition functions for CAs in order to compute elementary as well as advanced functions (e.g. parity, square, or prime number generation) [26]. Further problems were investigated in recent years [14, 18].

In addition to the computational tasks, various other (more general) benchmark problems have been investigated using cellular automata, e.g. including principles of self-organization, replication or pattern formation. For example, Basanta et al. used a genetic algorithm to evolve the rules of effector automata (a generalised variant of CA) to create microstructural patterns that are similar to crystal structures known from some materials [1]. An important aspect of this work was to investigate new materials with specific properties and their simulation using computers. Suzudo proposed an approach to the evolutionary design of 2D asynchronous CA for a specific formation of patterns in groups in order to better understand of the pattern-forming processes known from nature [23]. Elmenreich et al. proposed an original technique for growing self-organising structures in CA whose development is controlled by neural networks according to the internal cell states [9].

The proposed work represents an extended version of our recent study published in [4], the aim of which is to present a part of our wider research in the area of cellular automata, where representation techniques and automatic (evolutionary) methods for the design of complex multi-state cellular automata are investigated. The goal of this work is to design transition functions for cellular automata using evolutionary algorithms, which satisfy the given behaviour with respect to some specific initial and target conditions. In particular, it will be shown that the evolutionary algorithm can design various transition functions for uniform 1D CAs (that have never been seen before) to perform *generic* square calculations in the cellular space using just local interactions of cells. An additional analysis of the results demonstrates that various generic CA-based solutions of the squaring problem can be discovered, which substantially overcome the known solution regarding both the complexity of the transition functions and the number of steps (speed) of calculation. In order to show the abilities of the proposed method for designing CA using the concept of conditionally matching rules, some further experiments are presented regarding the evolution of 2D multi-state cellular automata in which the formation of some non-trivial patterns is treated as a case study. As cellular automata represent a platform potentially important for future technologies (see their utilisation in various emerging fields, e.g. [13, 21] or [19]), it is worth studying their design and behaviour on the elementary level as well (i.e. using various benchmark problems).

2 Cellular Automata for Square Calculations

For the purposes of developing algorithms for squaring natural numbers, 1D uniform cellular automata are treated with the following specification (target behaviour). The number of cell states is investigated for values 4, 6, 8 and 10 (this was chosen on the basis of the existing solution [26] that uses 8 states; moreover it is worth of determining whether less states will enable to design generic solutions and whether the EA will be able to find solutions in a huge search space induced by 10 cell states). The new state of a given cell depends on the states of its west neighbour (c_W), the cell itself (central cell, c_C) and its east neighbour (c_E), i.e. it is a case of 3-cell neighbourhood. A *step* of the CA will be considered as a synchronous update of state values of all its cells according to a given transition function. For the practical implementation purposes, cyclic boundary conditions are considered. However, it is important to note that CAs with sufficient sizes are used in order to avoid affecting the development by the finite number of cells.

The value of x is encoded in the initial CA state as a continuous sequence of cells in state 1, whose length (i.e. the number of cells in state 1) corresponds to x, the other cells possess state 0. For example, the state of a 12-cell CA, which encodes $x = 3$, can appear as 0000011100000. The result $y = x^2$, that will emerge from the initial state in a finite number of steps, is assumed as a stable state in which a continuous sequence of cells in non-zero states can be detected, the length of which equals the value of y, the other cells are required in state 0. For the aforementioned example, the result can appear as 002222222220 or even 023231323200 (there is a sequence of non-zero cells of length $3^2 = 9$). The concept of representing the input value x and the result y is graphically illustrated in Fig. 1. This is a generalised interpretation based on the idea presented in [26], p. 639. The goal is to discover transition functions for the CA, that are able to calculate the square of arbitrary number $x > 1$.

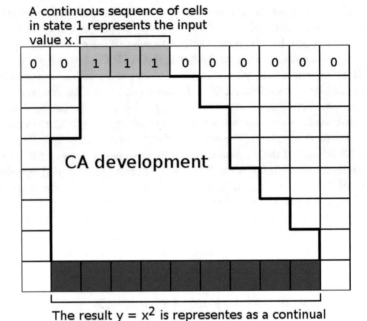

Fig. 1 Illustration of encoding integer values in a 1D cellular automaton. In this example $x = 3$, $y = 9$. Extracted from [4]

2.1 Conditionally Matching Rules

In order to represent the transition functions for CAs, the concept of Condition-ally Matching Rules (CMR), originally introduced in [6], will be applied. This technique showed as very promising for designing complex cellular automata [3, 5]. For the 1D CA working with 3-cell neighbourhood, a CMR is defined as $(cond_W\, s_W)(cond_C\, s_C)(cond_E\, s_E) \rightarrow s_{Cnew}$, where $cond_\star$ denotes a condition function and $s_\star$ denotes a state value. Each part $(cond_\star\, s_\star)$ on the left of the arrow is evaluated with respect to the state of a specific cell in the neighbourhood (in this case c_W, c_C and c_E respectively). For the experiments presented in this work the relation operators $=$, $\neq$, $\geq$ and $\leq$ are considered as the condition functions. A finite sequence of CMRs represents a transition function. In order to determine the new state of a cell, the CMRs are evaluated sequentially. If a rule is found in which all conditions are true (with respect to the states in the cell neighbourhood), s_{Cnew} from this rule is the new state of the central cell. Otherwise the cell state does not change. For example, consider a transition function that contains a CMR $(\neq 1)(\neq 2)(\leq 1) \rightarrow 1$. Let c_W, c_C, c_E be states of cells in a neighbourhood with values $2, 3, 0$ respectively, and a new state of the central cell ought to be calculated. According to the aforementioned rule, $c_W \neq s_W$ is true as $2 \neq 1$, similarly $c_C \neq s_C$ is true ($3 \neq 2$) and $c_E \leq s_E$ ($0 \leq 1$). Therefore, this CMR is said to match, i.e. $s_{Cnew} = 1$ on its right side will update the state of the central cell. Note that the same concept can also be applied to CA working with a wider cellular neighborhood. For example, a CMR for a 2D CA with 5-cell neighborhood would consist of 5 items for the conditional functions instead for 3 items.

The evolved CMRs can be transformed to the conventional table rules [5] without loss of functionality or violating the basic CA principles. In this work the transformation is performed as follows: (1) For every possible combination of states c_W c_C c_E in cellular neighborhood a new state s_{Cnew} is calculated using the CMR-based transition function. (2) If $c_C \neq s_{Cnew}$ (i.e. the cell state ought to be modified), then a table rule of the form c_W c_C c_E $\rightarrow$ s_{Cnew} is generated. Note that the combinations of states not included amongst the table rules do not change the state of the central cell, which is treated implicitly during the CA simulation. The number of such *generated rules* will represent a metrics indicating the complexity of the transition function.

In order to determine the complexity of the transition function with respect to a specific square calculation in CA, a set of *used rules* is created using the aforementioned principle whereas the combinations of states c_W c_C c_E are considered just occurring during the given square calculation in the CA. There metrics (together with the number of states and CA steps) will allow us to compare the solutions obtained by the evolution and to identify the best results with respect to their complexity and efficiency.

An evolutionary algorithm will be applied to search for suitable CMR-based transition functions as described in the following section.

3   Setup of the Evolutionary System

A custom evolutionary algorithm (EA) was utilised, which is a result of our long-term experimentation in this area. Note, however, that neither tuning of the EA nor in-depth analysis of the evolutionary process is a subject of this work. The EA is based on a simple genetic algorithm [10] with a tournament selection of base 4 and a custom mutation operator. Crossover is not used as it has not shown any improvement in success rate or efficiency of our experiments.

The EA utilises the following fixed-length representation of the conditionally matching rules in the genomes. For the purpose of encoding the condition functions $=$, $\neq$, $\geq$ and $\leq$, integer values 0, 1, 2 and 3 will be used respectively. Each part $(cond_\star$ $s_\star)$ of the CMR is encoded as a single integer $P_\star$ in the range from 0 to M where $M = 4 * S - 1$ (4 is the fixed number of condition functions considered and S is the number of cell states) and the part $\rightarrow$ s_{Cnew} is represented by an integer in the range from 0 to $S - 1$. In order to decode a specific condition and state value, the following operations are performed: $cond_\star = P_\star / S$, $s_\star = P_\star$ mod S (note that / is the integer division and *mod* is the modulo-division). This means that a CMR $(cond_W$ $s_W)(cond_C$ $s_C)(cond_E$ $s_E) \rightarrow s_{Cnew}$ can be represented by 4 integers;

if 20 CMRs ought to be encoded in the genome, then $4 * 20 = 80$ integers are needed. For example, consider $S = 3$ for which $M = 4 * 3 - 1 = 11$. If a 4-tuple of integers (2 9 11 2) representing a CMR in the genome ought to be decoded, then the integers are processed respectively as:

- $cond_W = 2/3 = 0$ which corresponds to the operator $=$, $s_W = 2 \bmod 3 = 2$,
- $cond_C = 9/3 = 3$ which corresponds to the operator $\leq$, $s_C = 9 \bmod 3 = 0$,
- $cond_E = 11/3 = 3$ which corresponds to the operator $\leq$, $s_E = 11 \bmod 3 = 2$,
- $s_{Cnew} = 2$ is directly represented by the 4th integer.

Therefore, a CMR of the form $(=2)(\leq 0)(\leq 2) \rightarrow 2$ has been decoded.

The following variants of the fitness functions are treated (note that the input x is set to the middle of the cellular array):

1. RESULT ANYWHERE (RA-fitness): The fitness is calculated with respect to any valid arrangement (position) of the result sequence in the CA. For example, $y = 4$ in an 8-cell CA may be $rrrr0000$, $0rrrr000$, $00rrrr00$, $000rrrr0$ or $0000rrrr$, where $r \neq 0$ represent the result states that may be generally different within the result sequence. A partial fitness value is calculated for every possible arrangement of the result sequence as the sum of the number of cells in the expected state for the given values of x. The final fitness is the highest of the partial fitness values.
2. SYMMETRIC RESULT (SR-fitness): The result is expected symmetrically with respect to the input. For example, if 0000011100000 corresponds to initial CA state for $x = 3$, then the result $y = 3^2$ is expected as a specific CA state $00rrrrrrrrr00$ (each r may be represented by any non-zero state). The fitness is the number of cells in the expected state.

The fitness evaluation of each genome is performed by simulating the CA for initial states with the values of x from 2 to 6. The result of the x^2 calculation is inspected after the 99th and 100th step of the CA, which allows to involve the state stability check into the evaluation. This approach was chosen on the basis of the maximal x evaluated during the fitness calculation and on the basis of the number of steps needed for the square calculation using the existing solution [26]. In particular, the fitness of a fully working solution evaluated for x from 2 to 6 in a 100-cell CA is given by $F_{max} = 5 * 2 * 100 = 1000$ (there are 5 different values of x for which the result x^2 is investigated in 2 successive CA states, each consisting of 100 cells). The evolved transition functions, satisfying the maximal fitness for the given range of x, are checked for the ability to work in larger CAs for up to $x = 25$ The solutions which pass this check are considered as generic.

The EA works with a population of 8 genomes initialised randomly at the beginning of evolution. After evaluating the genomes, four candidates are selected randomly, the candidate with the highest fitness becomes a parent. An offspring is created by mutating 2 randomly selected integers in the parent. The selection and mutation continue until a new population of the same size is created and the evolutionary process is repeated until 2 million generations are performed. If a solution with the maximal fitness is found, then the evolutionary run is considered as successful. If no such solution is found within the given generation limit, then the evolutionary run is terminated and regarded as unsuccessful.

4 Results of Square Calculations in 1D CA

The evolutionary design of CAs for the generic square calculation has been investigated for the following settings: the number of states 4, 6, 8 and 10, the transition functions consisting of 20, 30, 40 and 50 CMRs and two ways of the fitness calculation described in Sect. 3. For each setup, 100 independent evolutionary runs have been executed. The success rate and average number of generations needed to find a working solution were observed with respect to the evolutionary process. As regards the parameters of the CA, the minimal number of rules and steps needed to calculate the square of x were determined.

For the purposes of comparison of the results proposed in Sect. 4.3, the CA will be denominated by unique identifiers of the form CA–XX–YY, where XX and YY are integers distinguishing the sets of evolutionary experiments and the CA obtained.

4.1 Results for the RA-Fitness

For the RA-fitness, the statistical results are summarised in Table 1. The table also contains the total numbers of generic solutions discovered for the given state setups and parameters determined for these solutions. For every number of states considered, at least one generic solution was identified. For example, a transition function was discovered for the 4-state CA, which consists of 36 table rules (transformed from the CMR representation evolved). This solution can be optimised to 26 rules (by eliminating the rules not used during the square calculation) which represents the simplest CA for generic square calculations known so far (note that Wolfram's CA works with 8 states and 51 rules [26]). Moreover, for example, our solution needs 74 steps to calculate 6^2 whilst Wolfram's CA needs 112 steps, which also represents a substantial innovation discovered by the EA. The CA development corresponding to this solution is shown in Fig. 2.

Another result obtained using the RA-fitness is illustrated by the CA development in Fig. 3. In this case the CA works with 6 states and its transition function consists of 52 effective rules. The number of steps needed, for example, to calculate 6^2, is 46 (and compared to 112 steps of Wolframs CA, it is an improvement of the CA efficiency by more than 50%) which represents the best CA known so far for this operation and the best result obtained from our experiment.

One more example of evolved CA is shown in Fig. 4. This generic solution was obtained in the setup with 8-state CA, however, the transition function works with 6 different states only. There are 49 transition rules, the CA needs 68 steps to calculate 6^2. This means that the EA discovered a simpler solution (regarding the the number

Table 1 Statistics of the evolutionary experiments conducted using the RA-fitness (the upper part of the table) and the parameters of the generic solutions (in the lower part of the table). The parameters of the best results obtained are marked **bold**. Note that # denotes "the number of", the meaning of "generated rules", "used rules" and "steps" of the CA is defined in Sect. 2. Extracted from [4]

| | The number of states | | | | | | | | | | | | | | | |
| | 4 | | | | 6 | | | | 8 | | | | 10 | | | |
The num. of CMRs	succ. rate	avg. gen.	min. steps	min. rules	succ. rate	avg. gen.	min. steps	min. rules	succ. rate	avg. gen.	min. steps	min. rules	succ. rate	avg. gen.	min. steps	min. rules
20	3	844364	54	35	30	769440	45	120	45	570939	39	232	35	328210	47	569
30	3	620998	52	36	24	749837	40	120	38	595467	42	340	33	363360	45	663
40	2	1344286	77	46	19	629122	37	136	30	701612	41	365	29	244566	46	662
50	2	959689	73	43	20	813803	41	134	35	582342	39	348	38	373490	40	762

The number of generic solutions (#generic) obtained for the given number of states and parameters of the generic solutions: #generated rules/#used rules/#steps for 6^2)

#generic	1				5				6				3			
Parameters	**36/26/74**				**176/52/46**, 164/33/87, 152/49/78, 185/66/70, 175/52/69				435/49/68, 403/51/79, 422/39/65, 392/62/76, 423/41/68, 429/94/76				934/64/56, 835/61/79, 916/35/76			

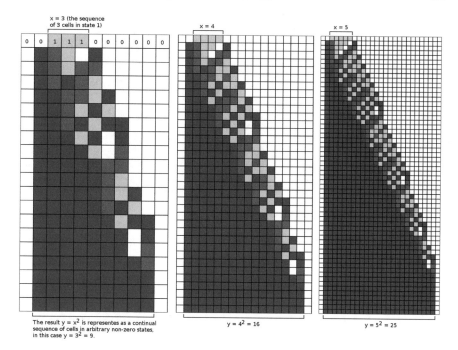

Fig. 2 Example of a 4-state squaring CA development for $x = 3, 4$ and 5 using our most compact transition function. This solution is denominated as CA–30–00 and its rules are: $0\,0\,1 \rightarrow 3, 0\,1\,1 \rightarrow 2,$ $0\,3\,0 \rightarrow 2, 1\,0\,0 \rightarrow 3, 1\,0\,2 \rightarrow 2, 1\,0\,3 \rightarrow 2, 1\,1\,0 \rightarrow 0, 1\,1\,2 \rightarrow 2, 1\,1\,3 \rightarrow 2, 1\,2\,1 \rightarrow 1,$ $1\,3\,0 \rightarrow 1, 1\,3\,1 \rightarrow 1, 1\,3\,2 \rightarrow 2, 1\,3\,3 \rightarrow 0, 2\,1\,2 \rightarrow 3, 2\,1\,3 \rightarrow 3, 2\,2\,0 \rightarrow 1, 2\,2\,1 \rightarrow 1,$ $2\,3\,1 \rightarrow 1, 2\,3\,2 \rightarrow 2, 3\,0\,2 \rightarrow 3, 3\,1\,0 \rightarrow 3, 3\,1\,1 \rightarrow 3, 3\,1\,3 \rightarrow 3, 3\,2\,0 \rightarrow 3, 3\,2\,3 \rightarrow 3.$ Extracted from [4]

of states and table rules) which is a part of the solution space of the 8-state CA. Again, this result exhibits generally better parameters compared to the known solution from [26]. The CA development, that was not observed in any other solution, is also interesting visually - as Fig. 4 shows, the CA generates a pattern with some "dead areas" (cells in state 0) within the cells that subsequently form the result sequence. The size of these areas is gradually reduced, which finally lead to derive the number of steps after which a stable state containing the correct result for the given x has emerged (illustrated by the right part of Fig. 4 for $x = 8$ whereas the CA needs 122 steps to produce the result).

4.2 Results for the SR-Fitness

Table 2 shows the statistics for the SR-fitness together with the total numbers of generic solutions discovered for the given state setups and parameters determined for

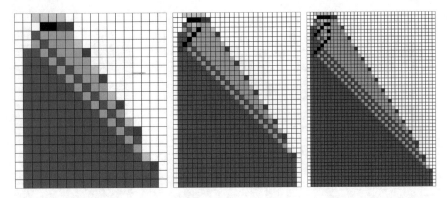

Fig. 3 Example of a 6-state CA development for $x = 4, 5$ and 6. This is the fastest CA-based (3-neighbourhood) solution known so far and the best result obtained from our experiments. This solution is denominated as CA–50–12 and its rules are: $0\,0\,4 \rightarrow 2, 0\,0\,5 \rightarrow 2, 0\,1\,1 \rightarrow 0, 0\,2\,3 \rightarrow 0,$ $0\,2\,4 \rightarrow 4, 0\,2\,5 \rightarrow 3, 0\,3\,2 \rightarrow 2, 0\,3\,3 \rightarrow 0, 0\,4\,0 \rightarrow 2, 0\,4\,2 \rightarrow 2, 0\,5\,3 \rightarrow 0, 0\,5\,5 \rightarrow 4,$ $1\,0\,0 \rightarrow 3, 1\,1\,0 \rightarrow 3, 1\,1\,1 \rightarrow 5, 1\,1\,4 \rightarrow 5, 1\,4\,4 \rightarrow 5, 2\,0\,4 \rightarrow 3, 2\,1\,0 \rightarrow 2, 2\,2\,5 \rightarrow 5,$ $2\,3\,0 \rightarrow 2, 2\,3\,4 \rightarrow 2, 2\,4\,0 \rightarrow 2, 2\,4\,1 \rightarrow 2, 2\,4\,2 \rightarrow 2, 2\,4\,3 \rightarrow 2, 2\,4\,4 \rightarrow 2, 2\,4\,5 \rightarrow 5,$ $2\,5\,2 \rightarrow 2, 2\,5\,4 \rightarrow 2, 3\,3\,0 \rightarrow 4, 3\,4\,4 \rightarrow 2, 4\,0\,0 \rightarrow 1, 4\,1\,0 \rightarrow 4, 4\,1\,1 \rightarrow 4, 4\,1\,4 \rightarrow 4,$ $4\,2\,0 \rightarrow 4, 4\,2\,1 \rightarrow 4, 4\,2\,3 \rightarrow 4, 4\,2\,4 \rightarrow 4, 4\,3\,0 \rightarrow 4, 4\,4\,5 \rightarrow 1, 4\,5\,1 \rightarrow 4, 4\,5\,4 \rightarrow 4,$ $4\,5\,5 \rightarrow 1, 5\,1\,1 \rightarrow 4, 5\,1\,4 \rightarrow 4, 5\,2\,4 \rightarrow 4, 5\,3\,3 \rightarrow 4, 5\,5\,3 \rightarrow 4, 5\,5\,4 \rightarrow 4, 5\,5\,5 \rightarrow 1.$ Extracted from [4]

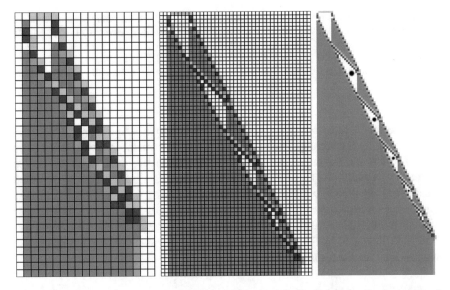

Fig. 4 Example of a 6-state squaring CA development (originally designed using 8-state setup) for $x = 4, 6$ and 8. This solution is denominated as CA–40–01. Its development shows a specific pattern evolved to derive the result of x^2, which was not observed in any other solution. The part on the right shows a complete global behaviour of this CA for $x = 8$ with some "dead areas" (marked by black spots) which lead to the correct stable result by progressively reducing the size of these areas (in this case the result of 8^2 is achieved after 122 steps). Extracted from [4]

these solutions. As evident, the success rates are generally lower compared to the RA-fitness which is expectable because the SR-fitness allows a single arrangement only of the result sequence in the CA. Moreover, just two generic CAs have been identified out of all the runs executed for this setup. However, the goal of this experiment was rather to determine whether solutions of this type ever exist for cellular automata and evaluate the ability of the EA to find them. As regards both generic solutions, their numbers of used rules and CA steps are significantly better in comparison with Wolfram's solution [26]. Specifically, Wolfram's solution uses are 51 rules and the calculation of 6^2 takes 112 steps, whilst the proposed results use 33, respective 36 rules and calculate 6^2 in 71, respective 78 steps. Moreover, one of them was discovered using a 4-state CA (Wolfram used 8 states), which belongs to the most compact solutions obtained herein and known so far.

Figure 5 shows examples of a CA (identified as generic) evolved using the SR-fitness. The transition function, originally obtained in 8-state CA setup, is represented by 36 used rules and works with 7 states only. Although this result cannot be considered as very efficient (for 6^2 the CA needs 78 steps), it exhibits one of the most complex emergent process obtained for the square calculation, the result of which is represented by a non-homogeneous state. The sample on the right of Fig. 5 shows a cutout of development for $x = 11$ in which the global behaviour can be observed. This result demonstrates that the EA can produce generic solutions to a non-trivial problem even for a single specific position of the result sequence required by the SR-fitness evaluation.

4.3 Analysis and Comparison of the Results

In this section an overall analysis and comparison of the results obtained for the generic square calculations is provided and some of further interesting CA are shown. Note that the data related to the CA behavior and visual samples of selected calculations were obtained using our experimental software (i.e. the evolutionary system and a dedicated CA simulator developed specifically for this purpose). There are in total 17 different CA obtained from our experiments and included in this evaluation.

In order to provide a direct comparison of computational efficiency of the CA, the number of steps needed to calculate the square was evaluated for the values of the input integer x from 2 to 16. We used the WolframAlpha computational knowledge engine[1] to generate expressions which allow us to determine the number of steps of a given CA for $x > 16$. Tables 3 and 4 summarize the analysis. The resulting CA are sorted from the best to worst regarding the number of steps for $x = 16$ as the sorting criterion.

For some CA the equations derived by WolframAlpha are specified for an independent integer variable $n > 0$ by means of which the number of steps can be determined for a given input value of x. For example, the number of steps a_n of the CA–50–

---

[1] https://www.wolframalpha.com/.

Table 2 Statistics of the evolutionary experiments conducted using the SR-fitness (the upper part of the table) and the parameters of the generic solutions (in the lower part of the table). The parameters of the best result obtained are marked **bold**. Note that # denotes "the number of", the meaning of "generated rules", "used rules" and "steps" of the CAs is defined in Sect. 2. Extracted from [4]

The num. of CMRs	The number of states															
	4				6				8				10			
	succ. rate	avg. gen.	min. steps	min. rules	succ. rate	avg. gen.	min. steps	min. rules	succ. rate	avg. gen.	min. steps	min. rules	succ. rate	avg. gen.	min. steps	min. rules
20	2	634948	71	38	4	734200	38	126	11	982446	34	234	18	855791	53	542
30	0	–	–	–	5	905278	48	150	17	934123	51	327	15	910269	35	742
40	1	1546681	79	45	4	928170	33	147	11	1033059	53	317	15	898314	52	748
50	0	–	–	–	3	989039	44	138	12	811686	32	380	17	861850	52	796
	The number of generic solutions (#generic) obtained for the given number of states and parameters of the generic solutions: #generated rules/#used rules/#steps for 6^2)															
#generic	1				0				1				0			
Parameters	**38/33/71**								234/36/78							

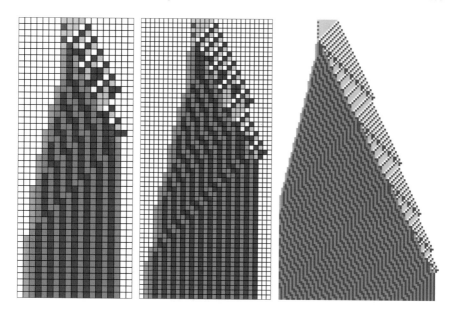

Fig. 5 Example of a 7-state CA controlled by a transition function evolved using the SR-fitness. This solution is denominated as CA–20–07. A complete development is shown for $x = 4$ and 5 (the left and middle sample respectively), the part on the right demonstrates a cutout of global behaviour of the CA for $x = 11$. Extracted from [4]

27 from Table 3 can be determined as $a_n = n(2n + 3)$ for all $x \geq 1$, i.e. in order to calculate the number of steps for $x = 7$, for example, then $n = x - 1 = 6$ and the substitution of this value to the equation gives $a_n = 6 * (2 * 6 + 3) = 90$ steps which corresponds to the value from Table 3 for $x = 7$ (observed for this CA using our CA simulator). The reason for taking $n = x - 1$ follows from the fact that the cellular automata work for $x \geq 2$.

For some CA WolframAlpha derived an iterative expression determining the number of steps a_{n+1} from the previous value a_n. For example, the CA–50–12 (the best one in this work) allows determining the number of steps for a given x as $a_{n+1} = n(3n + 7) - a_n$ for $n \geq 2$. Therefore, in order to calculate the number of steps e.g. for $x = 8$, it is needed to take $n = x - 2 = 8 - 2 = 6$, the value for the previous $x = 7$ must be known, i.e. $a_n = 64$ from Table 3, and substituting to the equation $a_{n+1} = 6 * (3 * 6 + 7) - 64 = 86$ which is the number of CA steps needed to calculate the result of 8^2 (as corresponds to the value from Table 3 for this CA for $x = 8$ observed in our CA simulator).

The aforementioned example also shows that it is not possible in some cases to express the number of CA steps for arbitrary $n \geq 1$ which means that the development of some CA for low values of x exhibit some anomaly in comparison with the development for larger input values. The reason for this behavior still remains in general an open question but our observations indicate that this is probably the case of CA exhibiting a complex (and mostly visually very attractive) pattern generated

Table 3 The number of steps of resulting CA needed to calculate the square for $x = 2, \ldots, 16$ together with expressions allowing to calculate the number of steps for given $n = x - 1$ (Part 1 of the CA comparison.)

							the input value of x							
2	3	4	5	6	7	8	9	10	11	12	13	14	15	16

the equation derived for the num. of steps of a given CA and

the num. of CA steps needed to finish the calculation of x^2

CA–50–12: $a_{n+1} = n(3n + 7) - a_n$ for $n \geq 2$

| 5 | 11 | 18 | 30 | 46 | 64 | 86 | 110 | 138 | 168 | 202 | 238 | 278 | 320 | 366 |

CA–30–08: $a_n = 2n^2 + n + 1, n \geq 1$

| 4 | 11 | 22 | 37 | 56 | 79 | 106 | 137 | 172 | 211 | 254 | 301 | 352 | 407 | 466 |

CA–50–27: $a_n = n(2n + 3), n \geq 1$

| 5 | 14 | 27 | 44 | 65 | 90 | 119 | 152 | 189 | 230 | 275 | 324 | 377 | 434 | 495 |

CA–40–01: $a_n = 2n^2 + 3n + 3, n \geq 1$

| 8 | 17 | 30 | 47 | 68 | 93 | 122 | 155 | 192 | 233 | 278 | 327 | 380 | 437 | 498 |

CA–40–14: $a_{n+1} = \frac{a_n(n-3)}{n-5} - \frac{2(11n+13)}{n-5}, n \geq 5$

| 8 | 16 | 29 | 46 | 68 | 94 | 124 | 158 | 196 | 238 | 284 | 334 | 388 | 446 | 508 |

CA–30–11: $a_{n+1} = -a_n + 4n^2 + 13n + 11$ for $n \geq 2$

| 3 | 17 | 35 | 51 | 76 | 100 | 133 | 165 | 206 | 246 | 295 | 343 | 400 | 456 | 521 |

CA–30–00: $a_{n+1} = \frac{a_n(n-1)}{n-3} + \frac{-15n-19}{n-3}, n \geq 3$

| 4 | 17 | 32 | 51 | 74 | 101 | 132 | 167 | 206 | 249 | 296 | 347 | 402 | 461 | 524 |

CA–50–22: $a_{n+1} = 2(2n^2 + 7n + 7) - a_n$ for $n \geq 1$

| 14 | 24 | 34 | 58 | 76 | 108 | 134 | 174 | 208 | 256 | 298 | 354 | 404 | 468 | 526 |

CA–20–07: $a_n = \frac{1}{4}(8n^2 + 22n - 5(-1)^n - 3), n \geq 1$

| 8 | 17 | 35 | 52 | 78 | 103 | 137 | 170 | 212 | 253 | 303 | 352 | 410 | 467 | 533 |

Table 4 The number of steps of resulting CA needed to calculate the square for $x = 2, \ldots, 16$ together with expressions allowing to calculate the number of steps for given $n = x - 1$ (Part 2 of the CA comparison.)

								the input value of x							
2	3	4	5	6	7	8	9	10	11	12	13	14	15		16

the equation derived for the num. of steps of a given CA and

the num. of CA steps needed to finish the calculation of x^2

CA–40–34: $a_n = \frac{1}{4}(8n^2 + 22n - 3(-1)^n + 3), n \geq 1$

9	19	36	54	79	105	138	172	213	255	304	354	411	469		534

CA–50–07: $a_n = \frac{1}{2}(5n^2 + 3n - 2), n \geq 1$

3	12	26	45	69	98	132	171	215	264	318	377	441	510		584

CA–20–00: $a_{n+1} = -a_n + 5n^2 + 9n + 2$ for $n \geq 1$

5	13	27	47	71	101	135	175	219	269	323	383	447	517		591

CA–30–31: $a_{n+1} = \frac{a_n(n^2 - 8n - 2)}{n^2 - 10n + 7} + \frac{-27n^2 + 13n + 9}{n^2 - 10n + 7}, n \geq 9$

2	16	33	54	79	108	141	178	219	271	331	397	469	547		631

CA–50–17: $a_n = 3n^2 + 1, n \geq 1$

4	13	28	49	76	109	148	193	244	301	364	433	508	589		676

CA–20–27: $a_{n+1} = \frac{a_n(n-1)}{n-3} - \frac{15(n+1)}{n-3}, n \geq 3$

7	15	30	51	78	111	150	195	246	303	366	435	510	591		678

CA–50–15: $a_n = 3n^2 + 2n + 2, n \geq 1$

7	18	35	58	87	122	163	210	263	322	387	458	535	618		707

Wolfram's CA: $a_n = 3n^2 + 7n + 2, n \geq 1$

12	28	50	78	112	152	198	250	308	372	442	518	600	688		782

by the development. For a low input value (e.g. $x < 5$ in case of the CA–40–14 from Table 3) the development does not need to involve all possible state transitions before reaching the result state, which would otherwise emerge for larger x. Therefore, the development for $x < 5$ is specific, leading to the numbers of steps that is not possible to express together with the numbers of steps for larger input values.

The overall comparison of some selected CA is shown in Fig. 6. As evident from this figure (and from the equations in Tables 3 and 4), the computational efficiency of the CA (i.e. the number of steps needed for given x) in all cases exhibit a quadratic form. However, the CA efficiency differ substantially between various solutions. Whilst Wolfram's CA exhibits the highest numbers of steps out of all CA that were available for the square calculations herein, our experiments showed that (1) this approach can be improved substantially and (2) various other solutions with a moderate efficiency exist for this task. Some of them are presented as visualisation of the appropriate CA development in Figs. 7 and 8. Specifically, the CA–30–08 from Fig. 7a represents an example of the second best result discovered from our evolutionary experiments. Its simple pattern exhibit a high degree of regularity which is probably the cause of very simple equation expressing the number of steps for given x (Table 3). More complex, less computationally efficient and visually interesting patterns are generated by CA–50–22 (Fig. 7b, Table 3) and CA–40–34 (Fig. 7c, Table 4). On the other hand, the CA–30–31 from Fig. 8a produces results of calculating x^2 that is composed of various (stable) state values and this CA also exhibit the most complex equation needed to express its number of steps for given x (see

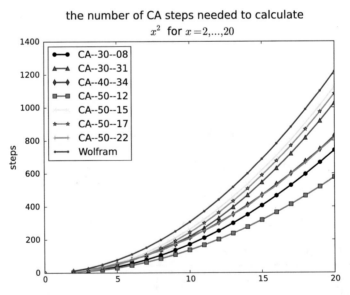

Fig. 6 Evaluation of the computational efficiency (i.e. the number of steps needed to achieve the result of x^2) of some selected CA whose development is also presented visually in various figures in this paper. A comparison with the existing Wolfram's CA [26] (the top-most curve) is included

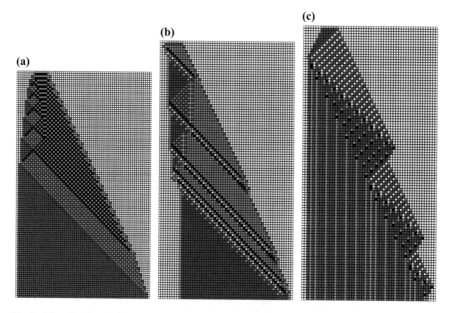

Fig. 7 Visualisation of the development of some selected squaring cellular automata obtained from our experiment: **a** CA–30–08, $x = 8$, **b** CA–50–22, $x = 7$, **c** CA–40–34, $x = 7$

Table 4). Together with CA from Fig. 8b, c it belongs to the least computationally efficient (i.e. requires many steps to produce the result – see the comparison in Fig. 6) but also can be viewed as solutions that demonstrate the variety of different styles of how the result of x^2 in CA can be achieved.

4.4 Discussion

In most cases of the experimental settings the EA was able to produce at least one generic solution for the CA-based square calculation. Despite the 2 million generation limit, the results from Tables 1 and 2 show that the average number of generations is mostly below 1 million, which indicates a potential of the EA to efficiently explore the search space. In comparison with the initial study of this problem proposed in [5], where 200,000 generations were performed, the significant increase of this parameter herein is important with respect to achieving a reasonable success rate and producing generic solutions (note that an initial comparison of various ranges for x evaluated in the fitness was proposed in [5], the result of which was considered in this work).

As regards the RA-fitness, which can be considered as the main technique proposed herein for the evolution of cellular automata, a more detailed analysis was performed with various multi-state CA. As the results in Table 1 show that the number of generic solutions increases for the number of states from 4 to 8, then for

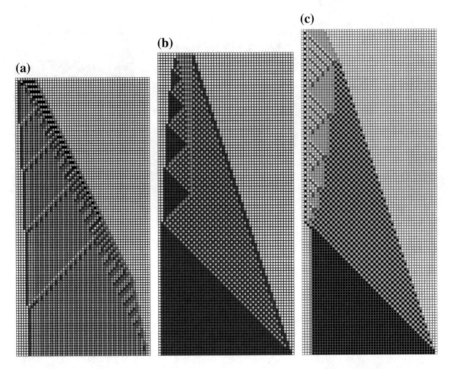

Fig. 8 Visualisation of the development of some selected squaring cellular automata obtained from our experiment: **a** CA–30–31, $x = 7$, **b** CA–50–17, $x = 7$, **c** CA–50–15, $x = 7$

10-state CAs a significant reduction can be observed. This is probably caused by the exponential increase of the search space depending on the number of states. The results indicate that the 8-state setup represents a very feasible value that may be considered as sufficient for this kind of problem (note that 6 generic solutions were obtained for this setup).

In both sets of experiments with the RA-fitness and SR-fitness, a phenomenon of a reduction of the number of states was observed. This is possible due to the identification of just the rules that are needed for the CA development to calculate the square out of all the rules generated from the evolved CMR-based transition function for every valid combination of states in the cellular neighbourhood. It was determined that the CAs in some cases do not need all the available cell states to perform the given operation.

5 Evolution of Complex 2D Cellular Automata

In order to provide a wider overview of what CA the proposed method can handle, some experiments dealing with the evolution of uniform multi-state 2D automata were conducted. The pattern development problem was chosen as a case study. In particular, some non-trivial and asymmetric patterns were chosen as shown in Fig. 9.

(a) **(b)**

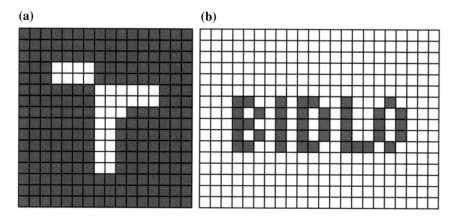

Fig. 9 Samples of selected patterns treated in the experiments for the pattern development problem in 2D CA: **a** the BUT logo (note that the T-like structure, composed of cells in state 1, is the subject of the CA development on the red background – cells in state 0), **b** a label containing the author's surname (composed of cells in state 1, the other cells are in state 0)

In order to evaluate a candidate CA, up to 40 development steps were performed and the steps 16–40 was assigned a partial fitness calculated as the number of cells in correct states with respect to the given pattern. The final fitness of the candidate CA was the maximum of the partial fitness values. The reason for this setup is to reduce the time needed for the evaluation because the patterns probably cannot be finished in less than 16 steps (this values was determined empirically). Therefore, no inspection of cell states is performed in steps 1–15. Moreover, no exact number of steps is known in which a pattern can be finished so that the aforementioned range of steps provides the evolution with a wider space to discover a solution. This setup does not eliminate a possibility of destroying the pattern that emerged at a certain step during the subsequent CA development. However, the development of a stable pattern was not a primary goal of these experiments.

It is difficult to provide statistical data from these experiments since only very few successful results have been obtained so far and there has been a research in progress regarding the optimization of evolutionary techniques used for the design of complex cellular automata. Therefore, only some selected results are presented in this section.

The first experiment — the development of the BUT logo — dealt with the CA working with 10 cell states. There are 10^5 various combinations in the 5-cell neighborhood which implies 10^{100000} possible transition functions. A single state-1 cell (a seed) was used as an initial state of the CA. The evolution managed to find a transition function (in the form of a sequence of CMRs that was converted to the table-based representation for the presentation purposes) that successfully develops the initial seed into the given pattern as shown in Fig. 9a. Although the pattern stability was not required, the pattern no longer changes by further CA development. A sample

(a)   **(b)**

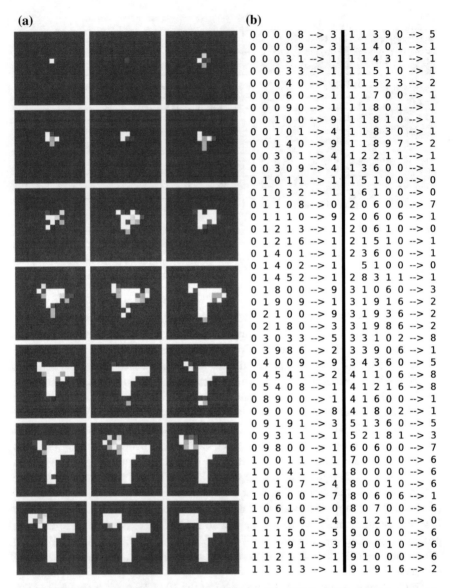

Fig. 10 Example of an evolved CA for the development of the BUT logo. The initial state consists of a single cell state 1, the other cells possess state 0. **a** The sequence of CA states producing the final pattern (ordered from left to right and top to bottom). **b** The appropriate transition rules for this CA

of the complete sequence of states leading to the emergence of this pattern (the BUT logo) is shown in Fig. 10 together with the appropriate transition function. The CA uses 90 transition rules and needs 20 steps to finish the pattern.

The second experiment — the development of the author's surname — dealt with the CA working with 12 cell states. There are 12^5 various combinations in the 5-cell neighborhood which implies 10^{248832} possible transition functions. In fact, this pattern requires to develop several (separate) structures (letters) of non-zero cells which form the complete label. A sample of the CA development together with the transition function is shown in Fig. 11. This is the only result obtained so far which is able to fully develop this pattern (from a triples of separate cells in states 1, 2 and 3 as an initial CA state – see the top-left corner of Fig. 11a). Similarly to the previous experiment, this pattern is also stable during further CA development which is especially interesting due to its increased complexity. The CA needs 32 steps to develop the target label from the initial state and the transition function consists of 161 rules (see Fig. 11b).

5.1 Discussion

Although the evolution of multi-state 2D CA is much more difficult than the 1D CA, the experiments provided some successful results with various target patterns. The results presented in this section probably represent the first case when complex 2D CA with at least 10 cell states were automatically designed using an evolutionary algorithm in a task of the non-trivial exact pattern development. In addition to the CA shown in Figs. 10 and 11, the evolution succeeded in searching other patterns as well (e.g. French flag, Czech flag, moving labels, replicating objects or multi-state gliders). This indicates that the proposed design method may be applicable in a wider area of cellular automata. A limitation for a higher success rate of the evolutionary experiments probably lies in the requirement of an exact pattern development. Our initial experiments suggest that promising areas for the CA applications may be those where approximate results are acceptable or the CA states allow us to tolerate some variations during the development. For example, the image processing, traffic prediction or design of approximative algorithms represent possible topics. Therefore, the future research will include modeling, simulation and optimization of such kinds of systems in cellular automata.

Even though the representation of the transition functions by means of conditional rules has proven a good applicability on various tasks, the optimization of the evolutionary algorithm used to search for the rules is probably still possible. This could not only improve the success rate of the evolutionary experiments but also reduce the computational effort and allow applying the concept of uniform computing platforms in real-world applications (e.g. with acceleration of the computations using modern reconfigurable technology).

(a)   **(b)**

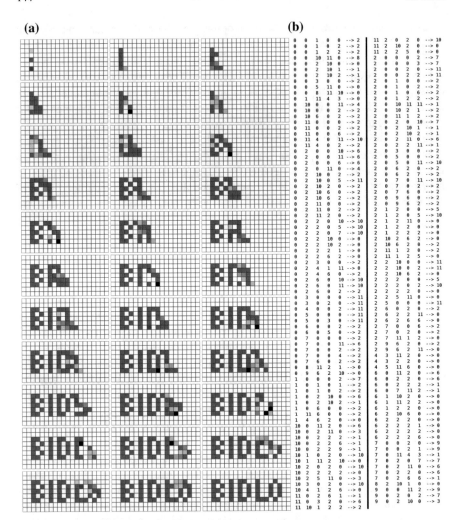

Fig. 11 Example of an evolved CA for the development of the author's surname. The initial state consists of three vertically aligned cells in states 1, 2 and 3 respectively with a single state-0 cells between them. **a** The sequence of CA states producing the final "BIDLO" pattern (ordered from left to right and top to bottom). **b** The appropriate transition rules for this CA

6 Conclusions

In this study we have presented some advanced topics related to the evolution of complex multi-state cellular automata. In particular, an analysis of CA for the generic square calculations has been proposed in the first case study. The results showed that some various algorithms to perform this task exist in CA, which differ both in the complexity of resulting transition functions and the efficiency of the computation (i.e. the number of steps of the CA needed to produce the result). Moreover, our best results presented herein have overcome the known solution (Wolfram's squaring CA), providing a reduction of the number of steps by approximately 50%.

The second case study has dealt with the non-trivial pattern development problem in two-dimensional CA. Several results have been presented that provide an exact and stable pattern developed from a simple initial CA state. Cellular automata working with 10 and 12 cell states have been treated, which induce search spaces of enormous sizes. Despite low success rates of the evolution, the results obtained have shown that the automatic design of such CA is possible even though our ongoing experiments indicate that the evolutionary algorithm still provides a space for further optimization.

In general, the proposed results probably represent the first case of the automatic design of exact behaviour in CA with more than 10 cell states. We believe that these pieces of knowledge will allow us to further improve our design method and to apply cellular automata for modeling, simulation and optimization of real-world problems.

Acknowledgements This work was supported by The Ministry of Education, Youth and Sports of the Czech Republic from the National Programme of Sustainability (NPU II), project IT4Innovations excellence in science – LQ1602, and from the Large Infrastructures for Research, Experimental Development and Innovations project "IT4Innovations National Supercomputing Center – LM2015070".

References

1. Basanta, D., Bentley, P., Miodownik, M., Holm, E.: Evolving cellular automata to grow microstructures. Genetic Programming. Lecture Notes in Computer Science, vol. 2610, pp. 1–10. Springer, Berlin Heidelberg (2003)
2. Berlekamp, E.R., Conway, J.H., Guy, R.K.: Winning Ways for Your Mathematical Plays, vol. 4, 2nd edn. A K Peters/CRC Press, Boca Raton (2004)
3. Bidlo, M.: Investigation of replicating tiles in cellular automata designed by evolution using conditionally matching rules. In: 2015 IEEE International Conference on Evolvable Systems (ICES), Proceedings of the 2015 IEEE Symposium Series on Computational Intelligence (SSCI), pp. 1506–1513. IEEE Computational Intelligence Society (2015)
4. Bidlo, M.: Evolution of generic square calculations in cellular automata. In: Proceedings of the 8th International Joint Conference on Computational Intelligence - Volume 3: ECTA, pp. 94–102. SciTePress - Science and Technology Publications (2016)
5. Bidlo, M.: On routine evolution of complex cellular automata. IEEE Trans. Evol. Comput. **20**(5), 742–754 (2016)
6. Bidlo, M., Vasicek, Z.: Evolution of cellular automata with conditionally matching rules. In: 2013 IEEE Congress on Evolutionary Computation (CEC 2013), pp. 1178–1185. IEEE Computer Society (2013)
7. Codd, E.F.: Cellular Automata. Academic Press, New York (1968)
8. Durand, B., Rka, Z.: The game of life: universality revisited. Mathematics and Its Applications. Cellular Automata, vol. 460, pp. 51–74. Springer, Netherlands (1999)
9. Elmenreich, W., Fehérvári, I.: Evolving self-organizing cellular automata based on neural network genotypes. In: Proceedings of the 5th International Conference on Self-organizing Systems, pp. 16–25. Springer (2011)
10. Holland, J.H.: Adaptation in Natural and Artificial Systems. University of Michigan Press, Ann Arbor (1975)

11. Ilachinski, A.: Cellular Automata: A Discrete Universe. World Scientific, Singapore (2001)
12. Lindgren, K., Nordahl, M.G.: Universal computation in simple one-dimensional cellular automata. Complex Syst. **4**(3), 299–318 (1990)
13. Mardiris, V., Sirakoulis, G., Karafyllidis, I.: Automated design architecture for 1-d cellular automata using quantum cellular automata. IEEE Trans. Comput. **64**(9), 2476–2489 (2015)
14. Ninagawa, S.: Solving the parity problem with rule 60 in array size of the power of two. J. Cell. Autom. **8**(3–4), 189–203 (2013)
15. Perrier, J.-Y., Sipper, M.: Toward a viable, self-reproducing universal computer. Phys. D **97**(4), 335–352 (1996)
16. Rendell, P.: A universal turing machine in Conway's game of life. In: 2011 International Conference on High Performance Computing and Simulation (HPCS), pp. 764–772 (2011)
17. Rendell, P.: A fully universal turing machine in Conway's game of life. J. Cell. Autom. **9**(1–2), 19–358 (2013)
18. Sahoo, S., Choudhury, P.P., Pal, A., Nayak, B.K.: Solutions on 1-d and 2-d density classification problem using programmable cellular automata. J. Cell. Autom. **9**(1), 59–88 (2014)
19. Sahu, S., Oono, H., Ghosh, S., Bandyopadhyay, A., Fujita, D., Peper, F., Isokawa, T., Pati, R.: Molecular implementations of cellular automata. In: Cellular Automata for Research and Industry. Lecture Notes in Computer Science, vol. 6350. Springer, Berlin (2010)
20. Sipper, M.: Quasi-uniform computation-universal cellular automata. Advances in Artificial Life. ECAL 1995. Lecture Notes in Computer Science, vol. 929, pp. 544–554. Springer, Berlin (1995)
21. Sridharan, K., Pudi, V.: Design of Arithmetic Circuits in Quantum Dot Cellular Automata Nanotechnology. Springer International Publishing, Switzerland (2015)
22. Stefano, G.D., Navarra, A.: Scintillae: how to approach computing systems by means of cellular automata. Cellular Automata for Research and Industry. Lecture Notes in Computer Science, vol. 7495, pp. 534–543. Springer, Berlin (2012)
23. Suzudo, T.: Searching for pattern-forming asynchronous cellular automata - an evolutionary approach. Cellular Automata. Lecture Notes in Computer Science, vol. 3305, pp. 151–160. Springer, Berlin (2004)
24. Tempesti, G.: A new self-reproducing cellular automaton capable of construction and computation. In: Advances in Artificial Life, Proceedings of the 3rd European Conference on Artificial Life. Lecture Notes in Artificial Intelligence, vol. 929, pp. 555–563. Springer, Berlin (1995)
25. von Neumann, J.: The Theory of Self-reproducing Automata (Burks, A.W. (ed.)). University of Illinois Press, Urbana (1966)
26. Wolfram, S.: A New Kind of Science. Wolfram Media, Champaign (2002)
27. Yuns, J.B.: Achieving universal computations on one-dimensional cellular automata. Cellular Automata for Research and Industry. Lecture Notes in Computer Science, vol. 6350, pp. 660–669. Springer, Berlin (2010)

Part II
Fuzzy Computation Theory and Applications

Fuzzy Modeling, Control and Prediction in Human-Robot Systems

Rainer Palm, Ravi Chadalavada and Achim J. Lilienthal

Abstract A safe and synchronized interaction between human agents and robots in shared areas requires both long distance prediction of their motions and an appropriate control policy for short distance reaction. In this connection recognition of mutual intentions in the prediction phase is crucial to improve the performance of short distance control. We suggest an approach for short distance control in which the expected human movements relative to the robot are being summarized in a so-called "compass dial" from which fuzzy control rules for the robot's reactions are derived. To predict possible collisions between robot and human at the earliest possible time, the travel times to predicted human-robot intersections are calculated and fed into a hybrid controller for collision avoidance. By applying the method of velocity obstacles, the relation between a change in robot's motion direction and its velocity during an interaction is optimized and a combination with fuzzy expert rules is used for a safe obstacle avoidance. For a prediction of human intentions to move to certain goals pedestrian tracks are modeled by fuzzy clustering, and trajectories of human and robot agents are extrapolated to avoid collisions at intersections. Examples with both simulated and real data show the applicability of the presented methods and the high performance of the results.

Keywords Fuzzy control · Fuzzy modeling · Prediction · Human-robot interaction · Human intentions · Obstacle avoidance · Velocity obstacles

Rainer Palm is adjunct professor at the AASS, Department of Technology, Orebro University.

R. Palm (✉) · R. Chadalavada · A. J. Lilienthal
AASS MRO Lab, School of Science and Technology, Orebro University, 70182 Orebro, Sweden
e-mail: rub.palm@t-online.de

R. Chadalavada
e-mail: ravi.chadalavada@oru.se

A. J. Lilienthal
e-mail: achim.lilienthal@oru.se

© Springer Nature Switzerland AG 2019
J. J. Merelo et al. (eds.), *Computational Intelligence*,
Studies in Computational Intelligence 792,
https://doi.org/10.1007/978-3-319-99283-9_8

149

1 Introduction

Interaction of humans and autonomous robots in common working areas is a challenging research field as to to system stability and performance and human safety. When human agents and robots share the same working area, both of them have to adapt their behavior, to either support their cooperation, or to enable them to do their own task separately. Research results on planning of mobile robot tasks, learning of repeated situations, navigation and obstacle avoidance have been published by [9, 15, 20, 22, 31]. In a scenario like this it is difficult to predict the behavior, motions and goals of a human agent. Even more it is important to predict the human behavior for a limited time horizon with a certain probability to enable the robot to perform adequate reactions. One class of solutions to this problem is the building of models of the human behavior by clustering methods [20, 24, 29]. Further research activities focus on Bayesian networks [12, 32], Hidden Markov Models (HMM) [2], Fuzzy logic or Fuzzy Cognitive Maps and reinforcement learning [5, 32]. Heinze addresses human intentions from the ontological point of view [13]. Another aspect is the (automatic) recognition of human intentions to aim at a certain goal. Some research on intention recognition describes human-robot interaction scenarios and the "philosophical and technical background for intention recognition" [32]. Further research deals with "Intention Recognition in Human-Robot Collaborative Systems" [1, 16], human-robot motions initiated by human intentions [10], and socially inspired planning [11]. In practice, an identification of a human intention needs to predict the direction of motion, the average velocity and parts of the future trajectory. In this connection, Bruce et al. address a planned human-robot rendezvous at an intersection zone [3]. Satake et al. [30] describe a social robot, that approaches a human agent in a way that is acceptable for humans. Further research on human intentions together with trajectory recognition and planning is presented by [4, 14]. Modeling of pedestrian trajectories using Gaussian processes is shown in [7]. In [19] fuzzy methods, probabilistic methods and HMM approaches for modeling of human trajectories are compared.

In addition to the recent research and as an extension of our work described in [25, 26], this paper concentrates on the recognition of human intentions to move on certain trajectories. The method is based on observations of early trajectory parts, plus a subsequent extrapolation. The here discussed control principles of interaction between human and robot mainly deal with trajectory planning and external sensor feedback on a higher level of the control hierarchy. Furthermore, the time schedule of the information flow and the kinematic relationship of a human-robot system in motion is considered. The observation of the human agent by the robot supplies motion data that are Kalman-filtered to cope with uncertainties and noise. This leads to an estimation of the velocity vector of the human relative to the robot which is depicted in a "compass dial". Because of the vague character of the problem to recognize a human intention, a set of fuzzy rules is extracted that consists of a reaction of the robot to a human motion either to prevent a collision or enable a cooperation. Our case is somehow different from the usual case of avoidance of moving obstacles due to the uncertainty to predict human intentions. A special issue is the case of *unchanged directions* of a motion both of the human and the robot that should be distinguished from a common obstacle avoidance methods [15]. This is done by a hybrid robot controller that computes the time for possible collisions at intersections. According to this knowledge the controller switches into a mode to change the robot's speed continuously to prevent from collisions at intersection of the planned paths. Due to uncertainties in the observations and to measurement noise the intersection points are extended to *intersection areas* which must not be reached at the same time neither by the human nor by the robot. Since this operation comprises only the first part of the human motion the subsequent extrapolations are updated at each time step in order to avoid larger errors. Another approach is the velocity obstacle method [8, 33] that has already been introduced in the early 20th century for ship navigation and mooring in harbors [21]. This method takes into account the size of contributing agents and the difference velocities between them, while moving in a shared working area. To take into account the size of another agent (human or robot) and the relative velocity between robot and human the so-called velocity obstacle method is applied where a compromise between the change of the direction and the speed of the robot is computed. Another essential issue is the identification of lanes from trajectories usually preferred by human agents in an open area or at a factory work floor. This is motivated by the need for an early reaction of the robot to the human's intention to walk, and either to avoid a possible collision or to plan a cooperation action. In this paper we concentrate on the fuzzy modeling of pedestrian tracks, the identification of lanes preferred by the human agents, and the identification of the membership of a pedestrian track (or parts of it) to a specific lane. Furthermore, the extrapolation of human and robot trajectories under certain measurements is addressed to avoid collisions between them at intersections.

The paper is organized as follows. In Sect. 2 we address the interaction between human and robot from the control structure point of view and the time schedule of the information flow. Section 3 deals with the kinematic and geometric relations between human and robot. A "compass dial" with the relative velocities and the corresponding fuzzy rules is presented in Sect. 4. Section 5 deals with avoidance strategies at intersections based on hybrid controllers, velocity obstacles and Mamdani fuzzy steering rules. Section 6 addresses the intention recognition and the prediction of human motions based on pedestrian data, the fuzzy modeling of sets (bundles) of pedestrian tracks and the identification of the membership of a single track to a certain lane. Finally, an extrapolation of human and robot trajectories in an early phase of motion predicts a possible collision at intersections which enables an obstacle avoidance. Section 7 presents simulation results based on simulated and experimental data. Section 8 ends with a discussion and conclusions.

2 Interaction Between Human and Robot

In a shared working environment, human and robot are constituted by a common control system shown in Fig. 1. Both human and robot are driven by their individual two dimensional goals (desired trajectories) x_{H_d} and x_{R_d}. Actions and reactions are represented by desired states $x_{H_d}(t_i)$ and $x_{R_d}(t_i)$. From the interaction of the two agents we obtain observable states $x_H(t_i)$ and $x_R(t_i)$. Intentions I_H and I_R are signals (e.g. information about parts of trajectories) transmitted to/observed by the other agent. The dynamic equations can formally be written as

$$
\begin{aligned}
\dot{x}_H &= f_H(x_H, u1_H, u2_H) \\
u1_H &= g1_H(x_{H_d}, x_H) \\
u2_H &= g2_H(I_R, x_H) \\
I_R &= h_R(x_R, x_H, x_{R_d})
\end{aligned}
\tag{1}
$$

$$
\begin{aligned}
\dot{x}_R &= f_R(x_R, u1_R, u2_R) \\
u1_R &= g1_R(x_{R_d}, x_R) \\
u2_R &= g2_R(I_H, x_R) \\
I_H &= h_H(x_H, x_R, x_{H_d})
\end{aligned}
\tag{2}
$$

The 1st two lines of (1) and (2), respectively, denote the individual dynamics plus internal control of human and robot, respectively, where the next two lines denote the 'crosswise' influence of intention and reaction between human and robot. The functions in (1) and (2) are highly nonlinear hybrid functions with continuous and switching attributes. A good example for modeling human-robot dynamics can be

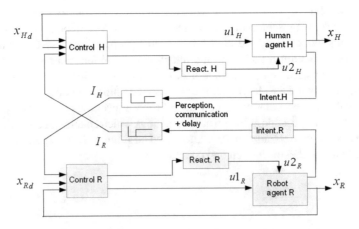

Fig. 1 Human-robot interaction, control scheme, extracted from [25]

found in [17]. Recall here that the here discussed control problems deal with the higher control level of external sensory and trajectory generation. Furthermore the feasibility of the desired trajectory x_{R_d} and its possible variations should be guaranteed because of the nonholonomic kinematics of the robot. In our case, intentions are functions of desired and actual states. The robot controllers $g1_R$ and $g2_R$ can be designed based on the knowledge about the system dynamics [27] whereas the human controllers $g1_H$ and $g2_H$, which have been introduced due to formal reasons, cannot be designed in the same way. The same is true for the formal description f_H of the human behavior which is usually only a rough approximation of the real behavior. Because of the difficult analytical description of the human's behavior and the nonlinear behavior of the robot, a modeling using local TS fuzzy models from motion data is recommended which leads to sets of rules like the following rule R_i for human and robot

$$\text{IF} \quad x_H = F_{x_H i} \quad \text{THEN} \quad \dot{x}_H = f_{H i}(x_H, u1_H, u2_H) \tag{3}$$
$$\text{IF} \quad x_R = F_{x_R i} \quad \text{THEN} \quad \dot{x}_R = f_{R i}(x_R, u1_R, u2_R)$$

where $x_H = [x_{H1}, x_{H1}]^T$, $x_R = [x_{R1}, x_{R1}]^T$ are state vectors.
$F_{x_H i} = [F_{x_{H1} i}, F_{x_{H2} i}]^T$ and $F_{x_R i} = [F_{x_{R1} i}, F_{x_{R2} i}]^T$ are sets of fuzzy sets with the convention
"$x_H = F_{x_H i}$" $\equiv$ "$x_{H1} = F_{x_H 1i}$ AND $x_{H2} = F_{x_H 2i}$".
$f_{H i}$ and $f_{R i}$ are local linear or nonlinear functions. For n rules each of the nonlinear functions of (1) and (2) are split up into n local functions

$$f_H = \sum_{i=1}^{n} w_i(x_H) \cdot f_{H i}(x_H, u1_H, u2_H) \tag{4}$$

$$f_R = \sum_{i=1}^{n} w_i(x_R) \cdot f_{R i}(x_R, u1_R, u2_R)$$

where

$$w_i(x_H) = \frac{\mu_i(x_H)}{\sum_{k=1}^{n} \mu_k(x_H)}; \qquad \mu_k(x_H) = \mu_{k1}(x_{H1}) \cdot \mu_{k2}(x_{H2}) \tag{5}$$

are weighting functions $w_i(x_H) \in [0, 1]$ with $\sum_{i=1}^{n} w_i(x_H) = 1$. $\mu_{km}(x_{Hm})$ is the degree of membership of x_{Hm} in $F_{x_H m i}$, $m = 1, 2$. The same is valid for the robot equation in (4). Let furthermore the robot controllers $u1_R$, $u2_R$ either be designed as weighted combinations of local controllers

$$u1_R = \sum_{i=1}^{n} w_i(x_R) \cdot g1_{R_i}(x_{R_d}, x_R) \tag{6}$$

$$u2_R = \sum_{i=1}^{n} w_i(x_R) \cdot g2_{R_i}(I_H, x_R)$$

or as Mamdani expert rules formulated in Sect. 4. On the other hand, human controllers $u1_H$, $u2_H$ are expressed as Mamdani expert rules which are the counterparts of the robot expert rules. Hybrid robot controllers for collision avoidance are presented in Sect. 5. An intention may become recognizable at an early part of the human trajectory $x_H(t_k | k = 1 \ldots m)$ where m is the time horizon on the basis of which the robot recognizes the human's intention. This information is sent to the robot with a delay T_{d_H}. After that the robot starts its intention recognition and starts to plan/compute a corresponding reaction $x_{R_d}(t_i)$. The intention to react is realized as a part of the trajectory of the robot $x_R(t_k | k = j \ldots n)$ where $(n - j)$ is the corresponding time horizon on the basis of which the human tries to recognize the robot's intention to move. Then this intention is transmitted to the human. Let the sampling time of the whole process be T_{tot}. Robot and human can control the mutual cooperation only if T_{tot} meets certain requirements on the time constants of the system. There are two time constants involved, the time constant τ_H of the human and the time constant τ_R of the robot. Let the total time constant of the process be the sum of the two

$$\tau_{tot} = \tau_H + \tau_R \tag{7}$$

A rule of the thumb says that the sample time T_{tot} should be $5 \ldots 30$ times shorter than the largest time constant of the process [18].

3 Kinematic Relations Between Human and Robot in a Common Working Area

The analysis of the mutual observations between human and robot requires the formulation of relative positions/orientations and velocities in the local coordinate systems C_H (human) and C_R (robot) and the global coordinate system C_B (basis). First, let us assume a sufficient knowledge of the positions/orientations of robot and human both in their local systems C_H, C_R and in the basis system C_B. The relation between two coordinate systems C_A and C_B is then defined by the transformation

$$T_{AB} = \begin{pmatrix} \cos(\phi_{AB}) & -\sin(\phi_{AB}) & x_A \\ \sin(\phi_{AB}) & \cos(\phi_{AB}) & y_A \\ 0 & 0 & 1 \end{pmatrix} \tag{8}$$

Fig. 2 Transformations
between frames, extracted
from [25]

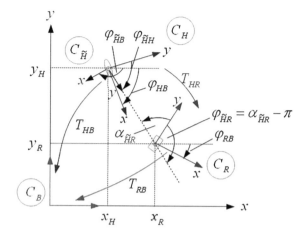

where $\mathbf{x} = (x_A, y_A)^T$, is a position vector in C_A and ϕ_{AB} is the orientation angle between C_A and C_B. Figure 2 shows the kinematic relations between the coordinate systems C_R, C_H, and C_B: T_{HB} between human and basis, T_{RB} between robot and basis, T_{HR} between human and robot. The orientation, of human and robot are chosen so that the y axis is pointing in the direction of motion. Next an additional coordinate system $C_{\tilde{H}}$ is defined whose y-axis points from the center of C_H to the center of C_R. This coordinate system is necessary for the formulation of the *heading angle* from human to robot. The distance between C_H (and $C_{\tilde{H}}$) and C_R is denoted by d_{HR}. In the following we assume parts of the *intended* trajectory $\mathbf{x}_{HR}(t_i)$ of the human to be measurable by the robot from which the velocity $\dot{\mathbf{x}}_{HR}(t_i)$, the orientation angle $\phi_{HR}(t_i)$ of the human in the robot system, and the transformation matrix $T_{HR}(t_i)$ can be estimated. Since the robot is assumed to know its own trajectory $\mathbf{x}_{RB}(t_i)$ and the transformation matrix $T_{RB}(t_i)$ we can compute the transformation matrix: $T_{HB}(t_i) = T_{RB}(t_i) \cdot T_{HR}(t_i)$.

Then we measure the distance d_{HR} between C_H and C_R, and the *relative angle* $\alpha_{\tilde{H}R}$ between the line $C_H - C_R$ and the y-axis of C_R. Finally we compute the *heading angle* $\phi_{\tilde{H}H} = \pi - (\alpha_{\tilde{H}R} + \phi_{HR})$ between $C_{\tilde{H}}$ and C_H which is necessary for the qualitative relation between human and robot. Now we have all information to formulate a set of qualitative fuzzy rules for human-robot interactions.

4   Fuzzy Rules for Human-Robot Interactions

In the center of $C_{\tilde{H}}$ a so-called compass dial is introduced that expresses the qualitative relationship of the human intentions seen from the robot. These comprise 8 human motions relative to the robot: 'APP=approach', 'AVR=avoid right', 'MOR=move right', 'RER=recede right', 'REC=recede', 'REL=recede left', 'MOL=move left', 'AVL=avoid left'.

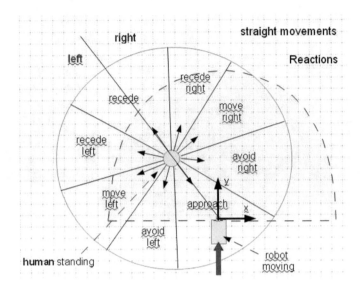

Fig. 3 Compass dial for human actions, extracted from [25]

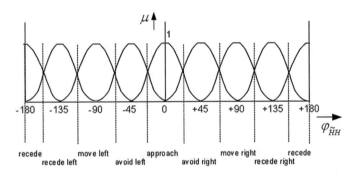

Fig. 4 Fuzzy sets for human motions, extracted from [25]

To identify the trajectory of the human relative to the robot it is enough to know the heading angle $\phi_{\tilde{H}H}$ and the relative velocity $\Delta \mathbf{v} = |\dot{\mathbf{x}}_{HB} - \dot{\mathbf{x}}_{RB}| = |\dot{\mathbf{x}}_{HR}|$. Since $\Delta \mathbf{v}$ is an invariant it becomes clear that it can be computed in any coordinate system. Ones the heading angle $\phi_{\tilde{H}H}$ is computed one can determine qualitative fuzzy relations between human and robot in form of fuzzy control rules according to the compass dial in Fig. 3. A fuzzy label is attached to each motion direction of the human agent. A crisp heading angle $\phi_{\tilde{H}H}$ is fuzzified with respect to the corresponding fuzzy sets for 'approach', 'avoid right' etc (see Fig. 4). From $\alpha_{\tilde{H}R}$, $\phi_{\tilde{H}H}$, $\Delta \mathbf{v}$ and the distances $|\Delta \mathbf{x}| = |\mathbf{x}_{HR}|$ the response of the robot to the human's intention is computed in an early state of the human action. However, because of the uncertain nature of the data (system/measurement noise) one obtains estimates of positions and velocities by an appropriate Kalman filter. Noise on velocity and measurement are filtered out leading to a smooth trajectory (or part of trajectory) from which the velocity vector is estimated. The variables to be processed are $\alpha_{\tilde{H}R}$, $\phi_{\tilde{H}H}$, the distance $|\Delta\mathbf{x}| = |\mathbf{x}_{HR}|$, and the relative velocities $|\Delta\mathbf{v}|$. For $|\mathbf{x}_{HR}|$ and $|\Delta\mathbf{v}|$ fuzzy sets 'Small', 'Medium', and 'Large' of Gaussian type are defined and appropriate fuzzy rules are formulated

$$
\begin{aligned}
IF \quad \alpha_{\tilde{H}R} \;&=\; A_i \;\; AND \;\; \phi_{\tilde{H}H} = P_i \;\; AND \\
|\Delta\mathbf{x}| \;&=\; DX_i \;\; AND \;\; |\Delta\mathbf{v}| = DV_i \\
&THEN \quad ACT_{rob}
\end{aligned} \tag{9}
$$

$$
ACT_{rob} : \quad \phi_{HR_{rob}} = PH_i \;\; AND \;\; |\mathbf{v}|_{rob} = VR_i
$$

i - rule number
$\alpha_{\tilde{H}R}$ - relative angle between $C_{\tilde{H}}$ and C_R
A_i - fuzzy set for the relative angle
$\phi_{\tilde{H}H}$ - heading angle between $C_{\tilde{H}}$ and C_H
P_i - fuzzy set for the heading angle
$|\Delta\mathbf{x}|$ - distance between C_H and C_R
DX_i - fuzzy set for the distance
$|\Delta\mathbf{v}|$ - relative velocity between C_H and C_R
DV_i - fuzzy set for the relative velocity
$\phi_{HR_{rob}}$ - orientation angle of human in robot system
PH_i - fuzzy set for the orientation angle
$|\mathbf{v}|_{rob}$ - desired velocity of the vehicle
VR_i - fuzzy set for the desired velocity

However, not every combination makes sense. Therefore "pruning" of the set of rules and intelligent hierarchization can solve this problem. A simple set of Mamdani-rules for $\alpha_{\tilde{H}R} > 0$ to avoid a collision between human and robot contains only the heading angle $\phi_{\tilde{H}H}$ and the orientation angle $\phi_{HR_{rob}}$ like:

$$
IF \quad \phi_{\tilde{H}H} = APP \;\; THEN \quad \phi_{HR_{rob}} = TUL \;\; AND \;\; |\mathbf{v}|_{rob} = SD \tag{10}
$$

where the robot actions are encoded by the following labels TR=turn right, TL=turn left, MS=move straight ahead, $MSSD$=move straight ahead/slow down. A whole set of rules is shown in (37) in Sect. 7.

5 Hybrid Controller for Collision Avoidance

5.1 Early Observation of Trajectories

From measured robot/human positions taken at an early point in time Kalman filtered sequences of robot and human positions $\mathbf{x}_{RB}(t_i)$ and $\mathbf{x}_{HR}(t_i)$ are gained from which the velocities $\mathbf{v}_H$ and $\mathbf{v}_R$ are estimated. Then the distance d_{HR} between C_H and C_R and the relative angle $\alpha_{\tilde{H}R}$ are measured and the angle ϕ_{HR} computed. Despite of existent traffic rules a collision between human agent and robot may occur especially in the cases AVR, MOR, RER or AVL, MOL, REL of the compass dial. Therefore at a certain distance a hybrid controller switches from the 'normal mode' to a 'prediction mode' to compute an 'area of intersection' and the time to reach it. After having reached this area the controller switches back to the 'normal mode' keeping its latest velocity constant.

5.2 Uncertainty in Measurements

Uncertainties in measurements and unexpected changes in directions and velocities lead to deviations in the calculations of possible crossing points. From simulations and experiments circular areas of possible collisions can be designed. Figure 5 shows the relations for the "1 human - 1 robot" case. Let P_H be the crossing point of human and robot and α the angle of uncertainty for both human and robot at a specific time t_i. A circle with the radius $r_1 = D_1 \cdot \sin \alpha/2$ describes the uncertainty area A_H of the human to be avoided by the robot. On the other hand, the robot has its own circular uncertainty area A_R that should not overlap with A_H. Let the distance between robot and crossing point be ΔD then the radius of A_R is $r_2 = (D_1 - \Delta D) \cdot \sin \alpha/2$. From the requirement $A_H \cap A_R = \emptyset$ we obtain $\Delta D \geq r_1 + r_2$. For $\Delta D = r_1 + r_2$ we obtain

$$r_2 = \frac{\sin \alpha/2}{(1 + \sin \alpha/2)}(D_2 - r_1) \tag{11}$$

and finally the velocity v_r for the robot to reach the distance point P_R in the time t_{HR} that the human would need to get to P_H (see (12a))

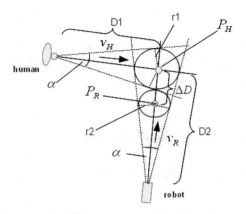

Fig. 5 Crossing areas, extracted from [25]

$$a: \quad v_{Ropt} = \frac{(D_2 - r_1)}{t_{HR}} \cdot \frac{1}{(1 + \sin \alpha/2)} \tag{12}$$

$$b: \quad v_{Ropt} = \frac{(D_2 + r_1)}{t_{HR}} \cdot \frac{1}{(1 - \sin \alpha/2)}$$

Equation (12a) is valid for the case when the human passes the crossing point before the robot. To be on the safe side one should require $v_R \leq v_{Ropt}$. For the case when the human passes the crossing point after the robot we get (12b) with $v_r \geq v_{Ropt}$.

5.3 Velocity Obstacles

Measurement errors or uncertainties of the geometrical model lead to crossing areas instead of crossing points that can be taken into account in advance. In addition, the size of an agent plays a particular role. An appropriate method is velocity obstacle method [8, 33] that considers size and difference velocity of an agent to other agents. Let two circular agents H and R with the radii r_H and r_R move in a common working area on straight trajectories with the velocities v_H and v_{R1} (see Fig. 6). The relative velocity between H and R is $v_{RH1} = v_{R1} - v_H$. Then we construct a larger agent H' at the location of H with a radius $r_{RH} = r_H + r_R$ and draw two tangents from R to this new agent H'. From this it becomes obvious that v_{RH1} lays inside a cone forming the "velocity obstacle" meaning that both agents will collide if the motion goes on in the same way. There are three options to bring the relative velocity outside the velocity obstacle.

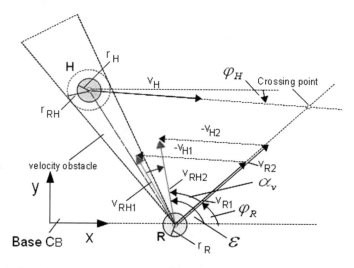

Fig. 6 Principle of velocity obstacles

(a) change $|v_R|$ and keep the direction of v_R constant
(b) change the direction of v_R and keep $|v_R|$ constant
(c) change $|v_R|$ and the direction of v_R.

In our case we choose option (c) which requires a compromise between the two other options. Bringing the difference velocity v_{RH} out of the velocity obstacle means changing the angle ϵ appropriately (see Fig. 6). $\epsilon = \alpha_v + \phi_R$ is a function $f(\phi_R, |v_R|, \phi_H, |v_H|)$, whereas only $\phi_R, |v_R|$ are control variables. Function f can be obtained as follows: Provided that $|v_R|, |v_H|$, and v_{RH} are available, we obtain

$$|v_H|^2 = |v_{RH}|^2 + |v_R|^2 - 2|v_R||v_{RH}|cos(\alpha_v) \tag{13}$$

and

$$\alpha_v = \arccos\left(\frac{|v_{RH}|^2 + |v_R|^2 - |v_H|^2}{2|v_R||v_{RH}|}\right) \tag{14}$$

$$\epsilon = f(\phi_R, |v_R|, \phi_H, |v_H|) = \alpha_v + \phi_R$$

A change of ϵ is defined as

$$d\epsilon = \frac{\partial f}{\partial \phi_R}d\phi_R + \frac{\partial f}{\partial |v_R|}d|v_R| \tag{15}$$

where $d\phi_R$ and $d|v_R|$ are the changes in the orientation angle ϕ_R and the velocity $|v_R|$, respectively. The partial derivatives in (15) are computed on the basis of given geometrical parameters and velocities. Let

$$N = |v_{RH}|^2 + |v_R|^2 - |v_H|^2$$

$$D = 2|v_R||v_{RH}|$$

$$\frac{\partial f}{\partial \phi_R} = 1 \tag{16}$$

$$\frac{\partial f}{\partial |v_R|} = \frac{-1}{\sqrt{1 - \frac{N^2}{D^2}}} \cdot \frac{|v_R|D - N|v_{RH}|}{2|v_R|^2|v_{RH}|^2}$$

Rewriting (15) together with (16) yields

$$d\epsilon = J \cdot d\omega \tag{17}$$

$$J = \left(\frac{\partial f}{\partial \phi_R}, \frac{\partial f}{\partial |v_R|}\right); \quad d\omega = (d\phi_R, d|v_R|)^T$$

To guarantee the above mentioned compromise between the orientation angle and the velocity of the robot a control law is formulated that leads to a desired angle ϵ_d outside the velocity obstacle. With the Lyapunov function

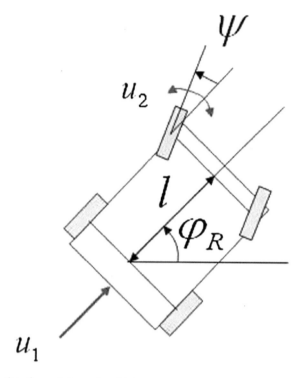

Fig. 7 Kinematics of a nonholonomic vehicle

$$V_\epsilon = 1/2 \cdot (\epsilon - \epsilon_d)^2 \tag{18}$$

we obtain a stable control law that leads the difference velocity v_{RH} away from the human velocity obstacle

$$d\omega = -K \cdot J^\dagger \cdot (\epsilon - \epsilon_d) \tag{19}$$

where $J^\dagger$ is the pseudo inverse of J and K is a gain matrix (see (17)).

To increase safety it is recommended to introduce the convention that both agents "human and robot" always turn to the same side seen from their local coordinate system. A combination of velocity obstacles and fuzzy rules leads to a "flat" hierarchy in which first a crisp decision the direction of avoidance is made and then a set of rules according to (10) decides on the orientation angle $\phi_{\tilde{H}H}$ and the velocity $|\mathbf{v}|_{rob}$.

Computation of the Steering Angle of a Car-like Vehicle. For car-like robots with nonholonomic kinematics the problem of computing the appropriate steering angle from the change in orientation and speed is left. Vehicles with nonholonomic kinematics are bound by constraints that are determined by the vehicle's geometry and the path, that it is up to follow. The kinematics of a nonholonomic rear-wheel driven vehicle *Rob* (see Fig. 7) is described by

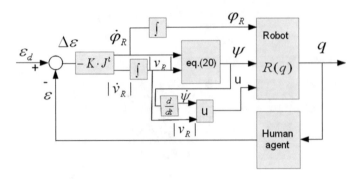

Fig. 8 Block scheme, velocity obstacle

$$\dot{q} = R(q) \cdot u$$
$$q = (x, y, \phi_R, \psi)^T \tag{20}$$
$$R(q) = \begin{pmatrix} \cos\phi_R & 0 \\ \sin\phi_R & 0 \\ \frac{1}{l} \cdot \tan\psi & 0 \\ 0 & 1 \end{pmatrix}$$

where $q \in \Re^4$ - state vector, $u = (u_1, u_2)^T$ - control vector, pushing/steering speed, $u_1 = |v_R|$; $u_2 = \dot{\phi}_R$; $x_R = (x, y)^T$ - position vector of *Rob*, ϕ_R - orientation angle, ψ - steering angle, l - length of vehicle,

From (19) we obtained the change in orientation angle $\dot{\phi}_R = d\phi_R/dt$. On the other hand we get from (20) $\dot{\phi}_R = |v_R| \cdot \frac{1}{l} \cdot \tan\psi$ from which we obtain the steering angle

$$\psi = \arctan\frac{l \cdot \dot{\phi}_R}{|v_R|} \tag{21}$$

After that, a change of the steering angle ψ leads to a change of the orientation angle ϕ_R and to a new ϵ which in a next step results in a new distribution of the change in orientation angle ϕ_R and robot speed v_R according to (19) (see Block scheme, Fig. 8).

6 Intention Recognition Based on Learning of Pedestrian Lanes

6.1 Fuzzy Modeling of Lanes

An option to identify/recognize human intention to aim at a specific goal is to learn from experience. The framework of "programming by demonstration" focuses on the recognition of human grasp pattern by fuzzy time clustering and modeling [24].

Fig. 9 Edinburgh pedestrian data, extracted from [25]

Likewise for the intention recognition it is a matter of pattern recognition, when a model is used that has been built on the basis of recorded pedestrian trajectories. For this purpose we used the "Edinburgh Informatics Forum Pedestrian Database" which consists of a large set of walking trajectories that has been measured over a period of months [6] (see Fig. 9). The idea is to identify lanes that people normally use in a specific working area. In our case the models are built by fuzzy time clustering as follows:

Assume m locations at a working area describing start and goal positions of pedestrian tracks.

1. Split a set of m_{tot} trajectories into m_k groups (bundles) each of them containing m_{k_l} trajectories with the same start/goal area.

2. Run a fuzzy time clustering for each trajectory separately with c time clusters $C_{i,k,l} \in R^2$, $i = 1 \ldots c$ - number of time cluster, $k = 1 \ldots m_k$ - number of group; $l = 1 \ldots m_{k_l}$ - number of trajectory in group k;

3. Compute the mean values of the ith time clusters in each group k: $C_{i,k} = 1/m_{k_l} \cdot \sum_{l=1}^{m_{k_l}} C_{i,k,l}$, m_{k_l} - number of trajectories in group k. $C_{i,k,l} = (c_x, c_y)_{i,k,l}^T$ are the x, y coordinates of the ith cluster center in the lth trajectory in the kth group. The time clusters $C_{i,k}$, $(i = 1 \ldots c, k = 1 \ldots m_k)$, represent finally the lanes. Figure 10 shows an example for $k = 3$ and $c = 10$.

6.2 Recognition of Intentions to Follow Certain Lanes

In order to recognize the intention of a human agent to aim at a certain goal, the membership of his trajectory (or part of it) to one of the lanes is to be computed. The

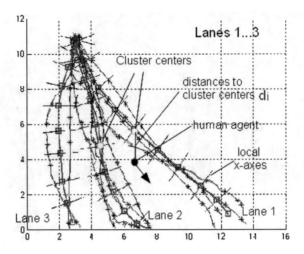

Fig. 10 Lanes 1 . . . 3, average trajectories, extracted from [25]

membership of a point $\mathbf{x} = (x, y)^T$ to a cluster center $C_{i,k}$ is here defined by

$$w_{i,k} = \frac{1}{\sum_{j=1}^{c} \left(\frac{d_{i,k}}{d_{j,k}}\right)^{\frac{2}{m_{proj}-1}}} \tag{22}$$

where $d_{i,k} = (\mathbf{x} - C_{i,k})^T (\mathbf{x} - C_{i,k})$, $C_{i,k}$ - ith cluster center in the kth lane, $m_{proj} > 1$ - fuzziness parameter [23, 28]. The algorithm works as follows:

1. Compute the closest distances $d_{i_{min},k} = min_j(|\mathbf{x} - C_{j,k}|)$ to the cluster centers $C_{j,k}$ $(j = 1 \ldots c, k = 1 \ldots m_k)$

2. Compare the membership functions $w_{i_{min},k}$ and select the lane with the highest membership:

$(w_{i_{min},k})_{max} = max(w_{i_{min},k}), k = 1 \ldots m_k$ or

$k_{max} = argmax(w_{i_{min},k})$.

However, only one data point is obviously not sufficient for intention recognition. Therefore moving averages $(\bar{w}_{i_{min},k})_{max}$ over a predefined number of time steps n are computed

$$(\bar{w}_{i_{min},k})_{max}(t_j) = \frac{1}{n} \sum_{i=0}^{n-1} (w_{i_{min},k})_{max}(t_j - i) \tag{23}$$

With this the influence of noise is further reduced and the degree of membership of a human trajectory to a particular lane is reliably identified.

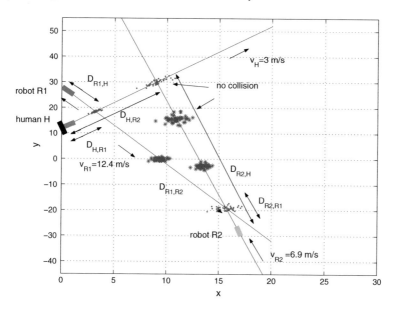

Fig. 11 Intersection points, one human, two robots

6.3 Prediction of Possible Collisions at Intersections

In a planning phase and for predefined paths one should concentrate on the avoidance of collisions at intersections. This can easily be handled for a combination "1 human - 1 robot" because it is sufficient to change the velocity of the robot to avoid a crash at the intersection. For more than two agents this problem is more involved due to potential conflicts in reaching local goals. This is illustrated by a "1 human - 2 robot" example (see Fig. 11). Let the human agent intend to move along a line with an estimated velocity v_H. Let robot 1 and 2 intend to move on lines with the velocities v_{R1} and v_{R2} such that all 3 lines cross at 3 different intersections $I_{H,R1}$, $I_{H,R2}$, and $I_{R1,R2}$ that can be computed in advance. The question is, which velocities v_{R1} and v_{R2} lead to a collision with the human agent provided that the velocity v_H of the human remains constant. Let the traveling times t_{ij} on the particular sections D_{ij} measured from start to intersection be

$$t_{ij} = D_{ij}/v_i; \qquad t_{ji} = D_{ji}/v_j \qquad (24)$$

Minimizing of the differences $\Delta t_{ij} = t_{ij} - t_{ji}$ by changing of the velocities v_{R1} and v_{R2} enables us to compute the time durations and respective velocities that would lead to collisions and from this to conclude how to avoid such collisions which is shown in the next steps.

Let $\Delta t_1 = \Delta t_{H,R1}$, $\Delta t_2 = \Delta t_{H,R2}$, $\Delta t_3 = \Delta t_{R1,R2}$, $v_1 = v_H$, $v_2 = v_{R1}$, $v_2 = v_{R2}$, $(v_i \geq 0)$ $\Delta \mathbf{t} = (\Delta t_1, \Delta t_2, \Delta t_3)^T$, $\mathbf{v} = (v_1, v_2, v_3)^T$. Define further the Lyapunov-function

$$V(\tau) = \frac{1}{2} \Delta\mathbf{t}(\tau)^T Q \Delta\mathbf{t}(\tau) \rightarrow min \qquad (25)$$

where $Q > 0$ is a diagonal weighting matrix. Differentiation of $V(\tau)$ by τ yields

$$\dot{V}(\tau) = \dot{\Delta\mathbf{t}}(\tau)^T Q \Delta\mathbf{t}(\tau) < 0 \qquad (26)$$

where $\dot{V} = \delta V / \delta \tau$ and

$$\dot{\Delta\mathbf{t}} = \frac{\delta \Delta\mathbf{t}}{\delta\mathbf{v}} \cdot \dot{\mathbf{v}} = J \cdot \dot{\mathbf{v}} \qquad (27)$$

$$J = \begin{pmatrix} \frac{\delta \Delta t_1}{\delta v_2} & \frac{\delta \Delta t_1}{\delta v_3} \\ \frac{\delta \Delta t_2}{\delta v_2} & \frac{\delta \Delta t_2}{\delta v_3} \\ \frac{\delta \Delta t_3}{\delta v_2} & \frac{\delta \Delta t_3}{\delta v_3} \end{pmatrix} \qquad (28)$$

which is not square due to $v_1 = v_H = const$. Choosing $\dot{\Delta\mathbf{t}}^T Q = -(\Delta\mathbf{t})^T$ leads to

$$\dot{V} = -\Delta\mathbf{t}^T \Delta\mathbf{t} < 0 \qquad (29)$$

From (27) we get

$$QJ \cdot \dot{\mathbf{v}} = -\Delta\mathbf{t} \qquad (30)$$

and finally

$$\delta\mathbf{v} = -K_{opt}(J^T J)^{-1} J^T Q^{-1} \Delta\mathbf{t} \qquad (31)$$

K_{opt} determines the length of the optimization step. Running (31) in an optimization loop

$$\mathbf{v}_{k+1} = \mathbf{v}_k + \delta\mathbf{v}_{k+1} \qquad (32)$$

the velocity vector converges to a velocity vector $\mathbf{v}_{opt}$ and a corresponding minimum of $\Delta\mathbf{t}$. In order to *avoid* a possible collision, Eq. (31) is changed in that a correction vector $C_{\Delta t}$ is introduced so that

$$V = \frac{1}{2}(\Delta\mathbf{t} + C_{\Delta t})^T Q(\Delta\mathbf{t} + C_{\Delta t}) \rightarrow min. \qquad (33)$$

From this follows that (31) changes to

Fig. 12 Motion before crossing, case 1, extracted from [25]

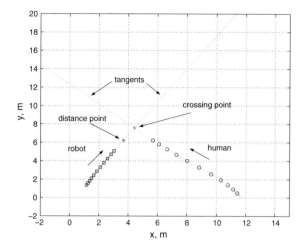

$$\delta \mathbf{v} = -K_{opt}(J^T J)^{-1} J^T Q^{-1}(\Delta \mathbf{t} + C_{\Delta t}) \tag{34}$$

This option leads to $|\Delta \mathbf{t}|_{|C_{\Delta t}|>0} > |\Delta \mathbf{t}|_{|C_{\Delta t}|=0}$ which prevents human and robot from a possible collision at some intersections.

Fuzziness due to measurement errors and uncertainties in kinematics and geometry have an impact on the generation of appropriate velocities to avoid collisions at the intersections. In the presence of given error rates in measurements for angles and distances, the resulting velocities lead to intersection areas (red scatter diagrams, see Fig. 11) instead of crossing points. The same is true for the results gained from the optimization outside the intersections (blue scatter diagrams). To cope with this problem numerically in the sense of fuzzy sets we run an example (1 human, 2 robots) several times (e.g. 20 times) and make a fuzzy clustering for the crossing points (red dots) and also for the optimized points (blue dots). The cluster centers are computed by the centers of gravity of each cluster. The clusters are modeled by Gaussian functions

$$w_{Gi} = \frac{1}{\sqrt{2\pi}\sigma} e^{-\left(\frac{d}{\sigma}\right)^2} \tag{35}$$

where w_i - membership of a point x to a cluster Cl_i, d - distance of a point x in the cluster from the cluster center x_{ci}, σ - standard deviation.

For obstacles with a diameter D it can easily be shown that a collision can be avoided if a single measurement satisfies

$$\tilde{D}_{ij} \geq 4\sigma + 2D \tag{36}$$

where $\tilde{D}_{ij}$ is the distance between a "red" and a "blue" measurement position.

Fig. 13 Time schedule
before/after crossing, case 1,
extracted from [25]

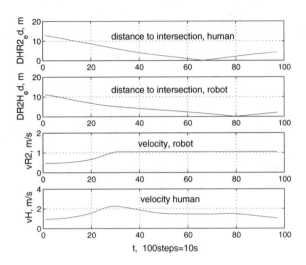

7 Simulation Results

7.1 Collision Avoidance with Hybrid Controller

In the 1st simulation the "1 human - 1 robot" case is considered where the estimations
of the crossing points are made during motion. This experiment is a combination of
real human walking data from the Edinburgh Data and a simulated robot. At each
time point the crossing points of the two tangents along the velocity vectors are
computed and the robot velocity will be adjusted according to Eq. (12a). Figure 12
shows the plot before the crossing point and Fig. 13 shows the time schedule before
and after the crossing point. In the case of stationarity the velocity is limited at
a predefined distance between robot and actual crossing point in order to prevent
from too high robot velocities. Figure 13 shows that after 3 s the robot has reached a
velocity that guarantees the human to pass the crossing point 1.5 s before the robot.
Case 2 (Eq. (12b)) is shown in Figs. 14 and 15. Here the robot passes the intersection
before the human with a time difference of 2.5 s.

7.2 Velocity Obstacles and Fuzzy Control

The simulation example includes both the analytical and the fuzzy solution. We simulated a robot with a length $i = 1.5$ m and a human agent with the goal positions
$x_{rob} = (0, 50)$ and $x_{hum} = (40, 10)$ (see Fig. 16). Figure 18 shows the whole evolution from $t = 0$–24 s. Figures 16 and 17 shows the velocity obstacles at the times
$t = 0$ s and $t = 18$ s, respectively. Figure 19 shows the time plots of the velocities,
orientation angles and steering angles of the robot and the human, respectively. The

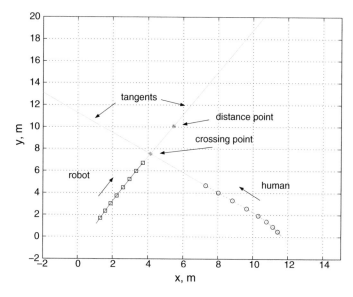

Fig. 14 Motion before crossing, case 2, extracted from [25]

Fig. 15 Time schedule
before/after crossing, case 2,
extracted from [25]

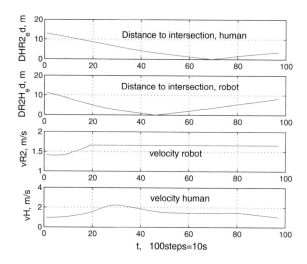

time distances between the squares/circles on the trajectories are 3 s. The robot is
controlled by the hybrid controller described in Sect. 5.3 whereas the human agent's
behavior follows the fuzzy controller described in Sect. 4. The complete set of rules
is shown in (37).

The time plots show a smooth avoidance procedure even in the case of the "fuzzy
- non-fuzzy" combination.

$$IF \quad \phi_{\tilde{H}H} = APP \quad THEN \quad \phi_{HR_{rob}} = TUL \quad AND \quad |\mathbf{v}|_{rob} = SD$$
$$IF \quad \phi_{\tilde{H}H} = AVR \quad THEN \quad \phi_{HR_{rob}} = AVL \quad AND \quad |\mathbf{v}|_{rob} = SD$$
$$IF \quad \phi_{\tilde{H}H} = MOR \quad THEN \quad \phi_{HR_{rob}} = MS \quad AND \quad |\mathbf{v}|_{rob} = NON$$
$$IF \quad \phi_{\tilde{H}H} = RER \quad THEN \quad \phi_{HR_{rob}} = MS \quad AND \quad |\mathbf{v}|_{rob} = NON$$
$$IF \quad \phi_{\tilde{H}H} = REC_R \quad THEN \quad \phi_{HR_{rob}} = MS \quad AND \quad |\mathbf{v}|_{rob} = SD \qquad (37)$$
$$IF \quad \phi_{\tilde{H}H} = AVL \quad THEN \quad \phi_{HR_{rob}} = AVR \quad AND \quad |\mathbf{v}|_{rob} = SD$$
$$IF \quad \phi_{\tilde{H}H} = MOL \quad THEN \quad \phi_{HR_{rob}} = AVL \quad AND \quad |\mathbf{v}|_{rob} = SD$$
$$IF \quad \phi_{\tilde{H}H} = REL \quad THEN \quad \phi_{HR_{rob}} = MS \quad AND \quad |\mathbf{v}|_{rob} = NON$$
$$IF \quad \phi_{\tilde{H}H} = REC_L \quad THEN \quad \phi_{HR_{rob}} = MS \quad AND \quad |\mathbf{v}|_{rob} = SD$$

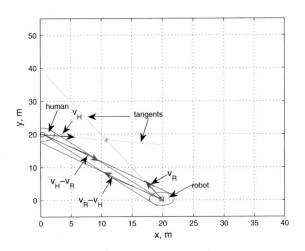

Fig. 16 Velocity obstacle, $t = 0$ s

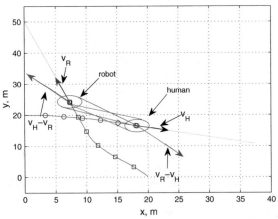

Fig. 17 Velocity obstacle, $t = 18$ s

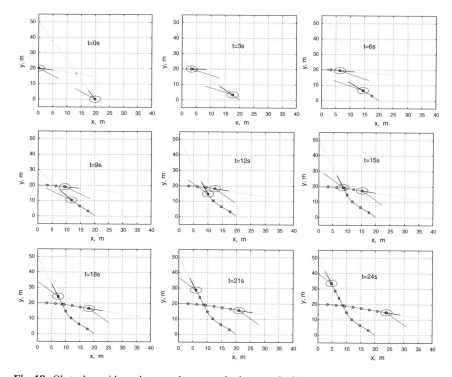

Fig. 18 Obstacle avoidance between human and robot, $t = 0-24$ s

The features of the controller are the following: Form of MsF: Gauss, fuzzifier= singleton, composition method=minmax, number of sets: $\phi_{\tilde{H}H} = 9$, $\phi_{HR_{rob}} = 5$, $|\mathbf{v}|_{rob} = 3$.

7.3 Recognition of Lanes While Tracking

From 11 trajectories of the Edinburgh-data 3 different lanes were identified used in both directions with different velocities. The modeling results are representative trajectories/paths of bundles of similar trajectories. In our example, 3 of 20 test trajectories (tracks) were selected from the Edinburgh-data which have not been used for modeling but whose entry/exit areas coincide with those of the modeled lanes (see Fig. 20). During motion the degrees of membership (see Fig. 21) for each track are computed according to (22) and (23) with a moving average about 10 time steps. Here, from the membership degrees in Figs. 22, 23 and 24 one can see, that from only a few samples the membership of a track to a lane can be recognized.

Fig. 19 Orientation angles, steering angles and velocities of robot and human

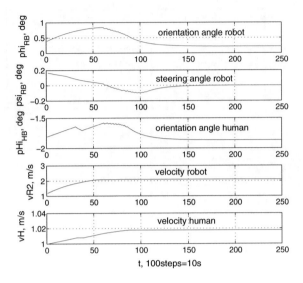

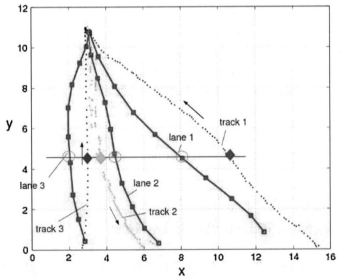

Fig. 20 Lanes 1–3, test tracks, extracted from [25]

Fig. 21 Membership functions, extracted from [25]

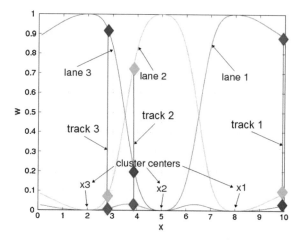

Fig. 22 Memberships track 1, extracted from [25]

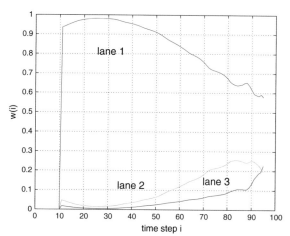

7.4 Prediction of Possible Collisions at Intersections

In this simulation a combination of 1 human agent and 3 robot agents is presented (see Fig. 25). The simulation shows sufficient distances between red and blue areas which shows that collisions between human agent and the robots and between robots can be avoided. In contrast to [26] the correction vector $C_{\Delta t}$ is velocity dependent which leads to better results than for $C_{\Delta t} = const$. In addition, an overlap of "blue areas" are tried to be avoided which results in a more consistent distribution of the "blue areas". The number of optimization steps and the resulting velocities (measured in m/s), respectively, are $N = 50$, $|v_H| = 5$, $|v_{R1}| = 20.11$, $|v_{R2}| = 20.07$, $|v_{R3}| = 22.63$.

Fig. 23 Memberships track
2, extracted from [25]

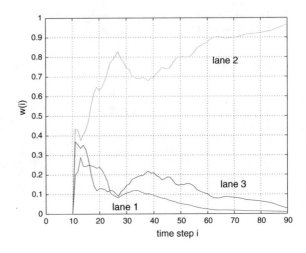

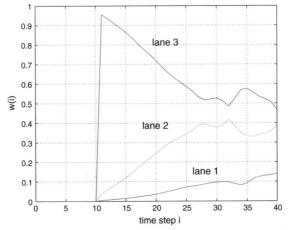

Fig. 24 Memberships track 3, extracted from [25]

8 Conclusions

The interaction between robots and humans in common working areas requires a safe,
reliable and stable harmonization of their motion, both for cooperation tasks and for
obstacle avoidance. This implies requirements for the short distance control task for
robot and human in addition to the usual goal to keep a certain distance between
the agents while considering certain constraints on the robot motion. In this article
we present a fuzzy rule approach based on the kinematic relations between robot
and human. Furthermore, for predefined tracks a hybrid controller is introduced that
changes the robot velocity depending on the direction and velocity of the motion of
the human. For obstacle avoidance the method of velocity obstacles is applied to the

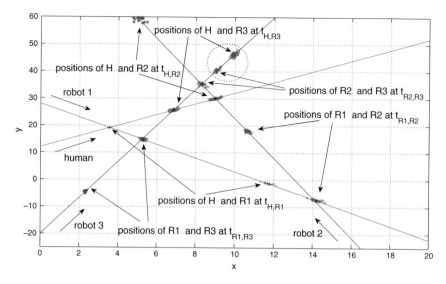

Fig. 25 Example with 1 human and 3 robots with correction

optimization of the changes of the robot's movements in direction and speed relative to humans. The latter method is further combined with Mamdani fuzzy expert rules. For longer distances it is crucial to predict the motion of humans and robots to avoid collisions at the intersections of predefined tracks or paths. This is done by fuzzy modeling of pedestrian tracks through identification of lanes preferred by human agents, and the identification of a membership of a pedestrian track to a specific lane. From this knowledge robots adjust their motion very early so that collisions can reliably be avoided. In the case of predefined paths the velocities of the robots are adjusted such that no collisions are to be expected at the intersections. Examples with both simulated and real data show the applicability of the presented methods and the high performance of the results.

Acknowledgements This research work has been supported by the AIR-project, Action and Intention Recognition in Human Interaction with Autonomous Systems.

References

1. Aarno, D.: Intention recognition in human machine collaborative systems. Licentiate thesis Stockholm, Sweden, pp. 1–104 (2007)
2. Bennewitz, M., Burgard, W., Cielniak, G., Thrun, S.: Learning motion patterns of people for compliant robot motion. Int. J. Robot. Res. **24**(1), 31–48 (2005)
3. Bruce, J., Wawer, J., Vaughan, R.: Human-robot rendezvous by co-operative trajectory signals, pp. 1–2 (2015)
4. Chadalavada, R.T., Andreasson, H., Krug, R., Lilienthal, A.: That's on my mind! robot to human intention communication through on-board projection on shared floor space. In: European Conference on Mobile Robots (ECMR) (2015)

5. Ciaramella, A., Cimino, M.G.C.A., Marcelloni, F., Straccia, U.: Combining fuzzy logic and semantic web to enable situation-awareness in service recommendation. Database and Expert Systems Applications. Lecture Notes in Computer Science, vol. 6261, pp. 31–45 (2010)

6. Edinburgh informatics forum pedestrian database (2010). http://homepages.inf.ed.ac.uk/rbf/FORUMTRACKING/

7. Ellis, T., Sommerlade, E., Reid, I.: Modelling pedestrian trajectory patterns with Gaussian processes. In: 2009 IEEE 12th International Conference on Computer Vision, Workshops (ICCV Workshops), pp. 1229–1234. IEEE (2009)

8. Fiorini, P., Shiller, Z.: Motion planning in dynamic environments using velocity obstacles. Int. J. Robot. Res. **17**(7), 1998 (1998)

9. Firl, J.: Probabilistic maneuver recognition in traffic scenarios. Doctoral dissertation, KIT Karlsruhe (2014)

10. Fraichard, T., Paulin, R., Reignier, P.: Human-robot motion: taking attention into account. Research report, RR-8487 (2014)

11. Gomez, J., Mavridis, N., Garrido, S.: Social path planning: generic human-robot interaction framework for robotic navigation tasks. Workshop of the IEEE/RSJ International Conference on Intelligent Robots and Systems (IROS'13) (2013)

12. Han, T.A., Pereira, L.: State of the art of intention recognition. AI Commun. **26**, 237–246 (2013)

13. Heinze, C.: Modelling intention recognition for intelligent agent systems. Research report, Air Operations Division (2004)

14. Johansson, A.F.: Data driven modeling of pedestrian crowds. Doctoral thesis, pp. 1–185 (2009)

15. Khatib, O.: Real-time obstacle avoidance for manipulators and mobile robots. In: IEEE International Conference on Robotics and Automatio, St. Loius, Missouri, p. 500505 (1985)

16. Krauthausen, P.: Learning dynamic systems for intention recognition in human-robot-cooperation. Doctoral dissertation University report, Karlsruhe (2012)

17. Leica, P., Toibero, M., Robert, F., Carelli, R.: Switched control to robot-human bilateral interaction for guiding people. J. Intell. Robot. Syst. Dordrecht, Argentina. **77**(1), 73–93 (2015)

18. Levine, W.: The control handbook, pp. 1–316 (1995)

19. Makris, D., Ellis, T.: Spatial and probabilistic modelling of pedestrian behaviour. In: Proceedings of the ICRA, pp. 3960–3965. IEEE (2010)

20. Mataric, M.: A distributed model for mobile robot environment-learning and navigation. Technical report, pp. 1–139 (1990)

21. Miller, F.S., Everett, A.F.: Instructions for the use of Martins mooring board and Battenbergs course indicator. Authority of the Lords of Commissioners of the Admirality (1903)

22. Palm, R., Bouguerra, A.: Particle swarm optimization of potential fields for obstacle avoidance. In: Proceedings of RARM 2013, Istanbul, Turkey. Volume: Scient. coop. Intern. Conf. in elect. and electr. eng. (2013)

23. Palm, R., Iliev, B.: Learning of grasp behaviors for an artificial hand by time clustering and Takagi-Sugeno modeling. In: Proceedings FUZZ-IEEE 2006 - IEEE International Conference on Fuzzy Systems. IEEE, Vancouver, BC, Canada, 16–21 July 2006

24. Palm, R., Iliev, B., Kadmiry, B.: Recognition of human grasps by time-clustering and fuzzy modeling. Robot. Auton. Syst. **57**(5), 484–495 (2009)

25. Palm, R., Chadalavada, R., Lilienthal, A.: Fuzzy modeling and control for intention recognition in human-robot systems. In: 7. IJCCI (FCTA) 2016, Porto, Portugal (2016)

26. Palm, R., Chadalavada, R., Lilienthal, A.: Recognition of human-robot motion intentions by trajectory observation. In: 9th International Conference on Human System Interaction, HSI2016. IEEE (2016)

27. Precup, R.-E., Preitl, S., Tomescu, M.L.: Fuzzy logic control system stability analysis based on Lyapunov's direct method. In: International Journal of Computers, Communications and Control, vol. 4(4), pp. 415–426 (2009)

28. Runkler, T., Palm, R.: Identification of nonlinear system using regular fuzzy c-elliptotype clustering. In: Proceedings of the FUZZIEEE 96, pp. 1026–1030. IEEE (1996)

29. Sadri, F., Wang, W., Xafi, A.: Intention recognition with clustering. Ambient Intelligence. Lecture Notes in Computer Science, vol. 7683, pp. 379–384 (2012)
30. Satake, S., Kanda, T., Glas, D., Imai, M., Ishiguro, H., Hagita, N.: How to approach humans?-strategies for social robots to initiate interaction. In: 13th International Symposium on Experimental Robotics, pp. 109–116 (2009)
31. Skoglund, A., Iliev, B., Kadmiry, B., Palm, R.: Programming by demonstration of pick-and-place tasks for industrial manipulators using task primitives. In: Proceedings Computational Intelligence in Robotics and Automation CIRA 2007. IEEE, Jacksonville, Florida, USA, 20–23 June 2007
32. Tahboub, K.A.: Intelligent human-machine interaction based on dynamic bayesian networks probabilistic intention recognition. J. Intell. Robot. Syst. **45**(1), 31–52 (2006)
33. van den Berg, J., Lin, M., Manocha, D.: Reciprocal velocity obstacles for real-time multi-agent navigation. In: Proceedings of the IEEE International Conference on Robotics and Automation (2008)

Input Value Skewness and Class Label Confusion in the NEFCLASS Neuro-Fuzzy System

Jamileh Yousefi and Andrew Hamilton-Wright

Abstract NEFCLASS is a common example of the construction of a NEURO-FUZZY system. The popular NEFCLASS classifier exhibits surprising behaviour when the feature values of the training and testing data sets exhibit significant skew. As skewed feature values are commonly observed in biological data sets, this is a topic that is of interest in terms of the applicability of such a classifier to these types of problems. This paper presents an approach to improve the classification accuracy of the NEFCLASS classifier, when data distribution exhibits positive skewness. The NEFCLASS classifier is extended to provide improved classification accuracy over the original NEFCLASS classifier when trained on skewed data. The proposed model uses two alternative discretization methods, MME and CAIM, to initialize fuzzy sets. From this study it is found that using the MME and CAIM discretization methods results in greater improvements in the classification accuracy of NEFCLASS as compared to using the original EQUAL-WIDTH technique NEFCLASS uses by default.

Keywords Fuzzy · Discretization · Neurofuzzy · Classification · Skewness
Nefclass

1 Introduction

Skewness is a numerical measure which indicates whether a data distribution is symmetric or not. Skewed feature values are commonly observed in biological and

J. Yousefi · A. Hamilton-Wright (✉)
School of Computer Science (SOCS), University of Guelph, Guelph, ON, Canada
e-mail: ahamiltonwright@mta.ca

J. Yousefi
e-mail: jyousefi@uoguelph.ca

A. Hamilton-Wright
Department of Mathematics and Computer Science,
Mount Allison University, Sackville, NB, Canada

medical datasets. Addressing skewness in medical diagnosis systems is vital for finding rare events, such as rare diseases [1].

Most machine learning algorithms perform poorly on skewed datasets [2, pp. 139–140]. Several studies [3–8] have examined the question of transforming the data when incorporating the distribution of input data into the classification system in order to more closely approximate normally distributed data. Very few studies [3, 4, 9, 10] addressed these issues while focusing on some alternative to a data transformation approach.

Data transformation is a common preprocessing step to treat the skewness and improve dataset normality. However, in the biological and biomedical domain, data transformation interferes with the transparency of the decision making process, and can lead to the exclusion of important information from the decision making process, and affect the system's ability to correctly classify the case. Therefore, rather than transformation of the data to achieve a more normally distributed input, in this paper we investigate using a non-transformation-based strategy to address the skewness problem.

This paper demonstrates that using an appropriate discretization method provides a considerable improvement in classification accuracy of NEFCLASS when trained on skewed data.

The choice of discretization technique is known to be one of the important factors that might affect the classification accuracy of a classifier. NEFCLASS classifiers use an EQUAL-WIDTH discretization method to divide the observed range of continuous values for a given feature into equally sized fuzzy intervals, overlapping by half of the interval width. EQUAL-WIDTH discretization does not take the class information into account, which, as we show here, results in a lower classification accuracy for NEFCLASS when the feature values of the training and testing data sets exhibit significant skew. Dealing with skewness without performing a transformation will provide greater clarity in interpretation, and by extension better classification transparency, as the data projection produced by the transformation does not need to be taken into account in interpretation.

We provide a study based on an easily reproducible synthetic data distribution, in order to allow deeper insights into the data analysis. Also, to show the pertinence of this analysis to real-world data problems, we ran all tests on a clinically applicable data set of quantitative electromyography (EMG) data.

We argue that the skewed data, in terms of feature value distribution, cause a higher misclassification percentage in classification learning algorithms. We further argue that distribution sensitive discretization methods such as CAIM and MME result in greater improvements in the classification accuracy of the NEFCLASS classifier as compared to using the original EQUAL-WIDTH technique. This is an expansion of the discussion presented in our paper at the 9th International Joint Conference on Computational Intelligence [11].

The next section of this paper contains a short review of the NEFCLASS classifier and three discretization methods that will be used to perform the classification task. Section 3 describes the methodology of our study, and in Sect. 4 the experimental results and statistical analysis are given. Finally, conclusions are presented.

2 Background

2.1 Discretization

A discretization process divides a continuous numerical range into a number of covering intervals where data falling into each discretized interval is treated as being describable by the same nominal value in a reduced complexity discrete event space. In fuzzy work, such intervals are then typically associated with the support of fuzzy sets, and the precise placement in the interval is mapped to the degree of membership in such a set.

Discretization methods are categorized into supervised and unsupervised algorithms. EQUAL-WIDTH intervals [12], EQUAL-FREQUENCY intervals, such as k-means clustering [13], and Marginal Maximum Entropy [14, 15] are examples of algorithms for unsupervised discretization. Maximum entropy [16], ChiMerge [17], CAIM [18], and URCAIM [19] are some examples of supervised algorithms that take class label assignment in a training data set into account when constructing discretization intervals.

In the following discussion, the three discretization methods, that we have chosen for the experiments are described. To demonstrate the partitioning produced by these algorithms we used a skewed dataset with 45 training instances and three classes, with sample discretizations shown in Fig. 1, which has been reproduced from [11]. The subfigures within Fig. 1 each show the same data, with the green, red and blue rows of dots (top, middle and bottom) within each figure describing the data for each class in the training data.

EQUAL-WIDTH

The EQUAL-WIDTH discretization algorithm divides the observed range of continuous values for a given feature into a set number of equally sized intervals, providing a simple mapping of the input space that is created independent of both the distribution of class and of the density of feature values within the input space [12, 17].

Figure 1a demonstrates the partitioning using EQUAL-WIDTH intervals; note that there is no clear relation between classes and intervals, and that the intervals shown have different numbers of data points within each (21, 19 and 5 in this case).

MME

Marginal Maximum Entropy, or MME, based discretization [14, 15] divides the dataset into a number of intervals for each feature, where the number of points is equal for all of the intervals, under the assumption that the information of each interval is expected to be equal. Figure 1b shows the MME intervals for the example three-class dataset. Note that the intervals in Fig. 1b do not cover the same fraction of the range of values (i.e., the widths differ), being the most dense in regions where there are more points. The same number of points (15) occur in each interval.

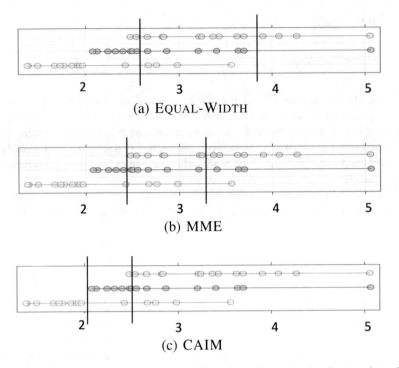

Fig. 1 Three discretization techniques result in different intervals produced on the same three-class data set. Figure extracted from [11]

CAIM

CAIM (class-attribute interdependence maximization) discretizes the observed range of a feature into a class-based number of intervals and maximizes the interdependency between class and feature values [18]. The CAIM discretization algorithm divides the data space into partitions, which leads to preserve the distribution of the original data [18], and obtain decisions less biased to the training data.

Figure 1c shows the three CAIM intervals for our sample data set. Note how the placement of the discretization boundaries is closely related to the points where the densest portion of the data for a particular class begins or ends.

2.2 NEFCLASS *Classifier*

NEFCLASS [20, 21] is a NEURO-FUZZY classifier that tunes a set of membership functions that describe input data and associates these through a rule set with a set of potential class labels. Training is done in a supervised fashion based on a training data set.

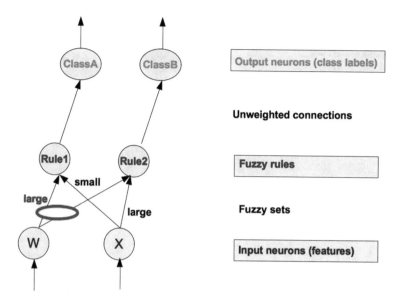

Fig. 2 A NEFCLASS model with two inputs, two rules, and two output classes. Figure extracted from [11]

Figure 2, which has been reproduced from [11] shows a NEFCLASS model that classifies input data with two features into two output classes by using three fuzzy sets and two fuzzy rules. Input features are supplied to the nodes at the bottom of the figure. These are then fuzzified, using a number of fuzzy sets. The sets used by a given rule are indicated by linkages between input nodes and rule nodes. If the same fuzzy set is used by multiple rules, these links are shown passing through an oval, such as the one marked "large" in Fig. 2. Rules directly imply an output classification, so these are shown by unweighted connections associating a rule with a given class. Multiple rules may support the same class, however that is not shown in this diagram. Rule weightings are computed based on the degree of association between an input fuzzy membership function and the rule (calculated based on the degree of association in the training data), as well as the level of activation of the fuzzy membership function, as is typical in a fuzzy system [22, Chapter 1].

Figure 3, reproduced from [11] describes the relationship between the support of the fuzzy intervals used in Nefclass and underlying class-related data points. In Fig. 3a, a set of initial fuzzy membership functions describing regions of the input space are shown, here for a two-dimensional problem in which the fuzzy sets are based on the initial discretization produced by the EQUAL-WIDTH algorithm. As will be demonstrated, NEFCLASS functions work best when these regions describe regions specific to each intended output class, as is shown here, and as is described in the presentation of a similar figure in the classic work describing this classifier [20, pp. 239].

Fig. 3 Fuzzy membership
functions before and after
training data based tuning
using the NEFCLASS
algorithm. Figure extracted
from [11]

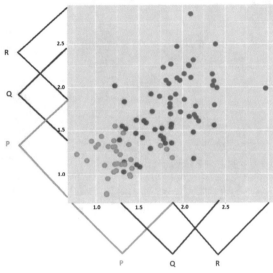

(a) Initial fuzzy set membership functions in NEFCLASS,
produced using EQUAL-WIDTH discretization

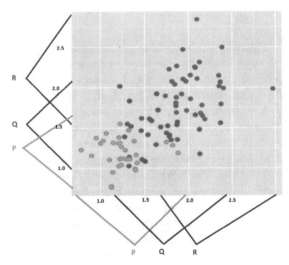

(b) Results of tuning the above membership functions to
better represent class/membership function information

As is described in the NEFCLASS overview paper [21, pp. 184], a relationship is constructed through training data based tuning to maximize the association of the support of a single fuzzy set with a single outcome class. This implies both that the number of fuzzy sets must match the number of outcome classes exactly, and in addition, that there is an assumption that overlapping classes will drive the fuzzy sets to overlap as well.

Figure 3a (Figure extracted from [11]) shows the input membership functions as they exist before membership function tuning, when the input space is partitioned into EQUAL-WIDTH fuzzy intervals.

Figure 3b (Figure extracted from [11]) demonstrates that during the fuzzy set tuning process, the membership function is shifted and the support is reduced or enlarged, in order to better match the coverage of the data points belonging to the associated class, however as we will see later, this process is strongly informed by the initial conditions set up by the discretization to produce the initial fuzzy membership functions.

3 Methodology

This paper has two objectives. The first is to characterize how the NEFCLASS classification accuracy degrades as data skewness increases. The second is to evaluate alternative discretization methods to counteract the performance problems in skewed data domains. To support this second goal we will evaluate maximum marginal entropy (MME) [14, 15] and the supervised, class-dependent discretization method CAIM [18].

We carried out two different set of experiments. In the first experiment, denoted as the effect of skewness of feature values, we evaluate unmodified NEFCLASS behaviour when dealing with different level of skewness.

In the second experiment, we investigate the classification accuracy of a modified NEFCLASS classifier upon employing the three different discretization techniques. Modification through alternative discretization methods takes into account an important difference between the discretization methods and their effects on the classifier's accuracy.

We implemented a modified NEFCLASS classifier, embedded with a choice of two alternative discretization methods, MME and CAIM. The MME and CAIM methods are not part of the standard NEFCLASS implementation, therefore we implemented two modified versions of NEFCLASS classifier, each utilizing one of these two discretization methods. For the EQUAL-WIDTH and MME methods, the number of fuzzy sets is defined as a parameter. For the CAIM method, the number of fuzzy sets is determined by the discretization algorithm.

Experiments were then performed on synthesized datasets with different levels of feature values skewness. Besides, we conducted a set of experiments to evaluate the effectiveness of our approaches for a real-world dataset, EMG data, which contains several highly skewed features. Results from the experiments are presented in terms of misclassification percentages, which are equal to the number of misclassified data instances divided by the total number of instances in the testing dataset.

3.1 Synthesized Datasets

Three synthesized datasets were used for experiments. All the synthesized data sets used describe classification problems within a 4-dimensional data space containing distributions of data from three separate classes.

Our data was produced by randomly generating numbers following the F-distribution with different degrees of freedom chosen to control skew. The F-distribution [23] has been chosen as the synthesis model because the degree of skew within an F-distribution is controlled by the pairs of degrees of freedom specified as a pair of distribution control parameters. This allows for a spectrum of skewed data distributions to be constructed. We designed the datasets to present different levels of skewness with increasing skew levels. Three pairs of degrees of freedom parameters have been used to generate datasets with different levels of skewness, defined here as low, medium, and high-skewed feature values. The specific degree of freedom values used were chosen after initial experimentation resulting in degree of freedom choices of $(100, 100)$ to provide data close to a normal distribution, $(100, 20)$ to provide moderate skew, and $(35, 8)$ to provide high.

A synthesized data set consisting of 1000 randomly generated examples consisting of four-feature (W, X, Y, Z) F-distribution data for each of three classes was created. The three classes (ClassA, ClassB and ClassC) overlap, and are skewed in the same direction.

The size of datasets were designed to explore the effect of skewness when enough data is available to clearly ascertain data set properties. Ten-fold cross validation was used to divide each dataset into training (2700) and testing (300 point) sets in which an equal number of each class is represented. We have taken care to ensure that all datasets used have a similar degree of overlap, and same degree of variability.

Figure 4 (also reproduced from [11]) shows the skewness of each dataset for each feature. From these figures one can see that the LOW-100, 100 data is relatively symmetric, while the MED-100, 20 and HIGH-35, 8 data show an increasing, and ultimately quite dramatic, skew.

3.2 Electromyography Dataset (EMG)

EMG is the study of the electrical potentials observed from contracting muscles as seen through the framework of quantitative measurement. EMG is used in research and diagnostic study [24]. The EMG datasets are known to contain features with highly skewed value distributions [25].

The EMG dataset used here contains seven features of MUP templates (Amplitude, Duration, Phases, Turns, AAR, SizeIndex, and MeanMUVoltage) observed on 791 examples representing three classes (Myopathy, Neuropathy, Healthy), collected through the number of contractions, and used in previous work [26]. Each example is based on measuring values from a MUP template extracted from an EMG signal, and it describes the contribution of all observed MUs that contributed to the obtained EMG signal.

Fig. 4 Skewness by label and feature for the three synthetic datasets. Figure extracted from [11]

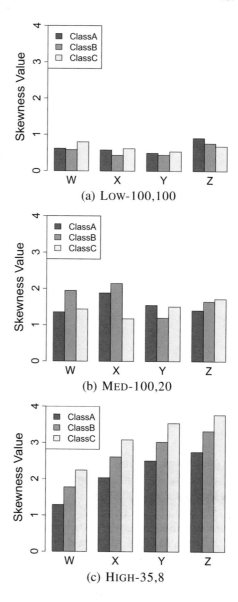

(a) Low-100,100

(b) Med-100,20

(c) High-35,8

Figure 5 shows the skewness values for all features with respect to each class as with the other figures in this paper, it is reproduced from our earlier paper [11]. As is shown in Fig. 5, the distribution of values for Amplitude and MeanMUVoltage are highly positively skewed, particularly in Myopathy cases. In contrast, Turns is highly skewed in Neuropathy and Normal cases. Also, Phases, AAR, and SizeIndex demonstrate relatively low skewness.

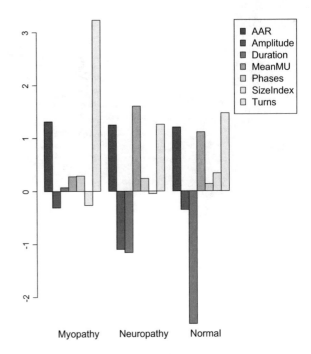

Fig. 5 Degree of skewness for the EMG dataset

4 Results and Discussion

4.1 The Effect of Feature Value Skewness on the NEFCLASS Classifier's Behaviour

In this section, analyses were performed to explore the relationship between different levels of skewness, misclassification percentages, and the number of rules obtained by the NEFCLASS classifier.

We begin by applying the NEFCLASS classifier to the synthesized datasets with our three different levels of skewness: low, medium, and high. Our null hypothesis is that the misclassification percentage observed for NEFCLASS are equal in all three datasets.

As mentioned earlier, NEFCLASS will attempt to tune the fuzzy membership functions provided by the initial EQUAL-WIDTH discretization to associate the support of each fuzzy set unambiguously with a single class. As our skewed data sets overlap substantially, this is a difficult task for the NEFCLASS classifier.

Figures 6 and 7 (Figures extracted from [11]) show the relationship between the distribution of input data values in feature X, and the placement of the final membership function by the NEFCLASS classifier when using the EQUAL-WIDTH discretization method for the three datasets LOW-100, 100, MED-100, 20 and HIGH-35, 8.

Figure 6a–c illustrate the density functions for the LOW-100, 100, MED-100, 20 and HIGH-35, 8 data sets, respectively. As can be seen in Fig. 6a, the feature values in

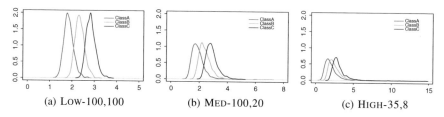

Fig. 6 Density plots for feature X for all datasets. Figure extracted from [11]

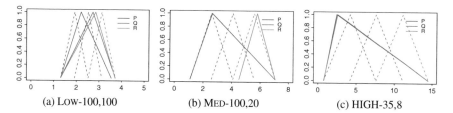

Fig. 7 Fuzzy sets and membership functions constructed by EQUAL-WIDTH for feature X of all datasets. Figure extracted from [11]

this case are centrally tended and generally symmetric, providing an approximation of a Normal distribution, though observable skew is still present, as shown in Fig. 4a. In Fig. 6b, tails are observable as the mean of feature values are pulled towards the right while the median values remain similar to those in Fig. 6a, for a moderately skewed distribution. In Fig. 6c, the feature values exhibit longer tails than normal, indicating a highly skewed distribution.

Figures 6 and 7 (also reproduced from [11]) allows comparison between a data distribution and the representation of fuzzy sets for this feature in this distribution, displayed immediately below. For example, Fig. 6a can be compared with Fig. 7a. Similar vertical comparisons are able to be made for Fig. 6b, c.

Figure 7a illustrates the placement of initial fuzzy sets and final membership functions with the EQUAL-WIDTH discretization method for dataset LOW-100, 100. As can be seen, the same number of fuzzy sets with equal support are constructed, and after tuning the membership functions are shifted and the supports are reduced or enlarged, in order to better match the distribution of class-specific feature values. The dotted line indicates the initial fuzzy sets and the solid line indicates the final placement of the fuzzy sets.

Fuzzy sets have been given the names "P", "Q" and "R" in these figures, rather than more traditional linguistic labels such as "small", "medium" and "large" because of the placement associations with the classes, and the assignment of class names to fuzzy sets lies within the NEFCLASS training algorithm, and is not under user control. For this reason, it is not possible to assign names a priori that have any linguistic meaning. The choices NEFCLASS makes in terms of which fuzzy set is used to represent a particular class is part of the underlying performance issue explored in this paper, as will be shown.

As one can see in comparing Fig. 7a with Fig. 6a, NEFCLASS has chosen to associate fuzzy set P with ClassB, set Q with ClassA, and fuzzy set R with ClassC based on this EQUAL-WIDTH initial discretization. The associations are not as clear with data exhibiting higher skew, as shown in Fig. 7b, c, in which NEFCLASS is attempting to set two, and then three, fuzzy membership functions to essentially the same support and with the same central peak. This, of course, limits or entirely destroys the information available to the rule based portion of the system, and is therefore unsurprisingly correlated with a higher number of classification errors.

The analysis of fuzzy sets and memberships for feature X in dataset LOW-100, 100 shows that all of three fuzzy sets are expanded in terms of their support. As shown in Fig. 7a, the support of fuzzy sets Q and R share identical support, as defined by the observed range of the training data. In addition, sets Q and R are defined by a nearly identical maximum point, rendering the distinction between them moot. Similarly, in Fig. 7b, we observe identical maxima and support for sets P and Q, however R is set apart with a different, but substantially overlapping, support and a clearly distinct maxima.

This confusion gets only more pronounced as the degree of skew increases, as is clear in Fig. 7c, representing the fuzzy sets produced for HIGH-35, 8, in which all fuzzy sets P, Q and R share identical support defined by the range of the observed data. In addition, all the fuzzy sets in this example are defined by a nearly identical maximum point.

It is perhaps surprising that difference in the proportion of data in the tails of the distribution are not represented more directly here, however this is largely due to the fact that the EQUAL-WIDTH discretization technique is insensitive to data density, concentrating instead purely on range.

Table 1 reports the misclassification percentages (as median $\pm$ IQR), as well as the number of rules (as median $\pm$ IQR) obtained by each classifier, using each discretization technique, and for each dataset. The results have been calculated over the

Table 1 Misclassification percentage (Median $\pm$ IQR) and Number of Rules (Median $\pm$ IQR) for each classifier trained on three synthesized dataset

	Discretization	Dataset		
		LOW-100, 100	MED-100, 20	HIGH-35, 8
Median $\pm$ IQR for misclassification percentages	EQUAL-WIDTH	22.66 ± 1.33	65.00 ± 5.51	66.67 ± 3.26
	MME	26.00 ± 1.50	34.16 ± 1.00	42.50 ± 1.00
	CAIM	24.33 ± 4.17	34.16 ± 2.60	41.67 ± 4.33
Median $\pm$ IQR for number of rules	EQUAL-WIDTH	49.00 ± 0.00	34.50 ± 2.00	15.00 ± 1.00
	MME	44.00 ± 1.50	50.00 ± 1.00	46.00 ± 1.00
	CAIM	43.50 ± 1.00	51.50 ± 2.50	45.00 ± 1.00

Table 2 M–W–W results comparing the misclassification percentages based on level of skew

Discretization	Low-100, 100 versus		MED-100, 20 versus
	MED-100, 20	HIGH-35, 8	HIGH-35, 8
EQUAL-WIDTH	$2.9 \times 10^{-11***}$	$2.9 \times 10^{-11***}$	0.9300
MME	$2.9 \times 10^{-11***}$	$2.9 \times 10^{-11***}$	0.0001***
CAIM	$2.9 \times 10^{-11***}$	$2.9 \times 10^{-11***}$	0.0001***

***Significant at 99.9% confidence ($p < 0.001$)

10 cross-validation trials. Figure 8 shows a graphical summary of the misclassification percentages and number of rules observed for each discretization method. In the figure, the magenta boxes represent the misclassification percentages, the green boxes represent the number of rules, and the black dots represent the outliers.

An exploration of the normality of the distribution of misclassification percentages using a Shapiro–Wilks test found that a non-parametric test was appropriate in all cases. To explore the statistical validity of the differences between observed misclassification percentages of NEFCLASS classifiers for different data sets and using different discretization techniques, Mann–Whitney–Wilcoxon (M–W–W) tests were performed.

By running a M–W–W test on the misclassification percentages for each pair of skewed data sets, the results shown in Table 2 were obtained. The test shows that there is a significant difference for almost all levels of skewness, the only exceptions being the MED-100, 20 and HIGH-35, 8 distributions when the EQUAL-WIDTH discretization method has been used. We therefore conclude that the NEFCLASS classification accuracy was significantly affected by feature value skewness in the majority of cases. As shown in Table 1, LOW-100, 100 achieved the lowest misclassification percentage. The decrease in the number of rules produced using the EQUAL-WIDTH method shows that less information is being captured about the data set as skewness increases. As there is no reason to assume that less information will be required to make a successful classification, this decrease in the number of rules is therefore an contributing cause for the increase in the misclassification percentage noted for EQUAL-WIDTH in Fig. 8a.

4.2 *Improving Accuracy by Using Alternative Discretization Methods*

In this section, we investigate how the choice of discretization method affects the classification accuracy of a NEFCLASS based classifier, when dealing with datasets with various degrees of skew. We compare the results for our new NEFCLASS implementations using MME and CAIM with the results of the default EQUAL-WIDTH discretization strategy. The null hypothesis is that there will be no difference in the observed misclassification percentages. Table 3 reports the M–W–W test results. The

Fig. 8 Summary of the
misclassification percentages
and number of rules for
EQUAL-WIDTH, MME and
CAIM

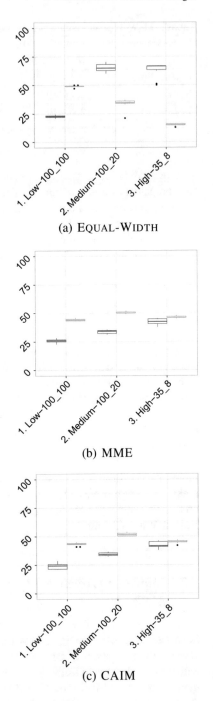

(a) EQUAL-WIDTH

(b) MME

(c) CAIM

Table 3 M–W–W results comparing the misclassification percentages based on discretization method

Dataset	Equal-Width versus		MME versus
	MME	CAIM	CAIM
Low-100, 100	0.0020***	0.2500	0.1600
Med-100, 20	0.0001***	0.0001***	0.5000
High-35, 8	0.0001***	0.0001***	0.9100

***Significant at 99.9% confidence ($p < 0.001$)

test identified the significance of the difference in misclassification percentages for Equal-Width versus MME and CAIM at medium and high skew, where very low p values are computed. In the case of Low-100, 100, a significant difference is observed when comparing Equal-Width to MME, however with a greater p value (0.002).

As shown in Table 1, MME and CAIM achieved lower misclassification percentages compared to Equal-Width. Note that there was no significant difference between MME and CAIM (also confirmed by the M–W–W test). The results also show considerably larger interquartile ranges of the misclassification percentages obtained by all classifiers when trained by High-35, 8, compared to other datasets.

Figure 9 (Figure extracted from [11]) illustrates the placement of initial fuzzy sets and final membership functions when using the MME discretization method for each of the three datasets. As can be seen in this figure, the same number of fuzzy sets with unequal support are constructed. Similarly, in Fig. 10 (Figure extracted from [11]), the fuzzy membership functions produced using CAIM discretization are

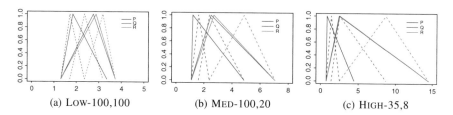

(a) Low-100,100   (b) Med-100,20   (c) High-35,8

Fig. 9 Fuzzy sets and membership functions constructed by MME for feature X of all datasets. Figure extracted from [11]

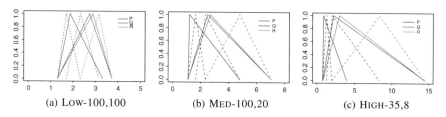

(a) Low-100,100   (b) Med-100,20   (c) High-35,8

Fig. 10 Fuzzy sets and membership functions constructed by CAIM for feature X of all datasets. Figure extracted from [11]

shown. The CAIM algorithm constructed three fuzzy sets for all features for our test data, however as the algorithm is free to choose a differing number of discretization intervals based on the observed data [18], for other applications this may vary from feature to feature. All fuzzy sets generated had distinct support, range, and a different centre value defining the triangular membership functions, though there was still a strong degree of similarity between some sets, as shown in the representative data displayed in Fig. 10.

As shown in Figs. 9 and 10, the final placement of support for fuzzy the first membership function, P, is different from the support for Q and R, in contrast with the very similar fuzzy set definitions of EQUAL-WIDTH shown in Fig. 7. This will preserve a greater degree of differentiation between the information captured in each fuzzy set. A comparison between the results for MME and CAIM for HIGH-35, 8 (not shown graphically in the paper) indicated that the choice of the triangular membership functions produced by CAIM for HIGH-35, 8 were slightly different, but the choice of the triangular membership functions produced by MME were nearly identical. As the difference between the support of the fuzzy membership functions and the centre points increases, the learning phase is more able to create meaningful new rules. This therefore leads to a lower number of classification errors. The larger number of rules generated by MME and CAIM for MED-100, 20 and HIGH-35, 8 is summarized in Table 1.

4.3   Experiments on the EMG Dataset

Table 4 depicts the misclassification percentages (as median $\pm$ IQR) and the number of rules (as median $\pm$ IQR) obtained by NEFCLASS using the EQUAL-WIDTH, MME and CAIM methods when trained on the EMG data. Table 5 reports the results of the M–W–W test.

As shown in Table 5, the test revealed a significant decrease in the misclassification percentages obtained by MME compared to EQUAL-WIDTH. The CAIM method did not achieve a significantly lower misclassification percentage. However, CAIM used significantly fewer rules without deterioration of accuracy.

Table 4 Misclassification percentages (Median $\pm$ IQR) and number of rules (Median $\pm$ IQR) obtained from EQUAL-WIDTH, MME and CAIM trained on the EMG dataset

Classifier	Misclassification percentage	Number of rules
EQUAL-WIDTH	54.30 $\pm$ 31.00	149.00 $\pm$ 4.00
MME	41.00 $\pm$ 22.00	90.00 $\pm$ 14.00
CAIM	50.00 $\pm$ 24.50	74.00 $\pm$ 11.00

Table 5 Results of one-tailed M–W–W to compare the discretization methods for the EMG dataset

Classifier	Misclassification percentage	Number of rules
EQUAL-WIDTH versus MME	**0.03**	0.84
EQUAL-WIDTH versus CAIM	0.11	0.0001***

***Significant at 95% confidence ($p < 0.05$)

It can be concluded that using MME led to a significant improvement in accuracy. Furthermore, CAIM achieved a comparable accuracy with fewer rules, compared to EQUAL-WIDTH.

5 Conclusions

The results of this study indicate that the NEFCLASS classifier performs increasingly poorly as data feature value skewness increases.

Further, this study indicates that the choice of initial discretization method affects the classification accuracy of NEFCLASS classifier, and that this effect is very strong in skewed data sets. Utilizing MME or CAIM discretization methods in the NEFCLASS classifier significantly improved classification accuracy for positively high skewed data.

Our non-transformation-based strategy using the MME or CAIM discretization methods can be useful in the medical diagnosis systems because it not only provides more accurate results but also can preserve the information carried by outliers that might be lost in data transformation. It also can be applied in other domains such as fraud detection, computer security, and finance where data exhibits skewness.

Acknowledgements The authors gratefully acknowledge the support of NSERC, the National Sciences and Engineering Research Council of Canada, for ongoing grant support.

References

1. Gao, J., Hu, W., Li, W., Zhang, Z., Wu, O.: Local outlier detection based on kernel regression. In: Proceedings of the 10th International Conference on Pattern Recognition, Washington, DC, USA, pp. 585–588. EEE Computer Society (2010)
2. Ben-Gal, I.: Outlier detection. In: Maimon, O., Rokach, L. (eds.) Data Mining and Knowledge Discovery Handbook, pp. 131–146. Springer Science & Business Media, Berlin (2010)
3. Mansoori, E., Zolghadri, M., Katebi, S.: A weighting function for improving fuzzy classification systems performance. Fuzzy Sets Syst. **158**, 588–591 (2007)
4. Liu, Y., Liu, X., Su, Z.: A new fuzzy approach for handling class labels in canonical correlation analysis. Neurocomputing **71**, 1785–1740 (2008)
5. Peker, N.E.S.: Exponential membership function evaluation based on frequency. Asian J. Math. Stat. **4**, 8–20 (2011)

6. Chittineni, S., Bhogapathi, R.B.: A study on the behavior of a neural network for grouping the data. Int. J. Comput. Sci. **9**, 228–234 (2012)
7. Changyong, F., Hongyue, W., Naiji, L., Tian, C., Hua, H., Ying, L., Xin, M.: Log-transformation and its implications for data analysis. Shanghai Arch Psychiatry **26**, 105–109 (2014)
8. Qiang, Q., Guillermo, S.: Learning transformations for clustering and classification. J. Mach. Learn. Res. **16**, 187–225 (2015)
9. Zadkarami, M.R., Rowhani, M.: Application of skew-normal in classification of satellite image. J. Data Sci. **8**, 597–606 (2010)
10. Hubert, M., Van der Veeken, S.: Robust classification for skewed data. Adv. Data Anal. Classif. **4**, 239–254 (2010)
11. Yousefi, J., Hamilton-Wright, A.: Classification confusion within NEFCLASS caused by feature value skewness in multi-dimensional datasets. In: Proceedings of the 9th International Joint Conference on Computational Intelligence, Porto, IJCCI-2016 (2016)
12. Chemielewski, M.R., Grzymala-Busse, J.W.: Global discretization of continuous attributes as preprocessing for machine learning. Int. J. Approx. Reason. **15**, 319–331 (1996)
13. Monti, S., Cooper, G.: A latent variable model for multivariate discretization. In: The Seventh International Workshop on Artificial Intelligence and Statistics, Fort Lauderdale, FL, pp. 249–254 (1999)
14. Chau, T.: Marginal maximum entropy partitioning yields asymptotically consistent probability density functions. IEEE Trans. Pattern Anal. Mach. Intell. **23**, 414–417 (2001)
15. Gokhale, D.V.: On joint and conditional entropies. Entropy **1**, 21–24 (1999)
16. Bertoluzza, C., Forte, B.: Mutual dependence of random variables and maximum discretized entropy. Ann. Probab. **13**, 630–637 (1985)
17. Kerber, R.: ChiMerge discretization of numeric attributes. In: Proceedings of AAAI-92, San Jose Convention Center, San Jose, California, pp. 123–128 (1992)
18. Kurgan, L.A., Cios, K.: CAIM discretization algorithm. IEEE Trans. Knowl. Data Eng. **16**, 145–153 (2004)
19. Cano, A., Nguyen, D.T., Ventura, S., Cios, K.J.: ur-CAIM: improved CAIM discretization for unbalanced and balanced data. Soft Comput. **33**, 173–188 (2016)
20. Nauck, D., Klawonn, F., Kruse, R.: Neuro-Fuzzy Systems. Wiley, New York (1996)
21. Nauck, D., Kruse, R.: NEFCLASS-X - a soft computing tool to build readable fuzzy classifiers. BT Technol. J. **16**, 180–190 (1998)
22. Mendel, J.M.: Uncertain Rule-Based Fuzzy Logic Systems. Prentice-Hall, Englewood Cliffs (2001)
23. Natrella, M.: NIST SEMATECH eHandbook of Statistical Methods. NIST (2003)
24. Stashuk, D.W., Brown, W.F.: Quantitative electromyography. In: Brown, W.F., Bolton, C.F., Aminoff, M.J. (eds.) Neuromuscular Function and Disease, vol. 1, pp. 311–348. W.B. Saunders, Philadelphia (2002)
25. Enoka, R., Fuglevand, A.: Motor unit physiology: some unresolved issues. Muscle & Nerve **24**, 4–17 (2001)
26. Varga, R., Matheson, S.M., Hamilton-Wright, A.: Aggregate features in multi-sample classification problems. IEEE Trans. Biomed. Health Inf. **99**, 1 (2014). (in press)

Hyperresolution for Propositional Product Logic with Truth Constants

Dušan Guller

Abstract In the paper, we generalise the well-known hyperresolution principle to the propositional product logic with explicit partial truth. We propose a hyperresolution calculus suitable for automated deduction in a useful expansion of the propositional product logic by intermediate truth constants and the equality, $=$, strict order, $\prec$, projection, Δ, operators. We expand the propositional product logic by a countable set of intermediate truth constants of the form $\bar{c}$, $c \in (0, 1)$. We propose translation of a formula to an equivalent satisfiable finite order clausal theory, which consists of order clauses - finite sets of order literals of the augmented form: $\varepsilon_1 \diamond \varepsilon_2$ where ε_i is either a truth constant $\bar{0}$ or $\bar{1}$, or a conjunction of powers of propositional atoms or intermediate truth constants, and $\diamond$ is a connective $=$ or $\prec$. $=$ and $\prec$ are interpreted by the standard equality and strict order on [0, 1], respectively. We shall investigate the canonical standard completeness, where the semantics of the propositional product logic is given by the standard $\mathbf{\Pi}$-algebra, and truth constants are interpreted by 'themselves'. The hyperresolution calculus over order clausal theories is refutation sound and complete for the finite case. We solve the deduction problem $T \models \phi$ of a formula ϕ from a finite theory T in this expansion.

Keywords Hyperresolution · The product logic · Intermediate truth constants · Automated deduction · Fuzzy logics

1 Introduction

Current research in many-valued logics is mainly concerned with t-norm based logics, including the important cases of continuous and left-continuous t-norm [18, 19]. The standard semantics of a t-norm based logic is formed by the unit interval of real

Partially supported by Grant APVV-16-0042.

D. Guller (✉)
Department of Applied Informatics, Comenius University, Mlynská dolina,
842 48 Bratislava, Slovakia
e-mail: guller@fmph.uniba.sk

© Springer Nature Switzerland AG 2019
J. J. Merelo et al. (eds.), *Computational Intelligence*,
Studies in Computational Intelligence 792,
https://doi.org/10.1007/978-3-319-99283-9_10

197

numbers $[0, 1]$ equipped with the standard order, supremum, infimum, the t-norm and its residuum. The condition of left-continuity ensures the existence of the unique residuum for a given t-norm. The basic logics of continuous and left-continuous t-norm are the BL (basic) [17] and MTL (monodial t-norm) [4] ones, respectively. By the Mostert–Shields theorem [22], the three fundamental logics of continuous t-norm are Gödel, Łukasiewicz, and Product ones. From a syntactical point of view, classical many-valued deduction calculi are widely studied, especially Hilbert-style ones. In addition, a perspective from automated deduction has received attractivity during the last two decades. The reason is its growing application potential in many fields, spanning from engineering to informatics, such as fuzzy control and optimisation of industrial processes, knowledge representation and reasoning, ontology languages, the Semantic Web, the Web Ontology Language (OWL), fuzzy description logics and ontologies, multi-step fuzzy (many-valued) inference, fuzzy knowledge/expert systems. A considerable effort has been made in development of SAT solvers for the problem of Boolean satisfiability. SAT solvers may exploit either complete solution methods (called complete or systematic SAT solvers) or incomplete or hybrid ones. Complete SAT solvers are mostly based on the Davis–Putnam–Logemann–Loveland procedure ($DPLL$) [2, 3] and its refinements (e.g. chronological backtracking is replaced with non-chronological one using conflict-driven clause learning ($CDCL$) [20, 32]), or on resolution proof methods [11, 27, 28] improved by various features [1, 30].

t-norm based logics are logics of comparative truth: the residuum of a t-norm satisfies, for all $x, y \in [0, 1]$, $x \to y = 1$ if and only if $x \leq y$. Since implication is interpreted by a residuum, in the propositional case, a formula of the form $\phi \to \psi$ is a consequence of a theory if $\|\phi\|^{\mathfrak{A}} \leq \|\psi\|^{\mathfrak{A}}$ for every model $\mathfrak{A}$ of the theory. Most explorations of t-norm based logics are focused on tautologies and deduction calculi with the only distinguished truth value 1 [17]. However, in many real-world applications, one may be interested in representation and inference with explicit partial truth; besides the truth constants $\bar{0}, \bar{1}$, intermediate truth constants are involved. In the literature, two main approaches to expansions with truth constants are described. Historically, the first one has been introduced in [25], where the propositional Łukasiewicz logic is augmented by truth constants $\bar{r}, r \in [0, 1]$, Pavelka's logic (PL). A formula of the form $\bar{r} \to \phi$ evaluated to 1 expresses that the truth value of ϕ is greater than or equal to r. In [24], further development of evaluated formulae, and in [17], Rational Pavelka's logic (RPL) - a simplification of PL exploiting book-keeping axioms, are described. Another approach relies on traditional algebraic semantics. Various completeness results for expansions of t-norm based logics with countably many truth constants are achieved, among others, in [5–10, 21, 29, 35, 36].

In recent years, our exploration of automated deduction in t-norm based logics also concerns the propositional product logic with the multiplication t-norm. We have introduced an extension of the $DPLL$ procedure [13, 15]. In [14], we have examined the resolution counterpart. Particularly, we have generalised the hyperresolution principle and devised a hyperresolution calculus. As another step, one may incorporate a countable set of intermediate truth constants of the form $\bar{c}, c \in (0, 1)$, to get a modification of the hyperresolution calculus suitable for automated deduction with explicit partial truth. We shall investigate the canonical standard completeness, where the semantics of the propositional product logic is given by the standard $\mathbf{\Pi}$-algebra, and truth constants are interpreted by 'themselves'. Note that the Hilbert-style calculus for the propositional product logic introduced in [17] is not suitable for expansion with intermediate truth constants. We have $\phi \vdash \psi$ if and only if $\phi \models \psi$ (wrt. $\mathbf{\Pi}$). However, that cannot be preserved after adding intermediate truth constants. Let $c \in (0, 1)$ and a be an atom. Then $\bar{c} \models a$ ($\bar{c}$ is unsatisfiable), however, $\not\models \bar{c}^n \to a$ for any n, $\not\vdash \bar{c}^n \to a$, $\bar{c} \not\vdash a$ (from the soundness and the deduction-detachment theorem for this calculus). We modify translation of a formula to an equivalent satisfiable finite order clausal theory, which consists of order clauses - finite sets of order literals of the augmented form: $\varepsilon_1 \diamond \varepsilon_2$ where ε_i is either a truth constant $\bar{0}$ or $\bar{1}$, or a conjunction of powers of propositional atoms or intermediate truth constants, and $\diamond$ is a connective $\approx$ or $\prec$. $\approx$ and $\prec$ are interpreted by the standard equality and strict order on $[0, 1]$, respectively. The modified hyperresolution calculus over order clausal theories is still refutation sound and complete for the finite case. We finally solve the deduction problem $T \models \phi$ of a formula ϕ from a finite theory T in this expansion of the propositional product logic.

The paper is arranged as follows. Section 2 recalls the propositional product logic. Section 3 presents modified translation to clausal form. Section 4 proposes a modified hyperresolution calculus. Section 5 brings conclusions.

2 Propositional Product Logic

$\mathbb{N}$, $\mathbb{Z}$, $\mathbb{R}$ designates the set of natural, integer, real numbers, and $=$, $\leq$, $<$ denotes the standard equality, order, strict order on $\mathbb{N}$, $\mathbb{Z}$, $\mathbb{R}$. We denote $\mathbb{R}_0^+ = \{c \mid 0 \leq c \in \mathbb{R}\}$, $\mathbb{R}^+ = \{c \mid 0 < c \in \mathbb{R}\}$, $[0, 1] = \{c \mid c \in \mathbb{R}, 0 \leq c \leq 1\}$; $[0, 1]$ is called the unit interval. Let X, Y, Z be sets and $f : X \longrightarrow Y$ a mapping. By $\|X\|$ we denote the set-theoretic cardinality of X. The relationship of X being a finite subset of Y is denoted as $X \subseteq_{\mathcal{F}} Y$. Let $Z \subseteq X$. We designate $f[Z] = \{f(z) \mid z \in Z\}$; $f[Z]$ is called the image of Z under f; $f|_Z = \{(z, f(z)) \mid z \in Z\}$; $f|_Z$ is called the restriction of f onto Z. Let $\gamma \leq \omega$. A sequence δ of X is a bijection $\delta : \gamma \longrightarrow X$. Recall that X is countable if and only if there exists a sequence of X. Let I be an index set, and $S_i \neq \emptyset$, $i \in I$, be sets. A selector $\mathcal{S}$ over $\{S_i \mid i \in I\}$ is a mapping $\mathcal{S} : I \longrightarrow \bigcup\{S_i \mid i \in I\}$ such that for all $i \in I$, $\mathcal{S}(i) \in S_i$. We denote $\mathcal{S}el(\{S_i \mid i \in I\}) = \{\mathcal{S} \mid \mathcal{S} \text{ is a selector over } \{S_i \mid i \in I\}\}$. Let $c \in \mathbb{R}^+$. $\log c$ denotes the binary logarithm of c. Let $f, g : \mathbb{N} \longrightarrow \mathbb{R}_0^+$. f is of the order of g, in symbols $f \in O(g)$, iff there exist n_0 and $c^* \in \mathbb{R}_0^+$ such that for all $n \geq n_0$, $f(n) \leq c^* \cdot g(n)$.

Throughout the paper, we shall use the common notions and notation of propositional logic. The set of propositional atoms of the product logic will be denoted as *PropAtom*. Let $\{0, 1\} \subseteq \mathbb{C} \subseteq [0, 1]$ be countable. We assume a countable set of truth constants $\overline{\mathbb{C}} = \{\bar{c} \mid c \in \mathbb{C}\}$; $\bar{0}$, $\bar{1}$ denotes the false, the true; $\bar{c}$, $0 < c < 1$, is called an intermediate truth constant. Let $x \in \overline{\mathbb{C}}$ and $X \subseteq \overline{\mathbb{C}}$. We denote $\underline{x} = c \in \mathbb{C}$ such that $\bar{c} = x$; and $\underline{X} = \{c \mid c \in \mathbb{C}, \bar{c} \in X\}$. By *PropForm* we designate the set of all propositional formulae of the product logic built up from *PropAtom* and $\overline{\mathbb{C}}$ using the connectives: ¬, negation, Δ, Delta, ∧, conjunction, &, strong conjunction, ∨, disjunction, →, implication, ↔, equivalence, ⩦, equality, and ≺, strict order.[1] In the paper, we shall assume that *PropAtom* is countably infinite; hence, so is *PropForm*. Let ε_i, $1 \le i \le n$, be either a formula or a set of formulae or a set of sets of formulae, in general. By $atoms(\varepsilon_1, \ldots, \varepsilon_n) \subseteq PropAtom$, $tcons(\varepsilon_1, \ldots, \varepsilon_n) \subseteq \overline{\mathbb{C}}$ we denote the set of all atoms, truth constants occurring in $\varepsilon_1, \ldots, \varepsilon_n$. Let $\phi \in PropForm$. We define the size of ϕ by recursion on the structure of ϕ as follows:

$$|\phi| = \begin{cases} 1 & \text{if } \phi \in PropAtom \cup \overline{\mathbb{C}}, \\ 1 + |\phi_1| & \text{if } \phi = \diamond\phi_1, \\ 1 + |\phi_1| + |\phi_2| & \text{if } \phi = \phi_1 \diamond \phi_2. \end{cases}$$

Let $T \subseteq PropForm$ be finite. We define the size of T as $|T| = \sum_{\phi \in T} |\phi|$.

The product logic is interpreted by the standard **Π**-algebra augmented by operators ⩦, ≺, **Δ** for the connectives ⩦, ≺, Δ, respectively.

$$\mathbf{\Pi} = ([0, 1], \le, \vee, \wedge, \cdot, \Rightarrow, \overline{}, \text{⩦}, \prec, \mathbf{\Delta}, 0, 1)$$

where ∨, ∧ denotes the supremum, infimum operator on $[0, 1]$;

$$a \Rightarrow b = \begin{cases} 1 & \text{if } a \le b, \\ \frac{b}{a} & \text{else}; \end{cases} \qquad \overline{a} = \begin{cases} 1 & \text{if } a = 0, \\ 0 & \text{else}; \end{cases}$$

$$a \text{⩦} b = \begin{cases} 1 & \text{if } a = b, \\ 0 & \text{else}; \end{cases} \qquad a \prec b = \begin{cases} 1 & \text{if } a < b, \\ 0 & \text{else}; \end{cases}$$

$$\mathbf{\Delta}a = \begin{cases} 1 & \text{if } a = 1, \\ 0 & \text{else}. \end{cases}$$

Recall that **Π** is a complete linearly ordered lattice algebra; ∨, ∧ is commutative, associative, idempotent, monotone; 0, 1 is its neutral element; · is commutative, associative, monotone; 1 is its neutral element; the residuum operator ⇒ of · satisfies the condition of residuation:

$$\text{for all } a, b, c \in \mathbf{\Pi}, \ a \cdot b \le c \iff a \le b \Rightarrow c; \tag{1}$$

Gödel negation $\overline{}$ satisfies the condition:

$$\text{for all } a \in \mathbf{\Pi}, \ \overline{a} = a \Rightarrow 0; \tag{2}$$

---

[1] We assume a decreasing connective precedence: ¬, Δ, &, ⩦, ≺, ∧, ∨, →, ↔.

$\mathbf{\Delta}$ satisfies the condition[2]:

$$\text{for all } a \in \mathbf{\Pi}, \ \mathbf{\Delta}a = a \rightleftharpoons 1. \tag{3}$$

A valuation $\mathcal{V}$ of propositional atoms is a mapping $\mathcal{V} : PropAtom \longrightarrow [0, 1]$. Let $\phi \in PropForm$ and $\mathcal{V}$ be a valuation. We define the truth value $\|\phi\|^{\mathcal{V}} \in [0, 1]$ of ϕ in $\mathcal{V}$ by recursion on the structure of ϕ as follows:

$\phi \in PropAtom, \qquad \|\phi\|^{\mathcal{V}} = \mathcal{V}(\phi);$

$\phi \in \overline{\mathbb{C}}, \qquad \|\phi\|^{\mathcal{V}} = \underline{\phi};$

$\phi = \neg\phi_1, \qquad \|\phi\|^{\mathcal{V}} = \overline{\|\phi_1\|^{\mathcal{V}}};$

$\phi = \mathbf{\Delta}\phi_1, \qquad \|\phi\|^{\mathcal{V}} = \mathbf{\Delta}\|\phi_1\|^{\mathcal{V}};$

$\phi = \phi_1 \diamond \phi_2, \qquad \|\phi\|^{\mathcal{V}} = \|\phi_1\|^{\mathcal{V}} \diamond \|\phi_2\|^{\mathcal{V}},$

$\qquad\qquad\qquad\qquad \diamond \in \{\wedge, \&, \vee, \rightarrow, \rightleftharpoons, \prec\};$

$\phi = \phi_1 \leftrightarrow \phi_2, \qquad \|\phi\|^{\mathcal{V}} = (\|\phi_1\|^{\mathcal{V}} \Rightarrow \|\phi_2\|^{\mathcal{V}}) \cdot (\|\phi_2\|^{\mathcal{V}} \Rightarrow \|\phi_1\|^{\mathcal{V}}).$

A theory is a set of formulae. Let $\phi, \phi' \in PropForm$ and $T \subseteq PropForm$. ϕ is true in $\mathcal{V}$, written as $\mathcal{V} \models \phi$, iff $\|\phi\|^{\mathcal{V}} = 1$. $\mathcal{V}$ is a model of T, in symbols $\mathcal{V} \models T$, iff, for all $\phi \in T$, $\mathcal{V} \models \phi$. ϕ is a tautology iff, for every valuation $\mathcal{V}$, $\mathcal{V} \models \phi$. ϕ is equivalent to ϕ', in symbols $\phi \equiv \phi'$, iff, for every valuation $\mathcal{V}$, $\|\phi\|^{\mathcal{V}} = \|\phi'\|^{\mathcal{V}}$.

3 Translation to Clausal Form

We firstly introduce a notion of power of propositional atom, truth constant, and a notion of conjunction of powers. Let $\varepsilon \in PropAtom \cup \overline{\mathbb{C}}$ and $n \geq 1$. An nth power of the propositional atom, truth constant ε, ε raised to the power of n, is a pair (ε, n), written as ε^n. A power ε^1 is denoted as ε; if it does not cause the ambiguity with the denotation of the single atom, truth constant ε in a given context. The set of all powers is designated as $PropPow$. Let $\varepsilon^n \in PropPow$. We define the size of ε^n as $|\varepsilon^n| = n \geq 1$. A conjunction Cn of powers is a non-empty finite set of powers such that for all $\varepsilon^m \neq v^n \in Cn$, $\varepsilon \neq v$. A conjunction $\{\varepsilon_0^{m_0}, \ldots, \varepsilon_n^{m_n}\}$ is written in the form $\varepsilon_0^{m_0} \& \cdots \& \varepsilon_n^{m_n}$. A conjunction $\{p\}$ is called unit and denoted as p; if it does not cause the ambiguity with the denotation of the single power p in a given context. The set of all conjunctions is designated as $PropConj$. Let $p \in PropPow$, $Cn, Cn_1, Cn_2 \in PropConj$, $\mathcal{V}$ be a valuation. The truth value $\|Cn\|^{\mathcal{V}} \in [0, 1]$ of $Cn = \varepsilon_0^{m_0} \& \cdots \& \varepsilon_n^{m_n}$ in $\mathcal{V}$ is defined by

---

[2] We assume a decreasing operator precedence: $^{-}, \mathbf{\Delta}, \cdot, \rightleftharpoons, \prec, \wedge, \vee, \Rightarrow$.

$$\|Cn\|^{\mathcal{V}} = \underbrace{\|\varepsilon_0\|^{\mathcal{V}} \cdot \ldots \cdot \|\varepsilon_0\|^{\mathcal{V}}}_{m_0} \cdot \ldots \cdot \underbrace{\|\varepsilon_n\|^{\mathcal{V}} \cdot \ldots \cdot \|\varepsilon_n\|^{\mathcal{V}}}_{m_n}.$$

We define the size of Cn as $|Cn| = \sum_{p \in Cn} |p| \geq 1$. By $p \,\&\, Cn$ we denote $\{p\} \cup Cn$ where $p \notin Cn$. Cn_1 is a subconjunction of Cn_2, in symbols $Cn_1 \sqsubseteq Cn_2$, iff, for all $\varepsilon^m \in Cn_1$, there exists $\varepsilon^n \in Cn_2$ such that $m \leq n$. Cn_1 is a proper subconjunction of Cn_2, in symbols $Cn_1 \sqsubset Cn_2$, iff $Cn_1 \sqsubseteq Cn_2$ and $Cn_1 \neq Cn_2$.

We finally introduce order clauses in the product logic. l is an order literal iff $l = \varepsilon_1 \diamond \varepsilon_2$, $\varepsilon_i \in \{\bar{0}, \bar{1}\} \cup PropConj$, $\diamond \in \{=, \prec\}$. The set of all order literals is designated as $OrdPropLit$. Let $l = \varepsilon_1 \diamond \varepsilon_2 \in OrdPropLit$. l is a pure order literal iff $\varepsilon_i \in PropConj$; l does not contain $\bar{0}, \bar{1}$. The set of all pure order literals is designated as $PurOrdPropLit$. Let $\mathcal{V}$ be a valuation. The truth value $\|l\|^{\mathcal{V}} \in [0, 1]$ of l in $\mathcal{V}$ is defined by $\|l\|^{\mathcal{V}} = \|\varepsilon_1\|^{\mathcal{V}} \diamond \|\varepsilon_2\|^{\mathcal{V}}$. Note that $\mathcal{V} \models l$ if and only if either $l = \varepsilon_1 = \varepsilon_2$, $\|\varepsilon_1 = \varepsilon_2\|^{\mathcal{V}} = 1$, $\|\varepsilon_1\|^{\mathcal{V}} = \|\varepsilon_2\|^{\mathcal{V}}$; or $l = \varepsilon_1 \prec \varepsilon_2$, $\|\varepsilon_1 \prec \varepsilon_2\|^{\mathcal{V}} = 1$, $\|\varepsilon_1\|^{\mathcal{V}} < \|\varepsilon_2\|^{\mathcal{V}}$. We define the size of l as $|l| = 1 + |\varepsilon_1| + |\varepsilon_2|$. An order clause is a finite set of order literals. A pure order clause is a finite set of pure order literals. Since $=$ is symmetric, $=$ is commutative; hence, for all $\varepsilon_1 = \varepsilon_2 \in OrdPropLit$, we identify $\varepsilon_1 = \varepsilon_2$ with $\varepsilon_2 = \varepsilon_1 \in OrdPropLit$ with respect to order clauses. An order clause $\{l_0, \ldots, l_n\} \neq \emptyset$ is written in the form $l_0 \vee \cdots \vee l_n$. The empty order clause $\emptyset$ is denoted as $\square$. An order clause $\{l\}$ is called unit and denoted as l; if it does not cause the ambiguity with the denotation of the single order literal l in a given context. We designate the set of all order clauses as $OrdPropCl$ and the set of all pure order clauses as $PurOrdPropCl$. Note that $PurOrdPropCl \subseteq OrdPropCl$. Let $l, l_0, \ldots, l_n \in OrdPropLit$ and $C, C' \in OrdPropCl$. We define the size of C as $|C| = \sum_{l \in C} |l|$. By $l_0 \vee \cdots \vee l_n \vee C$ we denote $\{l_0, \ldots, l_n\} \cup C$ where, for all $i, i' \leq n$ and $i \neq i'$, $l_i \notin C$, $l_i \neq l_{i'}$. By $C \vee C'$ we denote $C \cup C'$. C is a subclause of C', in symbols $C \sqsubseteq C'$, iff $C \subseteq C'$. An order clausal theory is a set of order clauses. A pure order clausal theory is a set of pure order clauses. A unit order clausal theory is a set of unit order clauses.

Let $\phi, \phi' \in PropForm$, $T, T' \subseteq PropForm$, $S, S' \subseteq OrdPropCl$, $\mathcal{V}$ be a valuation. C is true in $\mathcal{V}$, written as $\mathcal{V} \models C$, iff there exists $l^* \in C$ such that $\mathcal{V} \models l^*$. $\mathcal{V}$ is a model of S, in symbols $\mathcal{V} \models S$, iff, for all $C \in S$, $\mathcal{V} \models C$. Let $\varepsilon_1 \in \{\phi, T, C, S\}$ and $\varepsilon_2 \in \{\phi', T', C', S'\}$. ε_2 is a propositional consequence of ε_1, in symbols $\varepsilon_1 \models \varepsilon_2$, iff, for every valuation $\mathcal{V}$, if $\mathcal{V} \models \varepsilon_1$, then $\mathcal{V} \models \varepsilon_2$. ε_1 is satisfiable iff there exists a valuation $\mathcal{V}$ such that $\mathcal{V} \models \varepsilon_1$. Note that both $\square$ and $\square \in S$ are unsatisfiable. ε_1 is equisatisfiable to ε_2 iff ε_1 is satisfiable if and only if ε_2 is satisfiable. Let $S \subseteq_{\mathcal{F}} OrdPropCl$. We define the size of S as $|S| = \sum_{C \in S} |C|$.

Let $\mathbb{I} = \mathbb{N} \times \mathbb{N}$; a countably infinite index set. Since $PropAtom$ is countably infinite, there exist $\mathbb{O}, \tilde{\mathbb{A}} \subseteq PropAtom$ such that $\mathbb{O} \cup \tilde{\mathbb{A}} = PropAtom$, $\mathbb{O} \cap \tilde{\mathbb{A}} = \emptyset$, both are countably infinite, $\tilde{\mathbb{A}} = \{\tilde{a}_i \mid i \in \mathbb{I}\}$. Let $A \subseteq \tilde{\mathbb{A}}$. We denote $PropForm_A = \{\phi \mid \phi \in PropForm, atoms(\phi) \subseteq \mathbb{O} \cup A\}$ and $OrdPropCl_A = \{C \mid C \in OrdPropCl, atoms(C) \subseteq \mathbb{O} \cup A\}$.

From a computational point of view, the worst case time and space complexity will be estimated using the logarithmic cost measurement. Let $\mathcal{A}$ be an algorithm.

$\#\mathcal{O}_A(In) \geq 1$ denotes the number of all elementary operations executed by A on an input In.

Translation of a formula or theory to clausal form is based on the following lemma. A similar approach exploiting the renaming subformulae technique can be found in [16, 23, 26, 31, 33, 34].

Lemma 1 *Let* $n_\phi, n_0 \in \mathbb{N}$, $\phi \in PropForm_\emptyset$, $T \subseteq PropForm_\emptyset$.

(I) *There exist either* $J_\phi = \emptyset$, *or* n_{J_ϕ}, $J_\phi = \{(n_\phi, j) \mid j \leq n_{J_\phi}\}$, $J_\phi \subseteq \{(n_\phi, j) \mid j \in \mathbb{N}\} \subseteq \mathbb{I}$, *and* $S_\phi \subseteq_{\mathcal{F}} OrdPropCl_{\{\tilde{a}_j \mid j \in J_\phi\}}$ *such that*

   (a) $\|J_\phi\| \leq 2 \cdot |\phi|$;

   (b) *either* $J_\phi = \emptyset$, $S_\phi = \{\square\}$, *or* $J_\phi = S_\phi = \emptyset$, *or* $J_\phi \neq \emptyset$, $\square \notin S_\phi \neq \emptyset$;

   (c) *there exists a valuation* $\mathfrak{A}$ *and* $\mathfrak{A} \models \phi$ *if and only if there exists a valuation* $\mathfrak{A}'$ *and* $\mathfrak{A}' \models S_\phi$, *satisfying* $\mathfrak{A}|_\mathbb{O} = \mathfrak{A}'|_\mathbb{O}$;

   (d) $|S_\phi| \in O(|\phi|)$; *the number of all elementary operations of the translation of* ϕ *to* S_ϕ *is in* $O(|\phi|)$; *the time and space complexity of the translation of* ϕ *to* S_ϕ *is in* $O(|\phi| \cdot (\log(1 + n_\phi) + \log|\phi|))$;

   (e) *if* $S_\phi \neq \emptyset, \{\square\}$, *then* $J_\phi \neq \emptyset$, *for all* $C \in S_\phi$, $\emptyset \neq atoms(C) \cap \tilde{\mathbb{A}} \subseteq \{\tilde{a}_j \mid j \in J_\phi\}$;

   (f) $tcons(S_\phi) - \{\bar{0}, \bar{1}\} \subseteq tcons(\phi) - \{\bar{0}, \bar{1}\}$.

(II) *There exist* $J_T \subseteq \{(i, j) \mid i \geq n_0\} \subseteq \mathbb{I}$ *and* $S_T \subseteq OrdPropCl_{\{\tilde{a}_j \mid j \in J_T\}}$ *such that*

   (a) *either* $J_T = \emptyset$, $S_T = \{\square\}$, *or* $J_T = S_T = \emptyset$, *or* $J_T \neq \emptyset$, $\square \notin S_T \neq \emptyset$;

   (b) *there exists a valuation* $\mathfrak{A}$ *and* $\mathfrak{A} \models T$ *if and only if there exists a valuation* $\mathfrak{A}'$ *and* $\mathfrak{A}' \models S_T$, *satisfying* $\mathfrak{A}|_\mathbb{O} = \mathfrak{A}'|_\mathbb{O}$;

   (c) *if* $T \subseteq_{\mathcal{F}} PropForm_\emptyset$, *then* $J_T \subseteq_{\mathcal{F}} \{(i, j) \mid i \geq n_0\}$, $\|J_T\| \leq 2 \cdot |T|$, $S_T \subseteq_{\mathcal{F}} OrdPropCl_{\{\tilde{a}_j \mid j \in J_T\}}$, $|S_T| \in O(|T|)$; *the number of all elementary operations of the translation of* T *to* S_T *is in* $O(|T|)$; *the time and space complexity of the translation of* T *to* S_T *is in* $O(|T| \cdot \log(1 + n_0 + |T|))$;

   (d) *if* $S_T \neq \emptyset, \{\square\}$, *then* $J_T \neq \emptyset$, *for all* $C \in S_T$, $\emptyset \neq atoms(C) \cap \tilde{\mathbb{A}} \subseteq \{\tilde{a}_j \mid j \in J_T\}$;

   (e) $tcons(S_T) - \{\bar{0}, \bar{1}\} \subseteq tcons(T) - \{\bar{0}, \bar{1}\}$.

Proof It is straightforward to prove the following statements:

   Let $n_\theta \in \mathbb{N}$ and $\theta \in PropForm_\emptyset$. There exists $\theta' \in PropForm_\emptyset$ such that   (4)

   (a) $\theta' \equiv \theta$;

   (b) $|\theta'| \leq 2 \cdot |\theta|$; θ' can be built up from θ via a postorder traversal of θ with $\#\mathcal{O}(\theta) \in O(|\theta|)$ and the time, space complexity in $O(|\theta| \cdot (\log(1 + n_\theta) + \log|\theta|))$;

   (c) θ' does not contain $\neg$ and Δ;

   (d) $\theta' \in \overline{\mathbb{C}}$; or for every subformula of θ' of the form $\varepsilon_1 \diamond \varepsilon_2$, $\diamond \in \{\wedge, \vee\}$, $\varepsilon_i \neq \bar{0}, \bar{1}, \{\varepsilon_1, \varepsilon_2\} \nsubseteq \overline{\mathbb{C}}$; for every subformula of θ' of the form $\varepsilon_1 \diamond \varepsilon_2$, $\diamond \in \{\&, \leftrightarrow\}, \varepsilon_i \neq \bar{0}, \bar{1}$; for every subformula of θ' of the form $\varepsilon_1 \rightarrow \varepsilon_2$, $\varepsilon_1 \neq \bar{0}, \bar{1}, \varepsilon_2 \neq \bar{1}$; for every subformula of θ' of the form $\varepsilon_1 = \varepsilon_2$,

$\{\varepsilon_1, \varepsilon_2\} \not\subseteq \overline{\mathbb{C}}$; for every subformula of θ' of the form $\varepsilon_1 \prec \varepsilon_2$, $\varepsilon_1 \neq \bar{1}$,
$\varepsilon_2 \neq \bar{0}$, $\{\varepsilon_1, \varepsilon_2\} \not\subseteq \overline{\mathbb{C}}$;
(e) $tcons(\theta') - \{\bar{0}, \bar{1}\} \subseteq tcons(\theta) - \{\bar{0}, \bar{1}\}$.

The proof is by induction on the structure of θ.

Let $n_\theta \in \mathbb{N}$, $\theta \in PropForm_\emptyset - \{\bar{0}, \bar{1}\}$, (4c,d) hold for θ; $\mathbb{i} = (n_\theta, j_\mathbb{i}) \in$ \hfill (5)
$\{(n_\theta, j) \mid j \in \mathbb{N}\} \subseteq \mathbb{I}$, $\tilde{a}_\mathbb{i} \in \tilde{\mathbb{A}}$. There exist $n_J \geq j_\mathbb{i}$, $J = \{(n_\theta, j) \mid j_\mathbb{i} + 1$
$\leq j \leq n_J\} \subseteq \{(n_\theta, j) \mid j \in \mathbb{N}\} \subseteq \mathbb{I}$, $\mathbb{i} \notin J$, $S \subseteq_\mathcal{F} OrdPropCl_{\{\tilde{a}_\mathbb{i}\} \cup \{\tilde{a}_\mathbb{j} \mid \mathbb{j} \in J\}}$
such that

(a) $\|J\| \leq |\theta| - 1$;
(b) there exists a valuation $\mathfrak{A}$ and $\mathfrak{A} \models \tilde{a}_\mathbb{i} \leftrightarrow \theta \in PropForm_{\{\tilde{a}_\mathbb{i}\}}$ if and
    only if there exists a valuation $\mathfrak{A}'$ and $\mathfrak{A}' \models S$, satisfying $\mathfrak{A}|_{\mathbb{O} \cup \{\tilde{a}_\mathbb{i}\}} = \mathfrak{A}'|_{\mathbb{O} \cup \{\tilde{a}_\mathbb{i}\}}$;
(c) $|S| \leq 31 \cdot |\theta|$, S can be built up from θ via a preorder traversal of θ
    with $\#\mathcal{O}(\theta) \in O(|\theta|)$;
(d) for all $C \in S$, $\emptyset \neq atoms(C) \cap \tilde{\mathbb{A}} \subseteq \{\tilde{a}_\mathbb{i}\} \cup \{\tilde{a}_\mathbb{j} \mid \mathbb{j} \in J\}$, $\tilde{a}_\mathbb{i} \approx \bar{1}$, $\tilde{a}_\mathbb{i} \prec$
    $\bar{1} \notin S$;
(e) $tcons(S) - \{\bar{0}, \bar{1}\} = tcons(\theta) - \{\bar{0}, \bar{1}\}$.

The proof is by induction on the structure of θ using the interpolation rules in
Tables 1 and 2.
(I) By (4) for n_ϕ, ϕ, there exists $\phi' \in PropForm_\emptyset$ such that (4a–e) hold for n_ϕ, ϕ,
ϕ'. We get three cases for ϕ'.
Case 1: $\phi' \in \overline{\mathbb{C}} - \{\bar{1}\}$. We put $J_\phi = \emptyset \subseteq \{(n_\phi, j) \mid j \in \mathbb{N}\} \subseteq \mathbb{I}$ and $S_\phi = \{\square\} \subseteq_\mathcal{F}$
$OrdPropCl_\emptyset$.
Case 2: $\phi' = \bar{1}$. We put $J_\phi = \emptyset \subseteq \{(n_\phi, j) \mid j \in \mathbb{N}\} \subseteq \mathbb{I}$ and $S_\phi = \emptyset \subseteq_\mathcal{F}$
$OrdPropCl_\emptyset$.
Case 3: $\phi' \notin \overline{\mathbb{C}}$. We have (4c,d) hold for ϕ'. Then $\phi' \in PropForm_\emptyset - \{\bar{0}, \bar{1}\}$.
We put $j_\mathbb{i} = 0$ and $\mathbb{i} = (n_\phi, j_\mathbb{i}) \in \{(n_\phi, j) \mid j \in \mathbb{N}\} \subseteq \mathbb{I}$. Hence, $\tilde{a}_\mathbb{i} \in \tilde{\mathbb{A}}$. We get by (5)
for n_ϕ, ϕ' that there exist $n_J \geq j_\mathbb{i}$, $J = \{(n_\theta, j) \mid j_\mathbb{i} + 1 \leq j \leq n_J\} \subseteq \{(n_\theta, j) \mid j \in$
$\mathbb{N}\} \subseteq \mathbb{I}$, $\mathbb{i} \notin J$, $S \subseteq_\mathcal{F} OrdPropCl_{\{\tilde{a}_\mathbb{i}\} \cup \{\tilde{a}_\mathbb{j} \mid \mathbb{j} \in J\}}$, and (5a–e) hold for ϕ'.
We put $n_{J_\phi} = n_J$, $J_\phi = \{(n_\phi, j) \mid j \leq n_{J_\phi}\} \subseteq \{(n_\phi, j) \mid j \in \mathbb{N}\} \subseteq \mathbb{I}$, $S_\phi = \{\tilde{a}_\mathbb{i} \approx \bar{1}\} \cup$
$S \subseteq_\mathcal{F} OrdPropCl_{\{\tilde{a}_\mathbb{j} \mid \mathbb{j} \in J_\phi\}}$. (II) straightforwardly follows from (I). $\square$

Theorem 1 Let $n_0 \in \mathbb{N}$, $\phi \in PropForm_\emptyset$, $T \subseteq PropForm_\emptyset$. There exist
$J_T^\phi \subseteq \{(i, j) \mid i \geq n_0\} \subseteq \mathbb{I}$ and $S_T^\phi \subseteq OrdPropCl_{\{\tilde{a}_\mathbb{j} \mid \mathbb{j} \in J_T^\phi\}}$ such that

(i) there exists a valuation $\mathfrak{A}$, and $\mathfrak{A} \models T$, $\mathfrak{A} \not\models \phi$, if and only if there exists a
    valuation $\mathfrak{A}'$ and $\mathfrak{A}' \models S_T^\phi$, satisfying $\mathfrak{A}|_\mathbb{O} = \mathfrak{A}'|_\mathbb{O}$;
(ii) $T \models \phi$ if and only if S_T^ϕ is unsatisfiable;
(iii) if $T \subseteq_\mathcal{F} PropForm_\emptyset$, then $J_T^\phi \subseteq_\mathcal{F} \{(i, j) \mid i \geq n_0\}$, $\|J_T^\phi\| \in O(|T| + |\phi|)$,
    $S_T^\phi \subseteq_\mathcal{F} OrdPropCl_{\{\tilde{a}_\mathbb{j} \mid \mathbb{j} \in J_T^\phi\}}$, $|S_T^\phi| \in O(|T| + |\phi|)$; the number of all elemen-
    tary operations of the translation of T and ϕ to S_T^ϕ is in $O(|T| + |\phi|)$;

Table 1 Binary interpolation rules for $\wedge, \&, \vee, \rightarrow, \leftrightarrow, \equiv, \prec$

Case

$\theta = \theta_1 \wedge \theta_2$

$$\frac{\tilde{a}_{\mathbf{i}} \leftrightarrow (\theta_1 \wedge \theta_2)}{\left\{ \begin{array}{l} \tilde{a}_{\mathbf{i}_1} \prec \tilde{a}_{\mathbf{i}_2} \vee \tilde{a}_{\mathbf{i}_1} \equiv \tilde{a}_{\mathbf{i}_2} \vee \tilde{a}_{\mathbf{i}} \equiv \tilde{a}_{\mathbf{i}_2}, \tilde{a}_{\mathbf{i}_2} \prec \tilde{a}_{\mathbf{i}_1} \vee \tilde{a}_{\mathbf{i}} \equiv \tilde{a}_{\mathbf{i}_1}, \\ \tilde{a}_{\mathbf{i}_1} \leftrightarrow \theta_1, \tilde{a}_{\mathbf{i}_2} \leftrightarrow \theta_2 \end{array} \right\}} \tag{6}$$

$|\text{Consequent}| = 15 + |\tilde{a}_{\mathbf{i}_1} \leftrightarrow \theta_1| + |\tilde{a}_{\mathbf{i}_2} \leftrightarrow \theta_2| \leq 31 + |\tilde{a}_{\mathbf{i}_1} \leftrightarrow \theta_1| + |\tilde{a}_{\mathbf{i}_2} \leftrightarrow \theta_2|$

$\theta = \theta_1 \& \theta_2$

$$\frac{\tilde{a}_{\mathbf{i}} \leftrightarrow (\theta_1 \& \theta_2)}{\left\{ \tilde{a}_{\mathbf{i}} \equiv \tilde{a}_{\mathbf{i}_1} \& \tilde{a}_{\mathbf{i}_2}, \tilde{a}_{\mathbf{i}_1} \leftrightarrow \theta_1, \tilde{a}_{\mathbf{i}_2} \leftrightarrow \theta_2 \right\}} \tag{7}$$

$|\text{Consequent}| = 5 + |\tilde{a}_{\mathbf{i}_1} \leftrightarrow \theta_1| + |\tilde{a}_{\mathbf{i}_2} \leftrightarrow \theta_2| \leq 31 + |\tilde{a}_{\mathbf{i}_1} \leftrightarrow \theta_1| + |\tilde{a}_{\mathbf{i}_2} \leftrightarrow \theta_2|$

$\theta = \theta_1 \vee \theta_2$

$$\frac{\tilde{a}_{\mathbf{i}} \leftrightarrow (\theta_1 \vee \theta_2)}{\left\{ \begin{array}{l} \tilde{a}_{\mathbf{i}_1} \prec \tilde{a}_{\mathbf{i}_2} \vee \tilde{a}_{\mathbf{i}_1} \equiv \tilde{a}_{\mathbf{i}_2} \vee \tilde{a}_{\mathbf{i}} \equiv \tilde{a}_{\mathbf{i}_1}, \tilde{a}_{\mathbf{i}_2} \prec \tilde{a}_{\mathbf{i}_1} \vee \tilde{a}_{\mathbf{i}} \equiv \tilde{a}_{\mathbf{i}_2}, \\ \tilde{a}_{\mathbf{i}_1} \leftrightarrow \theta_1, \tilde{a}_{\mathbf{i}_2} \leftrightarrow \theta_2 \end{array} \right\}} \tag{8}$$

$|\text{Consequent}| = 15 + |\tilde{a}_{\mathbf{i}_1} \leftrightarrow \theta_1| + |\tilde{a}_{\mathbf{i}_2} \leftrightarrow \theta_2| \leq 31 + |\tilde{a}_{\mathbf{i}_1} \leftrightarrow \theta_1| + |\tilde{a}_{\mathbf{i}_2} \leftrightarrow \theta_2|$

$\theta = \theta_1 \rightarrow \theta_2, \theta_2 \neq \bar{0}$

$$\frac{\tilde{a}_{\mathbf{i}} \leftrightarrow (\theta_1 \rightarrow \theta_2)}{\left\{ \begin{array}{l} \tilde{a}_{\mathbf{i}_1} \prec \tilde{a}_{\mathbf{i}_2} \vee \tilde{a}_{\mathbf{i}_1} \equiv \tilde{a}_{\mathbf{i}_2} \vee \tilde{a}_{\mathbf{i}} \& \tilde{a}_{\mathbf{i}_1} \equiv \tilde{a}_{\mathbf{i}_2}, \tilde{a}_{\mathbf{i}_2} \prec \tilde{a}_{\mathbf{i}_1} \vee \tilde{a}_{\mathbf{i}} \equiv \bar{1}, \\ \tilde{a}_{\mathbf{i}_1} \leftrightarrow \theta_1, \tilde{a}_{\mathbf{i}_2} \leftrightarrow \theta_2 \end{array} \right\}} \tag{9}$$

$|\text{Consequent}| = 17 + |\tilde{a}_{\mathbf{i}_1} \leftrightarrow \theta_1| + |\tilde{a}_{\mathbf{i}_2} \leftrightarrow \theta_2| \leq 31 + |\tilde{a}_{\mathbf{i}_1} \leftrightarrow \theta_1| + |\tilde{a}_{\mathbf{i}_2} \leftrightarrow \theta_2|$

$\theta = \theta_1 \leftrightarrow \theta_2$

$$\frac{\tilde{a}_{\mathbf{i}} \leftrightarrow (\theta_1 \leftrightarrow \theta_2)}{\left\{ \begin{array}{l} \tilde{a}_{\mathbf{i}_1} \prec \tilde{a}_{\mathbf{i}_2} \vee \tilde{a}_{\mathbf{i}_1} \equiv \tilde{a}_{\mathbf{i}_2} \vee \tilde{a}_{\mathbf{i}} \& \tilde{a}_{\mathbf{i}_1} \equiv \tilde{a}_{\mathbf{i}_2}, \\ \tilde{a}_{\mathbf{i}_2} \prec \tilde{a}_{\mathbf{i}_1} \vee \tilde{a}_{\mathbf{i}_2} \equiv \tilde{a}_{\mathbf{i}_1} \vee \tilde{a}_{\mathbf{i}} \& \tilde{a}_{\mathbf{i}_2} \equiv \tilde{a}_{\mathbf{i}_1}, \\ \tilde{a}_{\mathbf{i}_1} \prec \tilde{a}_{\mathbf{i}_2} \vee \tilde{a}_{\mathbf{i}_2} \prec \tilde{a}_{\mathbf{i}_1} \vee \tilde{a}_{\mathbf{i}} \equiv \bar{1}, \tilde{a}_{\mathbf{i}_1} \leftrightarrow \theta_1, \tilde{a}_{\mathbf{i}_2} \leftrightarrow \theta_2 \end{array} \right\}} \tag{10}$$

$|\text{Consequent}| = 31 + |\tilde{a}_{\mathbf{i}_1} \leftrightarrow \theta_1| + |\tilde{a}_{\mathbf{i}_2} \leftrightarrow \theta_2| \leq 31 + |\tilde{a}_{\mathbf{i}_1} \leftrightarrow \theta_1| + |\tilde{a}_{\mathbf{i}_2} \leftrightarrow \theta_2|$

$\theta = \theta_1 \equiv \theta_2, \theta_i \neq \bar{0}, \bar{1}$

$$\frac{\tilde{a}_{\mathbf{i}} \leftrightarrow (\theta_1 \equiv \theta_2)}{\left\{ \begin{array}{l} \tilde{a}_{\mathbf{i}_1} \equiv \tilde{a}_{\mathbf{i}_2} \vee \tilde{a}_{\mathbf{i}} \equiv \bar{0}, \tilde{a}_{\mathbf{i}_1} \prec \tilde{a}_{\mathbf{i}_2} \vee \tilde{a}_{\mathbf{i}_2} \prec \tilde{a}_{\mathbf{i}_1} \vee \tilde{a}_{\mathbf{i}} \equiv \bar{1}, \\ \tilde{a}_{\mathbf{i}_1} \leftrightarrow \theta_1, \tilde{a}_{\mathbf{i}_2} \leftrightarrow \theta_2 \end{array} \right\}} \tag{11}$$

$|\text{Consequent}| = 15 + |\tilde{a}_{\mathbf{i}_1} \leftrightarrow \theta_1| + |\tilde{a}_{\mathbf{i}_2} \leftrightarrow \theta_2| \leq 31 + |\tilde{a}_{\mathbf{i}_1} \leftrightarrow \theta_1| + |\tilde{a}_{\mathbf{i}_2} \leftrightarrow \theta_2|$

$\theta = \theta_1 \prec \theta_2, \theta_1 \neq \bar{0}, \theta_2 \neq \bar{1}$

$$\frac{\tilde{a}_{\mathbf{i}} \leftrightarrow (\theta_1 \prec \theta_2)}{\left\{ \begin{array}{l} \tilde{a}_{\mathbf{i}_1} \prec \tilde{a}_{\mathbf{i}_2} \vee \tilde{a}_{\mathbf{i}} \equiv \bar{0}, \tilde{a}_{\mathbf{i}_2} \prec \tilde{a}_{\mathbf{i}_1} \vee \tilde{a}_{\mathbf{i}_2} \equiv \tilde{a}_{\mathbf{i}_1} \vee \tilde{a}_{\mathbf{i}} \equiv \bar{1}, \\ \tilde{a}_{\mathbf{i}_1} \leftrightarrow \theta_1, \tilde{a}_{\mathbf{i}_2} \leftrightarrow \theta_2 \end{array} \right\}} \tag{12}$$

$|\text{Consequent}| = 15 + |\tilde{a}_{\mathbf{i}_1} \leftrightarrow \theta_1| + |\tilde{a}_{\mathbf{i}_2} \leftrightarrow \theta_2| \leq 31 + |\tilde{a}_{\mathbf{i}_1} \leftrightarrow \theta_1| + |\tilde{a}_{\mathbf{i}_2} \leftrightarrow \theta_2|$

Table 2 Unary interpolation rules for $\to$, $\eqcirc$, $\prec$

Case	

$\theta = \theta_1 \to \bar{0}$

$$\frac{\tilde{a}_{\mathfrak{i}} \leftrightarrow (\theta_1 \to \bar{0})}{\{\tilde{a}_{\mathfrak{i}_1} \eqcirc \bar{0} \vee \tilde{a}_{\mathfrak{i}} \eqcirc \bar{0}, \bar{0} \prec \tilde{a}_{\mathfrak{i}_1} \vee \tilde{a}_{\mathfrak{i}} \eqcirc \bar{1}, \tilde{a}_{\mathfrak{i}_1} \leftrightarrow \theta_1\}} \tag{13}$$

$|\text{Consequent}| = 12 + |\tilde{a}_{\mathfrak{i}_1} \leftrightarrow \theta_1| \leq 31 + |\tilde{a}_{\mathfrak{i}_1} \leftrightarrow \theta_1|$

$\theta = \theta_1 \eqcirc \bar{0}$

$$\frac{\tilde{a}_{\mathfrak{i}} \leftrightarrow (\theta_1 \eqcirc \bar{0})}{\{\tilde{a}_{\mathfrak{i}_1} \eqcirc \bar{0} \vee \tilde{a}_{\mathfrak{i}} \eqcirc \bar{0}, \bar{0} \prec \tilde{a}_{\mathfrak{i}_1} \vee \tilde{a}_{\mathfrak{i}} \eqcirc \bar{1}, \tilde{a}_{\mathfrak{i}_1} \leftrightarrow \theta_1\}} \tag{14}$$

$|\text{Consequent}| = 12 + |\tilde{a}_{\mathfrak{i}_1} \leftrightarrow \theta_1| \leq 31 + |\tilde{a}_{\mathfrak{i}_1} \leftrightarrow \theta_1|$

$\theta = \theta_1 \eqcirc \bar{1}$

$$\frac{\tilde{a}_{\mathfrak{i}} \leftrightarrow (\theta_1 \eqcirc \bar{1})}{\{\tilde{a}_{\mathfrak{i}_1} \eqcirc \bar{1} \vee \tilde{a}_{\mathfrak{i}} \eqcirc \bar{0}, \tilde{a}_{\mathfrak{i}_1} \prec \bar{1} \vee \tilde{a}_{\mathfrak{i}} \eqcirc \bar{1}, \tilde{a}_{\mathfrak{i}_1} \leftrightarrow \theta_1\}} \tag{15}$$

$|\text{Consequent}| = 12 + |\tilde{a}_{\mathfrak{i}_1} \leftrightarrow \theta_1| \leq 31 + |\tilde{a}_{\mathfrak{i}_1} \leftrightarrow \theta_1|$

$\theta = \bar{0} \prec \theta_1$

$$\frac{\tilde{a}_{\mathfrak{i}} \leftrightarrow (\bar{0} \prec \theta_1)}{\{\bar{0} \prec \tilde{a}_{\mathfrak{i}_1} \vee \tilde{a}_{\mathfrak{i}} \eqcirc \bar{0}, \tilde{a}_{\mathfrak{i}_1} \eqcirc \bar{0} \vee \tilde{a}_{\mathfrak{i}} \eqcirc \bar{1}, \tilde{a}_{\mathfrak{i}_1} \leftrightarrow \theta_1\}} \tag{16}$$

$|\text{Consequent}| = 12 + |\tilde{a}_{\mathfrak{i}_1} \leftrightarrow \theta_1| \leq 31 + |\tilde{a}_{\mathfrak{i}_1} \leftrightarrow \theta_1|$

$\theta = \theta_1 \prec \bar{1}$

$$\frac{\tilde{a}_{\mathfrak{i}} \leftrightarrow (\theta_1 \prec \bar{1})}{\{\tilde{a}_{\mathfrak{i}_1} \prec \bar{1} \vee \tilde{a}_{\mathfrak{i}} \eqcirc \bar{0}, \tilde{a}_{\mathfrak{i}_1} \eqcirc \bar{1} \vee \tilde{a}_{\mathfrak{i}} \eqcirc \bar{1}, \tilde{a}_{\mathfrak{i}_1} \leftrightarrow \theta_1\}} \tag{17}$$

$|\text{Consequent}| = 12 + |\tilde{a}_{\mathfrak{i}_1} \leftrightarrow \theta_1| \leq 31 + |\tilde{a}_{\mathfrak{i}_1} \leftrightarrow \theta_1|$

the time and space complexity of the translation of T and ϕ to S_T^ϕ is in $O(|T| \cdot \log(1 + n_0 + |T|) + |\phi| \cdot (\log(1 + n_0) + \log|\phi|))$;

(iv) $tcons(S_T^\phi) - \{\bar{0}, \bar{1}\} \subseteq (tcons(\phi) \cup tcons(T)) - \{\bar{0}, \bar{1}\}$.

Proof We get by Lemma 1(II) for $n_0 + 1$ that there exist $J_T \subseteq \{(i, j) \mid i \geq n_0 + 1\} \subseteq \mathbb{I}$, $S_T \subseteq OrdPropCl_{\{\tilde{a}_{\mathfrak{j}} \mid \mathfrak{j} \in J_T\}}$, and Lemma 1(II a–e) hold for $n_0 + 1$. By (4) for n_0, ϕ, there exists $\phi' \in PropForm_\emptyset$ such that (4a–e) hold for n_0, ϕ, ϕ'. We get three cases for ϕ'.

Case 1: $\phi' \in \overline{\mathbb{C}} - \{\bar{1}\}$. We put $J_T^\phi = J_T \subseteq \{(i, j) \mid i \geq n_0 + 1\} \subseteq \{(i, j) \mid i \geq n_0\} \subseteq \mathbb{I}$ and $S_T^\phi = S_T \subseteq OrdPropCl_{\{\tilde{a}_{\mathfrak{j}} \mid \mathfrak{j} \in J_T^\phi\}}$.

Case 2: $\phi' = \bar{1}$. We put $J_T^\phi = \emptyset \subseteq \{(i, j) \mid i \geq n_0\} \subseteq \mathbb{I}$ and $S_T^\phi = \{\square\} \subseteq OrdPropCl_\emptyset$.

Case 3: $\phi' \notin \overline{\mathbb{C}}$. We have (4c,d) hold for ϕ'. Then $\phi' \in PropForm_\emptyset - \{\bar{0}, \bar{1}\}$. We put $\mathfrak{j}_{\mathfrak{i}} = 0$ and $\mathfrak{i} = (n_0, j_{\mathfrak{i}}) \in \{(n_0, j) \mid j \in \mathbb{N}\} \subseteq \mathbb{I}$. Hence, $\tilde{a}_{\mathfrak{i}} \in \tilde{\mathbb{A}}$. We get by (5) for n_0,

ϕ' that there exist $n_J \geq j_{\mathring{\imath}}$, $J = \{(n_0, j) \mid j_{\mathring{\imath}} + 1 \leq j \leq n_J\} \subseteq \{(n_0, j) \mid j \in \mathbb{N}\} \subseteq \mathbb{I}$, $\mathring{\imath} \notin J$, $S \subseteq_{\mathcal{F}} OrdPropCl_{\{\tilde{a}_{\mathring{\imath}}\} \cup \{\tilde{a}_j \mid j \in J\}}$, and (5a–e) hold for ϕ'. We put $J_T^{\phi} = J_T \cup \{\mathring{\imath}\} \cup J \subseteq \{(i, j) \mid i \geq n_0\} \subseteq \mathbb{I}$ and $S_T^{\phi} = S_T \cup \{\tilde{a}_{\mathring{\imath}} \prec \bar{1}\} \cup S \subseteq OrdPropCl_{\{\tilde{a}_j \mid j \in J_T^{\phi}\}}$. $\square$

As an example, we translate $\phi = (\bar{0} \prec c) \,\&\, (a \,\&\, c \prec \overline{0.5} \,\&\, c) \to a \prec \overline{0.5} \in$ *PropForm* to an equisatisfiable order clausal theory S^{ϕ}.

$$\phi = (\bar{0} \prec c) \,\&\, (a \,\&\, c \prec \overline{0.5} \,\&\, c) \to a \prec \overline{0.5}$$

$$\left\{ \tilde{a}_0 = \bar{1}, \tilde{a}_0 \leftrightarrow \Big(\underbrace{(\bar{0} \prec c) \,\&\, (a \,\&\, c \prec \overline{0.5} \,\&\, c)}_{\tilde{a}_1} \to \underbrace{a \prec \overline{0.5}}_{\tilde{a}_2} \Big) \right\} \tag{9}$$

$$\left\{ \tilde{a}_0 = \bar{1}, \tilde{a}_1 \prec \tilde{a}_2 \vee \tilde{a}_1 = \tilde{a}_2 \vee \tilde{a}_0 \,\&\, \tilde{a}_1 = \tilde{a}_2, \tilde{a}_2 \prec \tilde{a}_1 \vee \tilde{a}_0 = \bar{1}, \right.$$
$$\left. \tilde{a}_1 \leftrightarrow \underbrace{(\bar{0} \prec c)}_{\tilde{a}_3} \,\&\, \underbrace{(a \,\&\, c \prec \overline{0.5} \,\&\, c)}_{\tilde{a}_4}, \tilde{a}_2 \leftrightarrow \underbrace{a}_{\tilde{a}_5} \prec \underbrace{\overline{0.5}}_{\tilde{a}_6} \right\} \tag{7), (12}$$

$$\left\{ \tilde{a}_0 = \bar{1}, \tilde{a}_1 \prec \tilde{a}_2 \vee \tilde{a}_1 = \tilde{a}_2 \vee \tilde{a}_0 \,\&\, \tilde{a}_1 = \tilde{a}_2, \tilde{a}_2 \prec \tilde{a}_1 \vee \tilde{a}_0 = \bar{1}, \right.$$
$$\tilde{a}_1 = \tilde{a}_3 \,\&\, \tilde{a}_4, \tilde{a}_3 \leftrightarrow \bar{0} \prec \underbrace{c}_{\tilde{a}_7}, \tilde{a}_4 \leftrightarrow \underbrace{a \,\&\, c}_{\tilde{a}_8} \prec \underbrace{\overline{0.5} \,\&\, c}_{\tilde{a}_9},$$
$$\left. \tilde{a}_5 \prec \tilde{a}_6 \vee \tilde{a}_2 = \bar{0}, \tilde{a}_6 \prec \tilde{a}_5 \vee \tilde{a}_6 = \tilde{a}_5 \,\&\, \tilde{a}_2 = \bar{1}, \tilde{a}_5 = a, \tilde{a}_6 = \overline{0.5} \right\} \tag{16), (12}$$

$$\left\{ \tilde{a}_0 = \bar{1}, \tilde{a}_1 \prec \tilde{a}_2 \vee \tilde{a}_1 = \tilde{a}_2 \vee \tilde{a}_0 \,\&\, \tilde{a}_1 = \tilde{a}_2, \tilde{a}_2 \prec \tilde{a}_1 \vee \tilde{a}_0 = \bar{1}, \right.$$
$$\tilde{a}_1 = \tilde{a}_3 \,\&\, \tilde{a}_4, \bar{0} \prec \tilde{a}_7 \vee \tilde{a}_3 = \bar{0}, \tilde{a}_7 = \bar{0} \vee \tilde{a}_3 = \bar{1}, \tilde{a}_7 = c,$$
$$\tilde{a}_8 \prec \tilde{a}_9 \vee \tilde{a}_4 = \bar{0}, \tilde{a}_9 \prec \tilde{a}_8 \vee \tilde{a}_9 = \tilde{a}_8 \vee \tilde{a}_4 = \bar{1},$$
$$\tilde{a}_8 \leftrightarrow \underbrace{a}_{\tilde{a}_{10}} \,\&\, \underbrace{c}_{\tilde{a}_{11}}, \tilde{a}_9 \leftrightarrow \underbrace{\overline{0.5}}_{\tilde{a}_{12}} \,\&\, \underbrace{c}_{\tilde{a}_{13}},$$
$$\left. \tilde{a}_5 \prec \tilde{a}_6 \vee \tilde{a}_2 = \bar{0}, \tilde{a}_6 \prec \tilde{a}_5 \vee \tilde{a}_6 = \tilde{a}_5 \vee \tilde{a}_2 = \bar{1}, \tilde{a}_5 = a, \tilde{a}_6 = \overline{0.5} \right\} \tag{7}$$

$$S^{\phi} = \left\{ \tilde{a}_0 = \bar{1}, \tilde{a}_1 \prec \tilde{a}_2 \vee \tilde{a}_1 = \tilde{a}_2 \vee \tilde{a}_0 \,\&\, \tilde{a}_1 = \tilde{a}_2, \tilde{a}_2 \prec \tilde{a}_1 \vee \tilde{a}_0 = \bar{1}, \right.$$
$$\tilde{a}_1 = \tilde{a}_3 \,\&\, \tilde{a}_4, \bar{0} \prec \tilde{a}_7 \vee \tilde{a}_3 = \bar{0}, \tilde{a}_7 = \bar{0} \vee \tilde{a}_3 = \bar{1}, \tilde{a}_7 = c,$$
$$\tilde{a}_8 \prec \tilde{a}_9 \vee \tilde{a}_4 = \bar{0}, \tilde{a}_9 \prec \tilde{a}_8 \vee \tilde{a}_9 = \tilde{a}_8 \vee \tilde{a}_4 = \bar{1},$$
$$\tilde{a}_8 = \tilde{a}_{10} \,\&\, \tilde{a}_{11}, \tilde{a}_{10} = a, \tilde{a}_{11} = c, \tilde{a}_9 = \tilde{a}_{12} \,\&\, \tilde{a}_{13}, \tilde{a}_{12} = \overline{0.5}, \tilde{a}_{13} = c,$$
$$\left. \tilde{a}_5 \prec \tilde{a}_6 \vee \tilde{a}_2 = \bar{0}, \tilde{a}_6 \prec \tilde{a}_5 \vee \tilde{a}_6 = \tilde{a}_5 \vee \tilde{a}_2 = \bar{1}, \tilde{a}_5 = a, \tilde{a}_6 = \overline{0.5} \right\}$$

4 Hyperresolution over Order Clauses

In this section, we propose an order hyperresolution calculus operating over order clausal theories and prove its refutational soundness, completeness. At first, we introduce some basic notions and notation. Let $l \in OrdPropLit$. l is a contradiction iff

$l = \bar{0} \Rightarrow \bar{1}$ or $l = \varepsilon \prec \bar{0}$ or $l = \bar{1} \prec \varepsilon$ or $l = \varepsilon \prec \varepsilon$. l is a tautology iff either $l = \varepsilon \Rightarrow \varepsilon$ or $l = \bar{0} \prec \bar{1}$. Let $Cn \in PropConj$ and $C \in OrdPropCl$. We define an auxiliary function $simplify : (\{\bar{0}, \bar{1}\} \cup PropConj \cup OrdPropLit \cup OrdPropCl) \times PropAtom \times \{\bar{0}, \bar{1}\} \longrightarrow \{\bar{0}, \bar{1}\} \cup PropConj \cup OrdPropLit \cup OrdPropCl$ as follows:

$$simplify(\bar{0}, a, v) = \bar{0};$$

$$simplify(\bar{1}, a, v) = \bar{1};$$

$$simplify(Cn, a, \bar{0}) = \begin{cases} \bar{0} & \text{if } a \in atoms(Cn), \\ Cn & \text{else}; \end{cases}$$

$$simplify(Cn, a, \bar{1}) = \begin{cases} \bar{1} & \text{if } \exists n^* \, Cn = a^{n^*}, \\ Cn - a^{n^*} & \text{if } \exists n^* \, a^{n^*} \in Cn \neq a^{n^*}, \\ Cn & \text{else}; \end{cases}$$

$$simplify(l, a, v) = simplify(\varepsilon_1, a, v) \diamond simplify(\varepsilon_2, a, v) \text{ if } l = \varepsilon_1 \diamond \varepsilon_2;$$

$$simplify(C, a, v) = \{simplify(l, a, v) \mid l \in C\}.$$

For an input expression, atom, truth constant, $simplify$ replaces every occurrence of the atom by the truth constant in the expression and returns a simplified expression according to laws holding in $\mathbf{\Pi}$. Let $Cn_1, Cn_2 \in PropConj$ and $l_1, l_2 \in OrdPropLit$. Another auxiliary function $\odot : (\{\bar{0}, \bar{1}\} \cup PropConj) \times (\{\bar{0}, \bar{1}\} \cup PropConj) \longrightarrow \{\bar{0}, \bar{1}\} \cup PropConj$ is defined as follows:

$$\bar{0} \odot \varepsilon = \varepsilon \odot \bar{0} = \bar{0};$$

$$\bar{1} \odot \varepsilon = \varepsilon \odot \bar{1} = \varepsilon;$$

$$Cn_1 \odot Cn_2 = \{\varepsilon^{m+n} \mid \varepsilon^m \in Cn_1, \varepsilon^n \in Cn_2\} \cup$$
$$\{\varepsilon^n \mid \varepsilon^n \in Cn_1, \varepsilon \notin atoms(Cn_2) \cup tcons(Cn_2)\} \cup$$
$$\{\varepsilon^n \mid \varepsilon^n \in Cn_2, \varepsilon \notin atoms(Cn_1) \cup tcons(Cn_1)\}.$$

For two input expressions, $\odot$ returns the product of them. It can be extended to $\{\bar{0}, \bar{1}\} \cup OrdPropLit$ component-wisely. $\odot : (\{\bar{0}, \bar{1}\} \cup OrdPropLit) \times (\{\bar{0}, \bar{1}\} \cup OrdPropLit) \longrightarrow \{\bar{0}, \bar{1}\} \cup OrdPropLit$ is defined as follows:

$$\bar{0} \odot \varepsilon = \varepsilon \odot \bar{0} = \bar{0};$$

$$\bar{1} \odot \varepsilon = \varepsilon \odot \bar{1} = \varepsilon;$$

$$l_1 \odot l_2 = (\varepsilon_1 \odot \varepsilon_2) \diamond (v_1 \odot v_2) \text{ if } l_i = \varepsilon_i \diamond_i v_i,$$

$$\diamond = \begin{cases} \Rightarrow & \text{if } \diamond_1 = \diamond_2 = \Rightarrow, \\ \prec & \text{else}. \end{cases}$$

Note that $\odot$ is a binary commutative and associative operator. We denote $l^n = \underbrace{l \odot \cdots \odot l}_{n}, n \geq 1$, and say that l^n is an nth power of l. Let $I \subseteq_{\mathcal{F}} \mathbb{N}, l_i \in OrdPropLit$, $\alpha_i \geq 1, i \in I$. We define by recursion on I:

$$\bigodot_{i \in I} l_i^{\alpha_i} = \begin{cases} \bar{1} & \text{if } I = \emptyset, \\ l_{i^*}^{\alpha_{i^*}} \odot \left(\bigodot_{i \in I - \{i^*\}} l_i^{\alpha_i} \right) & \text{if } \exists i^* \in I, \end{cases}$$

$$\in \{\bar{1}\} \cup OrdPropLit.$$

Let $a \in PropAtom$. C is a guard iff either $C = a \rightleftharpoons \bar{0}$ or $C = \bar{0} \prec a$ or $C = a \prec \bar{1}$ or $C = a \rightleftharpoons \bar{1}$. Let $S \subseteq OrdPropCl$. We denote $guards(a) = \{a \rightleftharpoons \bar{0}, \bar{0} \prec a, a \prec \bar{1}, a \rightleftharpoons \bar{1}\} \subseteq OrdPropCl$, $guards(S) = \{C \mid C \in S \text{ is a guard}\}$, $guards(S, a) = S \cap guards(a)$, $ordtcons(S) = \{\bar{0} \prec c \mid c \in tcons(S) - \{\bar{0}, \bar{1}\}\} \cup \{c \prec \bar{1} \mid c \in tcons(S) - \{\bar{0}, \bar{1}\}\} \cup \{c_1 \prec c_2 \mid c_1, c_2 \in tcons(S) - \{\bar{0}, \bar{1}\}, \underline{c_1 < c_2}\} \subseteq OrdPropCl$. Note that $guards(S, a) \subseteq guards(S)$. a is guarded in S iff either $guards(S, a) = \{a \rightleftharpoons \bar{0}\}$ or $guards(S, a) = \{\bar{0} \prec a, a \prec \bar{1}\}$ or $guards(S, a) = \{a \rightleftharpoons \bar{1}\}$. Note that if a is guarded in S, then $a \in atoms(guards(S, a)) \subseteq atoms(S)$. S is a guarded order clausal theory iff, for all $a \in atoms(S)$, a is guarded in S.

The basic order hyperresolution calculus is defined as follows. Let $\kappa \geq 1$ and $S_{\kappa-1}, S_\kappa \subseteq OrdPropCl$. Order hyperresolution rules are defined with respect to κ, S, $S_{\kappa-1}$, S_κ. The first rule is the central order hyperresolution one.

(*Order hyperresolution rule*) (18)

$$\frac{\begin{array}{l} \bar{0} \prec a_1, \ldots, \bar{0} \prec a_r, a_1 \prec \bar{1}, \ldots, a_r \prec \bar{1} \in guards(S_{\kappa-1}), \\ \bar{0} \prec c_1, \ldots, \bar{0} \prec c_s, c_1 \prec \bar{1}, \ldots, c_s \prec \bar{1} \in ordtcons(S), \\ l_0 \vee C_0, \ldots, l_n \vee C_n \in S_{\kappa-1} \end{array}}{\displaystyle\bigvee_{i=0}^{n} C_i \in S_\kappa};$$

$atoms(l_0, \ldots, l_n) = \{a_1, \ldots, a_r\}$, $tcons(l_0, \ldots, l_n) = \{c_1, \ldots, c_s\}$,
$l_i \in PurOrdPropLit$,
there exist $\alpha_i^* \geq 1, i = 0, \ldots, n$, $J^* \subseteq \{j \mid 1 \leq j \leq r\}$, $\beta_j^* \geq 1, j \in J^*$,
$K^* \subseteq \{k \mid 1 \leq k \leq s\}$, $\gamma_k^* \geq 1, k \in K^*$, *such that*
$\left(\bigodot_{i=0}^{n} l_i^{\alpha_i^*} \right) \odot \left(\bigodot_{j \in J^*} (a_j \prec \bar{1})^{\beta_j^*} \right) \odot \left(\bigodot_{k \in K^*} (c_k \prec \bar{1})^{\gamma_k^*} \right)$ *is a contradiction.*

If there exists a product of powers of the input pure order literals $l_0, \ldots, l_n$, of powers of some guards $a_j \prec \bar{1}, j \in J^*$, and of powers of some literals $c_k \prec \bar{1}$ from $ordtcons(S), k \in K^*$, which is a contradiction of the form $\varepsilon \prec \varepsilon$, then we can derive an output order clause $\bigvee_{i=0}^{n} C_i$ consisting of the remainder order clauses $C_0, \ldots, C_n$. We say that $\bigvee_{i=0}^{n} C_i$ is an order hyperresolvent of $\bar{0} \prec a_1, \ldots, \bar{0} \prec a_r, a_1 \prec \bar{1}, \ldots, a_r \prec \bar{1}, \bar{0} \prec c_1, \ldots, \bar{0} \prec c_s, c_1 \prec \bar{1}, \ldots, c_s \prec \bar{1}, l_0 \vee C_0, \ldots, l_n \vee C_n$.

(*Order contradiction rule*) (19)

$$\frac{l \vee C \in S_{\kappa-1}}{C \in S_\kappa};$$

l is a contradiction.

If the order literal l is a contradiction, then it can be removed from the input order clause $l \vee C$. C is an order contradiction resolvent of $l \vee C$.

(Order $\bar{0}$-simplification rule) (20)

$$\frac{a \doteq \bar{0} \in guards(S_{\kappa-1}), C \in S_{\kappa-1}}{simplify(C, a, \bar{0}) \in S_{\kappa}};$$
$$a \in atoms(C), a \doteq \bar{0} \neq C.$$

If the guard $a \doteq \bar{0}$ is in the input order clausal theory $S_{\kappa-1}$, and the input order clause C contains the atom a, then C can be simplified using the auxiliary function *simplify*. $simplify(C, a, \bar{0})$ is an order $\bar{0}$-simplification resolvent of $a \doteq \bar{0}$ and C. Analogously, C can be simplified with respect to the guard $a \doteq \bar{1}$.

(Order $\bar{1}$-simplification rule) (21)

$$\frac{a \doteq \bar{1} \in guards(S_{\kappa-1}), C \in S_{\kappa-1}}{simplify(C, a, \bar{1}) \in S_{\kappa}};$$
$$a \in atoms(C), a \doteq \bar{1} \neq C.$$

$simplify(C, a, \bar{1})$ is an order $\bar{1}$-simplification resolvent of $a \doteq \bar{1}$ and C.

(Order $\bar{0}$-contradiction rule) (22)

$$\frac{\bar{0} \prec a_0, \ldots, \bar{0} \prec a_n \in guards(S_{\kappa-1}), a_0^{\alpha_0} \, \& \, \cdots \, \& \, a_n^{\alpha_n} \doteq \bar{0} \vee C \in S_{\kappa-1}}{C \in S_{\kappa}}.$$

C is an order $\bar{0}$-contradiction resolvent of $\bar{0} \prec a_0, \ldots, \bar{0} \prec a_n, a_0^{\alpha_0} \, \& \, \cdots \, \& \, a_n^{\alpha_n} \doteq \bar{0} \vee C$.

(Order $\bar{1}$-contradiction rule) (23)

$$\frac{a_i \prec \bar{1} \in guards(S_{\kappa-1}), a_0^{\alpha_0} \, \& \, \cdots \, \& \, a_n^{\alpha_n} \doteq \bar{1} \vee C \in S_{\kappa-1}}{C \in S_{\kappa}};$$
$$i \leq n.$$

C is an order $\bar{1}$-contradiction resolvent of $a_i \prec \bar{1}$ and $a_0^{\alpha_0} \, \& \, \cdots \, \& \, a_n^{\alpha_n} \doteq \bar{1} \vee C$. The last two rules detect a contradictory set of order literals of the form either $\{\bar{0} \prec a_0, \ldots, \bar{0} \prec a_n, a_0^{\alpha_0} \, \& \, \cdots \, \& \, a_n^{\alpha_n} \doteq \bar{0}\}$ or $\{a_i \prec \bar{1}, a_0^{\alpha_0} \, \& \, \cdots \, \& \, a_n^{\alpha_n} \doteq \bar{1}\}, i \leq n$. In either case, the remainder order clause C can be derived. Note that all the rules are sound; for every rule, the consequent order clausal theory is a propositional consequence of the antecedent one.

Let $S_0 = \emptyset \subseteq OrdPropCl$. Let $\mathcal{D} = C_1, \ldots, C_n, C_\kappa \in OrdPropCl, n \geq 1$. $\mathcal{D}$ is a deduction of C_n from S by order hyperresolution iff, for all $1 \leq \kappa \leq n, C_\kappa \in ordtcons(S) \cup S$, or there exist $1 \leq j_k^* \leq \kappa - 1, k = 0, \ldots, m$, such that C_κ is an order resolvent of $C_{j_0^*}, \ldots, C_{j_m^*} \in S_{\kappa-1}$ using Rule (18)–(23) with respect to $S_{\kappa-1}$; S_κ is defined by recursion on $1 \leq \kappa \leq n$ as follows:

$$S_\kappa = S_{\kappa-1} \cup \{C_\kappa\} \subseteq OrdPropCl.$$

$\mathcal{D}$ is a refutation of S iff $C_n = \square$. We denote

$$clo^{\mathcal{H}}(S) = \{C \mid there\ exists\ a\ deduction\ of\ C\ from\ S\ by\ order\ hyperresolution\}$$
$$\subseteq OrdPropCl.$$

Lemma 2 *Let $S \subseteq_{\mathcal{F}} OrdPropCl.$ $clo^{\mathcal{H}}(S) \subseteq_{\mathcal{F}} OrdPropCl.$*

Proof Straightforward. $\qquad\square$

Lemma 3 *Let $S \subseteq OrdPropCl.$ $tcons(clo^{\mathcal{H}}(S)) - \{\bar{0}, \bar{1}\} = tcons(S) - \{\bar{0}, \bar{1}\}.$*

Proof Straightforward. $\qquad\square$

Lemma 4 *Let $A = \{a_i \mid 1 \leq i \leq m\} \subseteq PropAtom,$ $S_1 = \{\bar{0} \prec a_i \mid 1 \leq i \leq m\} \cup \{a_i \prec \bar{1} \mid 1 \leq i \leq m\} \subseteq OrdPropCl,$ $S_2 = \{l_i \mid l_i \in PurOrdPropLit, 1 \leq i \leq n\} \subseteq PurOrdPropCl,$ $atoms(S_2) \subseteq A,$ $tcons(S_2) - \{\bar{0}, \bar{1}\} = \{c_i \mid 1 \leq i \leq s\},$ $S = S_1 \cup ordtcons(S_2) \cup S_2 \subseteq OrdPropCl,$ there not exist an application of Rule (18) with respect to S. S is satisfiable.*

Proof S is unit. Note that an application of Rule (18) with respect to S would derive $\square$; S would be unsatisfiable. We denote

$$PropConj_A = \{Cn \mid Cn \in PropConj, atoms(Cn) \subseteq A\}.$$

Let $Cn_1, Cn_2 \in PropConj_A$ and $Cn_2 \sqsubseteq Cn_1$. We define

$$cancel(Cn_1, Cn_2) = \{\varepsilon^{r-s} \mid \varepsilon^r \in Cn_1, \varepsilon^s \in Cn_2, r > s\} \cup$$
$$\{\varepsilon^r \mid \varepsilon^r \in Cn_1, \varepsilon \notin atoms(Cn_2) \cup tcons(Cn_2)\} \in PropConj_A.$$

We further denote

$$gen = \Big\{ Cn_1 \eqcirc Cn_2 \mid Cn_i \in PropConj_A,$$
$$there\ exist\ \emptyset \neq I^* \subseteq \{i \mid 1 \leq i \leq n\}, \alpha_i^* \geq 1, i \in I^*,$$
$$Cn_1 \eqcirc Cn_2 = \bigodot_{i \in I^*} l_i^{\alpha_i^*} \Big\} \cup$$
$$\Big\{ Cn_1 \prec Cn_2 \mid Cn_i \in PropConj_A,$$
$$there\ exist\ \emptyset \neq I^* \subseteq \{i \mid 1 \leq i \leq n\}, \alpha_i^* \geq 1, i \in I^*,$$
$$J^* \subseteq \{j \mid 1 \leq j \leq m\}, \beta_j^* \geq 1, j \in J^*,$$
$$K^* \subseteq \{k \mid 1 \leq k \leq s\}, \gamma_k^* \geq 1, k \in K^*,$$

$$Cn_1 \prec Cn_2 = \Big(\underset{i \in I^*}{\odot} l_i^{\alpha_i^*} \Big) \odot \Big(\underset{j \in J^*}{\odot} (a_j \prec \bar{1})^{\beta_j^*} \Big)$$

$$\odot \Big(\underset{k \in K^*}{\odot} (c_k \prec \bar{1})^{\gamma_k^*} \Big) \}$$

$\subseteq PurOrdPropLit,$

$$cnl = \big\{ Cn_1 \diamond Cn_2 \mid Cn_i \in PropConj_A, \text{ there exist}$$

$$Cn_1^* \diamond Cn_2^* \in gen, Cn^* \in PropConj_A,$$

$$Cn^* \sqsubset Cn_i^*, Cn_i = cancel(Cn_i^*, Cn^*) \big\} \subseteq PurOrdPropLit,$$

$clo = gen \cup cnl \subseteq PurOrdPropLit.$

Then $S_2 \subseteq gen \subseteq clo$.

$$\text{For all } Cn \in PropConj_A, Cn \prec Cn \notin gen, clo. \qquad (24)$$

The proof is straightforward; we have that there does not exist an application of Rule (18) with respect to S.

Let $tcons(S_2) \cup \{\bar{0}, \bar{1}\} \subseteq X \subseteq tcons(S_2) \cup \{\bar{0}, \bar{1}\} \cup A$. A partial valuation $\mathcal{V}$ is a mapping $\mathcal{V} : X \longrightarrow [0, 1]$ such that for all $c \in tcons(S_2) \cup \{\bar{0}, \bar{1}\}$, $\mathcal{V}(c) = \underline{c}$. We denote $dom(\mathcal{V}) = X$, $tcons(S_2) \cup \{\bar{0}, \bar{1}\} \subseteq dom(\mathcal{V}) \subseteq tcons(S_2) \cup \{\bar{0}, \bar{1}\} \cup A$. We define a partial valuation $\mathcal{V}_\iota$ by recursion on $\iota \leq m$ as follows:

$$\mathcal{V}_0 = \{(c, \underline{c}) \mid c \in tcons(S_2) \cup \{\bar{0}, \bar{1}\}\};$$
$$\mathcal{V}_\iota = \mathcal{V}_{\iota-1} \cup \{(a_\iota, \lambda_\iota)\} \quad (1 \leq \iota \leq m),$$

$$\mathbb{E}_{\iota-1} = \left\{ \left(\frac{\|Cn_2\|^{\mathcal{V}_{\iota-1}}}{\|Cn_1\|^{\mathcal{V}_{\iota-1}}} \right)^{\frac{1}{\alpha}} \, \middle| \, \begin{array}{l} Cn_1 \,\&\, a_\iota^\alpha \approx Cn_2 \in clo, \\ atoms(Cn_i) \subseteq dom(\mathcal{V}_{\iota-1}) \end{array} \right\} \cup$$

$$\left\{ (\|Cn_2\|^{\mathcal{V}_{\iota-1}})^{\frac{1}{\alpha}} \, \middle| \, \begin{array}{l} a_\iota^\alpha \approx Cn_2 \in clo, \\ atoms(Cn_2) \subseteq dom(\mathcal{V}_{\iota-1}) \end{array} \right\},$$

$$\mathbb{D}_{\iota-1} = \left\{ \left(\frac{\|Cn_2\|^{\mathcal{V}_{\iota-1}}}{\|Cn_1\|^{\mathcal{V}_{\iota-1}}} \right)^{\frac{1}{\alpha}} \, \middle| \, \begin{array}{l} Cn_2 \prec Cn_1 \,\&\, a_\iota^\alpha \in clo, \\ atoms(Cn_i) \subseteq dom(\mathcal{V}_{\iota-1}) \end{array} \right\} \cup$$

$$\left\{ (\|Cn_2\|^{\mathcal{V}_{\iota-1}})^{\frac{1}{\alpha}} \, \middle| \, \begin{array}{l} Cn_2 \prec a_\iota^\alpha \in clo, \\ atoms(Cn_2) \subseteq dom(\mathcal{V}_{\iota-1}) \end{array} \right\},$$

$$\mathbb{U}_{\iota-1} = \left\{ \left(\frac{\|Cn_2\|^{\mathcal{V}_{\iota-1}}}{\|Cn_1\|^{\mathcal{V}_{\iota-1}}} \right)^{\frac{1}{\alpha}} \;\middle|\; \begin{array}{l} Cn_1 \mathbin{\&} a_\iota^\alpha \prec Cn_2 \in clo, \\ atoms(Cn_i) \subseteq dom(\mathcal{V}_{\iota-1}) \end{array} \right\} \cup$$

$$\left\{ \left(\|Cn_2\|^{\mathcal{V}_{\iota-1}} \right)^{\frac{1}{\alpha}} \;\middle|\; \begin{array}{l} a_\iota^\alpha \prec Cn_2 \in clo, \\ atoms(Cn_2) \subseteq dom(\mathcal{V}_{\iota-1}) \end{array} \right\},$$

$$\lambda_\iota = \begin{cases} \dfrac{\bigvee \mathbb{D}_{\iota-1} + \bigwedge \mathbb{U}_{\iota-1}}{2} & \text{if } \mathbb{E}_{\iota-1} = \emptyset, \\ \bigvee \mathbb{E}_{\iota-1} & \text{else.} \end{cases}$$

For all $\iota \le \iota' \le m$, $\mathcal{V}_\iota$ is a partial valuation, $dom(\mathcal{V}_\iota) = tcons(S_2) \cup$ (25)
$\{\bar{0}, \bar{1}\} \cup \{a_1, \ldots, a_\iota\}$, $\mathcal{V}_\iota \subseteq \mathcal{V}_{\iota'}$.

The proof is by induction on $\iota \le m$.

For all $\iota \le m$, for all $a \in dom(\mathcal{V}_\iota) - (tcons(S_2) \cup \{\bar{0}, \bar{1}\})$, (26)
$Cn_1, Cn_2 \in PropConj_A$ and $atoms(Cn_i) \subseteq dom(\mathcal{V}_\iota)$,
$0 < \mathcal{V}_\iota(a) < 1$;
if $Cn_1 \equiv Cn_2 \in clo$, then $\|Cn_1\|^{\mathcal{V}_\iota} = \|Cn_2\|^{\mathcal{V}_\iota}$;
if $Cn_1 \prec Cn_2 \in clo$, then $\|Cn_1\|^{\mathcal{V}_\iota} < \|Cn_2\|^{\mathcal{V}_\iota}$.

The proof is by induction on $\iota \le m$.

We have $atoms(S_2) \subseteq A$. Then $atoms(S_1) = A$, $atoms(ordtcons(S_2)) = \emptyset$, $atoms(S) = atoms(S_1) \cup atoms(ordtcons(S_2)) \cup atoms(S_2) = A$. We put $\mathcal{V} = \mathcal{V}_m$, $dom(\mathcal{V}) \overset{(25)}{=} tcons(S_2) \cup \{\bar{0}, \bar{1}\} \cup A = tcons(S_2) \cup \{\bar{0}, \bar{1}\} \cup atoms(S)$.

For all $a \in A$, $Cn_1, Cn_2 \in PropConj_A$, (27)
$0 < \mathcal{V}(a) < 1$;
if $Cn_1 \equiv Cn_2 \in clo$, then $\|Cn_1\|^{\mathcal{V}} = \|Cn_2\|^{\mathcal{V}}$;
if $Cn_1 \prec Cn_2 \in clo$, then $\|Cn_1\|^{\mathcal{V}} < \|Cn_2\|^{\mathcal{V}}$.

The proof is by (26) for m.

We put $\mathfrak{A} = \mathcal{V}|_A \cup \{(a, 0) \mid a \in PropAtom - A\}$; $\mathfrak{A}$ is a valuation. Let $l \in S$. We have that S is unit. Then $l \in OrdPropLit$ and $atoms(l) \subseteq atoms(S) \subseteq dom(\mathcal{V})$. We get three cases for l.

Case 1: $l \in S_1$, either $l = \bar{0} \prec a$ or $l = a \prec \bar{1}$. Hence, $a \in A$, $\{\bar{0}, \bar{1}\} \subseteq dom(\mathcal{V})$, either $\|\bar{0}\|^{\mathfrak{A}} = \mathcal{V}(\bar{0}) = 0 \underset{(27)}{<} \mathcal{V}(a) = \mathfrak{A}(a)$, $\|l\|^{\mathfrak{A}} = \|\bar{0} \prec a\|^{\mathfrak{A}} = \|\bar{0}\|^{\mathfrak{A}} \prec \mathfrak{A}(a) = 1$, or $\mathfrak{A}(a) = \mathcal{V}(a) \underset{(27)}{<} 1 = \mathcal{V}(\bar{1}) = \|\bar{1}\|^{\mathfrak{A}}$, $\|l\|^{\mathfrak{A}} = \|a \prec \bar{1}\|^{\mathfrak{A}} = \mathfrak{A}(a) \prec \|\bar{1}\|^{\mathfrak{A}} = 1$, $\mathfrak{A} \models l$.

Case 2: $l \in ordtcons(S_2)$, $l = c_1 \prec c_2$, $c_i \in tcons(S_2) \cup \{\bar{0}, \bar{1}\}$, $\underline{c_1} < \underline{c_2}$. Hence, $\|c_1\|^{\mathfrak{A}} = \underline{c_1} < \underline{c_2} = \|c_2\|^{\mathfrak{A}}$, $\|l\|^{\mathfrak{A}} = \|c_1 \prec c_2\|^{\mathfrak{A}} = \|c_1\|^{\mathfrak{A}} \prec \|c_2\|^{\mathfrak{A}} = 1$, $\mathfrak{A} \models l$.

Case 3: $l \in S_2$, $l \in PurOrdPropLit$, either $l = Cn_1 \rightleftharpoons Cn_2$ or $l = Cn_1 \prec Cn_2$. Hence, $l \in S_2 \subseteq clo$, either $Cn_1 \rightleftharpoons Cn_2 \in clo$ or $Cn_1 \prec Cn_2 \in clo$, $Cn_1, Cn_2 \in PropConj_A$, either $\|Cn_1\|^{\mathfrak{A}} = \|Cn_1\|^{\mathcal{V}} \overset{(27)}{=} \|Cn_2\|^{\mathcal{V}} = \|Cn_2\|^{\mathfrak{A}}$, $\|l\|^{\mathfrak{A}} = \|Cn_1 \rightleftharpoons Cn_2\|^{\mathfrak{A}} = \|Cn_1\|^{\mathfrak{A}} \rightleftharpoons \|Cn_2\|^{\mathfrak{A}} = 1$, or $\|Cn_1\|^{\mathfrak{A}} = \|Cn_1\|^{\mathcal{V}} \underset{(27)}{\leq} \|Cn_2\|^{\mathcal{V}} = \|Cn_2\|^{\mathfrak{A}}$,

$\|l\|^{\mathfrak{A}} = \|Cn_1 \prec Cn_2\|^{\mathfrak{A}} = \|Cn_1\|^{\mathfrak{A}} \prec \|Cn_2\|^{\mathfrak{A}} = 1$, $\mathfrak{A} \models l$.

So, in all Cases 1–3, $\mathfrak{A} \models l$; $\mathfrak{A} \models S$; S is satisfiable.   □

Lemma 5 (Reduction Lemma) *Let* $A = \{a_i \mid i \leq m\} \subseteq PropAtom$, $S_1 = \{\bar{0} \prec a_i \mid i \leq m\} \cup \{a_i \prec \bar{1} \mid i \leq m\} \subseteq OrdPropCl$, $S_2 = \{(\bigvee_{j=0}^{k_i} l_j^i) \vee C_i \mid l_j^i \in PurOrdPropLit, i \leq n\} \subseteq PurOrdPropCl$, $atoms(S_2) \subseteq A$, $S = S_1 \cup ordtcons(S_2) \cup S_2 \subseteq OrdPropCl$ such that for all $S \in Sel(\{\{j \mid j \leq k_i\}_i \mid i \leq n\})$, there exists an application of Rule (18) with respect to $S_1 \cup ordtcons(S_2) \cup \{l_{S(i)}^i \mid i \leq n\} \subseteq OrdPropCl$. There exists $\emptyset \neq I^* \subseteq \{i \mid i \leq n\}$ such that $\bigvee_{i \in I^*} C_i \in clo^{\mathcal{H}}(S)$.*

Proof Analogous to the one of Proposition 2, [12].   □

Lemma 6 (Normalisation Lemma) *Let* $S \subseteq_{\mathcal{F}} OrdPropCl$ *be guarded. There exists* $S^* \subseteq_{\mathcal{F}} clo^{\mathcal{H}}(S)$ *such that there exist* $A = \{a_i \mid 1 \leq i \leq m\} \subseteq PropAtom$, $S_1 = \{\bar{0} \prec a_i \mid 1 \leq i \leq m\} \cup \{a_i \prec \bar{1} \mid 1 \leq i \leq m\} \subseteq OrdPropCl$, $S_2 = \{\bigvee_{j=1}^{k_i} l_j^i \mid l_j^i \in PurOrdPropLit, 1 \leq i \leq n\} \subseteq PurOrdPropCl$; and $atoms(S_2) = A$, $S^* = S_1 \cup ordtcons(S_2) \cup S_2$, $guards(S^*) = S_1$, S^* is guarded; S^* is equisatisfiable to S.*

Proof Let $B_{\bar{0}} = \{b \mid b \rightleftharpoons \bar{0} \in guards(S)\} \subseteq atoms(S)$ and $B_{\bar{1}} = \{b \mid b \rightleftharpoons \bar{1} \in guards(S)\} \subseteq atoms(S)$. Then, for all $b \in B_{\bar{0}}$, $clo^{\mathcal{H}}(S)$ is closed with respect to applications of Rule (20); for all $b \in B_{\bar{1}}$, $clo^{\mathcal{H}}(S)$ is closed with respect to applications of Rule (21); $clo^{\mathcal{H}}(S)$ is closed with respect to applications of Rule (19); $clo^{\mathcal{H}}(S)$ is closed with respect to applications of Rule (22); $clo^{\mathcal{H}}(S)$ is closed with respect to applications of Rule (23); the order clausal theory in the antecedent is equisatisfiable to the one in the consequent of every Rule (19), (20)–(23). By Lemma 2, $clo^{\mathcal{H}}(S) \subseteq_{\mathcal{F}} OrdPropCl$. We put $S_2 = \{C \mid C = \bigvee_{j=1}^{k} l_j \in clo^{\mathcal{H}}(S), l_j \in PurOrdPropLit, atoms(C) \cap (B_{\bar{0}} \cup B_{\bar{1}}) = \emptyset\} \subseteq PurOrdPropCl$, $A = atoms(S_2) \subseteq PropAtom$, $S_1 = \{\bar{0} \prec a \mid a \in A, \bar{0} \prec a \in guards(S)\} \cup \{a \prec \bar{1} \mid a \in A, a \prec \bar{1} \in guards(S)\} \subseteq S \subseteq clo^{\mathcal{H}}(S), tcons(S_2) - \{\bar{0}, \bar{1}\} \subseteq tcons(clo^{\mathcal{H}}(S)) - \{\bar{0}, \bar{1}\} \overset{\text{Lemma 3}}{=\!=\!=\!=} tcons(S) - \{\bar{0}, \bar{1}\}, ordtcons(S_2) \subseteq ordtcons(S) \subseteq clo^{\mathcal{H}}(S), S^* = S_1 \cup ordtcons(S_2) \cup S_2 \subseteq_{\mathcal{F}} clo^{\mathcal{H}}(S)$. Hence, $guards(S^*) = S_1$, S^* is guarded; S^* is equisatisfiable to S.   □

Theorem 2 (Refutational Soundness and Completeness) *Let* $S \subseteq_{\mathcal{F}} OrdPropCl$ *be guarded.* $\square \in clo^{\mathcal{H}}(S)$ *if and only if* S *is unsatisfiable.*

Proof ($\Longrightarrow$) Let $\mathfrak{A}$ be a model of S and $C \in clo^{\mathcal{H}}(S)$. Then $\mathfrak{A} \models C$. The proof is by complete induction on the length of a deduction of C from S by order hyperresolution. Let $\square \in clo^{\mathcal{H}}(S)$ and $\mathfrak{A}$ be a model of S. Hence, $\mathfrak{A} \models \square$, which is a contradiction; S is unsatisfiable.

($\Longleftarrow$) Let $\square \notin clo^{\mathcal{H}}(S)$. Then, by Lemma 6, there exists $S^* \subseteq_{\mathcal{F}} clo^{\mathcal{H}}(S)$ such that there exist $A = \{a_i \mid 1 \leq i \leq m\} \subseteq PropAtom$, $S_1 = \{\bar{0} \prec a_i \mid 1 \leq i \leq m\} \cup \{a_i \prec$

$\bar{1} \mid 1 \leq i \leq m\} \subseteq OrdPropCl$, $S_2 = \{\bigvee_{j=1}^{k_i} l_j^i \mid l_j^i \in PurOrdPropLit, 1 \leq i \leq n\} \subseteq$ $PurOrdPropCl$, and $atoms(S_2) = A$, $S^* = S_1 \cup ordtcons(S_2) \cup S_2$, S^* is equisatisfiable to S; $\square \notin clo^{\mathcal{H}}(S^*) \subseteq clo^{\mathcal{H}}(S)$. We get two cases for S^*.

Case 1: $S^* = \emptyset$. Then S^* is satisfiable, and S is satisfiable.

Case 2: $S^* \neq \emptyset$. Then $m, n \geq 1$, for all $1 \leq i \leq n$, $k_i \geq 1$, by Lemma 5 for S^*, there exists $\mathcal{S}^* \in Sel(\{\{j \mid 1 \leq j \leq k_i\}_i \mid 1 \leq i \leq n\})$ such that there does not exist an application of Rule (18) with respect to $S_1 \cup ordtcons(S_2) \cup \{l_{\mathcal{S}^*(i)}^i \mid 1 \leq i \leq n\} \subseteq OrdPropCl$. We put $S_2' = \{l_{\mathcal{S}^*(i)}^i \mid l_{\mathcal{S}^*(i)}^i \in PurOrdPropLit, 1 \leq i \leq n\} \subseteq$ $PurOrdPropCl$, $S' = S_1 \cup ordtcons(S_2') \cup S_2' \subseteq OrdPropCl$. Hence, $atoms(S_2') \subseteq$ $atoms(S_2) = A$, $tcons(S_2') \subseteq tcons(S_2)$, $ordtcons(S_2') \subseteq ordtcons(S_2)$, $S' = S_1 \cup$ $ordtcons(S_2') \cup S_2' \subseteq S_1 \cup ordtcons(S_2) \cup S_2$, there does not exist an application of Rule (18) with respect to S'; by Lemma 4 for S_2', S', S' is satisfiable; S^* is satisfiable; S is satisfiable.

So, in both Cases 1 and 2, S is satisfiable. $\qquad\square$

Let $S \subseteq S' \subseteq OrdPropCl$. S' is a guarded extension of S iff S' is guarded and minimal with respect to $\subseteq$.

Theorem 3 (Satisfiability Problem) *Let* $S \subseteq_{\mathcal{F}} OrdPropCl$. *$S$ is satisfiable if and only if there exists a guarded extension* $S' \subseteq_{\mathcal{F}} OrdPropCl$ *of S which is satisfiable.*

Proof ($\Longrightarrow$) Let S be satisfiable, and $\mathfrak{A}$ be a model of S. Then $atoms(S) \subseteq_{\mathcal{F}}$ $PropAtom$. We put $S_1 = \{a \doteq \bar{0} \mid a \in atoms(S), \mathfrak{A}(a) = 0\} \cup \{\bar{0} \prec a \mid a \in atoms(S), 0 < \mathfrak{A}(a) < 1\} \cup \{a \prec \bar{1} \mid a \in atoms(S), 0 < \mathfrak{A}(a) < 1\} \cup \{a \doteq \bar{1} \mid a \in atoms(S), \mathfrak{A}(a) = 1\} \subseteq_{\mathcal{F}} OrdPropCl$ and $S' = S_1 \cup S \subseteq_{\mathcal{F}} OrdPropCl$. Hence, S' is a guarded extension of S, for all $l \in S_1$, $\mathfrak{A} \models l$; $\mathfrak{A} \models S_1$; $\mathfrak{A} \models S'$; S' is satisfiable.

($\Longleftarrow$) Let there exist a guarded extension $S' \subseteq_{\mathcal{F}} OrdPropCl$ of S which is satisfiable. Then $S \subseteq S'$ is satisfiable. $\qquad\square$

Corollary 1 *Let* $n_0 \in \mathbb{N}$, $\phi \in PropForm_\emptyset$, $T \subseteq_{\mathcal{F}} PropForm_\emptyset$. *There exist* $J_T^\phi \subseteq_{\mathcal{F}}$ $\{(i, j) \mid i \geq n_0\} \subseteq \mathbb{I}$ *and* $S_T^\phi \subseteq_{\mathcal{F}} OrdPropCl_{\{\tilde{a}_j \mid j \in J_T^\phi\}}$ *such that* $T \models \phi$ *if and only if, for every guarded extension* $S' \subseteq_{\mathcal{F}} OrdPropCl$ *of* S_T^ϕ, $\square \in clo^{\mathcal{H}}(S')$.

Proof An immediate consequence of Theorems 1, 2, and 3. $\qquad\square$

We illustrate the solution to the deduction problem with an example. We show that $\phi = (\overline{0.81} \doteq \overline{0.9} \& \overline{0.9}) \& (a \& a \doteq \overline{0.81}) \to a \doteq \overline{0.9} \in PropForm$ is a tautology using the proposed translation to clausal form and the order hyperresolution calculus.

$$\phi = (\overline{0.81} \doteq \overline{0.9} \& \overline{0.9}) \& (a \& a \doteq \overline{0.81}) \to a \doteq \overline{0.9}$$

$$\left\{\tilde{a}_0 \prec \bar{1}, \tilde{a}_0 \leftrightarrow \left(\underbrace{(\overline{0.81} \doteq \overline{0.9} \& \overline{0.9}) \& (a \& a \doteq \overline{0.81})}_{\tilde{a}_1} \to \underbrace{a \doteq \overline{0.9}}_{\tilde{a}_2}\right)\right\} \qquad (9)$$

$$\left\{\tilde{a}_0 \prec \bar{1}, \tilde{a}_1 \prec \tilde{a}_2 \vee \tilde{a}_1 \doteq \tilde{a}_2 \vee \tilde{a}_0 \& \tilde{a}_1 \doteq \tilde{a}_2, \tilde{a}_2 \prec \tilde{a}_1 \vee \tilde{a}_0 \doteq \bar{1},\right.$$

$$\tilde{a}_1 \leftrightarrow \underbrace{(\overline{0.81} = \overline{0.9} \,\&\, \overline{0.9})}_{\tilde{a}_3} \,\&\, \underbrace{(a \,\&\, a = \overline{0.81})}_{\tilde{a}_4}, \tilde{a}_2 \leftrightarrow \underbrace{a}_{\tilde{a}_5} = \underbrace{\overline{0.9}}_{\tilde{a}_6} \Big\} \qquad (7), (11)$$

$$\Big\{ \tilde{a}_0 \prec \bar{1}, \tilde{a}_1 \prec \tilde{a}_2 \vee \tilde{a}_1 = \tilde{a}_2 \vee \tilde{a}_0 \,\&\, \tilde{a}_1 = \tilde{a}_2, \tilde{a}_2 \prec \tilde{a}_1 \vee \tilde{a}_0 = \bar{1},$$

$$\tilde{a}_1 = \tilde{a}_3 \,\&\, \tilde{a}_4, \tilde{a}_3 \leftrightarrow \underbrace{\overline{0.81}}_{\tilde{a}_7} = \underbrace{\overline{0.9} \,\&\, \overline{0.9}}_{\tilde{a}_8}, \tilde{a}_4 \leftrightarrow \underbrace{a \,\&\, a}_{\tilde{a}_9} = \underbrace{\overline{0.81}}_{\tilde{a}_{10}},$$

$$\tilde{a}_5 = \tilde{a}_6 \vee \tilde{a}_2 = \bar{0}, \tilde{a}_5 \prec \tilde{a}_6 \vee \tilde{a}_6 \prec \tilde{a}_5 \vee \tilde{a}_2 = \bar{1}, \tilde{a}_5 = a, \tilde{a}_6 = \overline{0.9} \Big\} \qquad (11)$$

$$\Big\{ \tilde{a}_0 \prec \bar{1}, \tilde{a}_1 \prec \tilde{a}_2 \vee \tilde{a}_1 = \tilde{a}_2 \vee \tilde{a}_0 \,\&\, \tilde{a}_1 = \tilde{a}_2, \tilde{a}_2 \prec \tilde{a}_1 \vee \tilde{a}_0 = \bar{1},$$

$$\tilde{a}_1 = \tilde{a}_3 \,\&\, \tilde{a}_4,$$

$$\tilde{a}_7 = \tilde{a}_8 \vee \tilde{a}_3 = \bar{0}, \tilde{a}_7 \prec \tilde{a}_8 \vee \tilde{a}_8 \prec \tilde{a}_7 \vee \tilde{a}_3 = \bar{1}, \tilde{a}_7 = \overline{0.81}, \tilde{a}_8 \leftrightarrow \underbrace{\overline{0.9}}_{\tilde{a}_{11}} \,\&\, \underbrace{\overline{0.9}}_{\tilde{a}_{12}},$$

$$\tilde{a}_9 = \tilde{a}_{10} \vee \tilde{a}_4 = \bar{0}, \tilde{a}_9 \prec \tilde{a}_{10} \vee \tilde{a}_{10} \prec \tilde{a}_9 \vee \tilde{a}_4 = \bar{1},$$

$$\tilde{a}_9 \leftrightarrow \underbrace{a}_{\tilde{a}_{13}} \,\&\, \underbrace{a}_{\tilde{a}_{14}}, \tilde{a}_{10} = \overline{0.81},$$

$$\tilde{a}_5 = \tilde{a}_6 \vee \tilde{a}_2 = \bar{0}, \tilde{a}_5 \prec \tilde{a}_6 \vee \tilde{a}_6 \prec \tilde{a}_5 \vee \tilde{a}_2 = \bar{1}, \tilde{a}_5 = a, \tilde{a}_6 = \overline{0.9} \Big\} \qquad (7)$$

$$S^\phi = \Big\{ \boxed{\tilde{a}_0 \prec \bar{1}} \qquad\qquad\qquad [1]$$

$$\tilde{a}_1 \prec \tilde{a}_2 \vee \tilde{a}_1 = \tilde{a}_2 \vee \tilde{a}_0 \,\&\, \tilde{a}_1 = \tilde{a}_2 \qquad [2]$$

$$\tilde{a}_2 \prec \tilde{a}_1 \vee \boxed{\tilde{a}_0 = \bar{1}} \qquad\qquad [3]$$

$$\boxed{\tilde{a}_1 = \tilde{a}_3 \,\&\, \tilde{a}_4} \qquad\qquad\qquad [4]$$

$$\tilde{a}_7 = \tilde{a}_8 \vee \boxed{\tilde{a}_3 = \bar{0}} \qquad\qquad [5]$$

$$\tilde{a}_7 \prec \tilde{a}_8 \vee \tilde{a}_8 \prec \tilde{a}_7 \vee \tilde{a}_3 = \bar{1} \qquad [6]$$

$$\boxed{\tilde{a}_7 = \overline{0.81}} \qquad\qquad\qquad [7]$$

$$\boxed{\tilde{a}_8 = \tilde{a}_{11} \,\&\, \tilde{a}_{12}} \qquad\qquad [8]$$

$$\boxed{\tilde{a}_{11} = \overline{0.9}} \qquad\qquad\qquad [9]$$

$$\boxed{\tilde{a}_{12} = \overline{0.9}} \qquad\qquad\qquad [10]$$

$$\tilde{a}_9 = \tilde{a}_{10} \vee \boxed{\tilde{a}_4 = \bar{0}} \qquad\qquad [11]$$

$$\tilde{a}_9 \prec \tilde{a}_{10} \vee \tilde{a}_{10} \prec \tilde{a}_9 \vee \tilde{a}_4 = \bar{1} \qquad [12]$$

$$\boxed{\tilde{a}_9 = \tilde{a}_{13} \,\&\, \tilde{a}_{14}} \qquad\qquad [13]$$

$$\boxed{\tilde{a}_{13} = a} \qquad\qquad\qquad [14]$$

$$\boxed{\tilde{a}_{14} = a} \qquad\qquad\qquad [15]$$

$$\boxed{\tilde{a}_{10} = \overline{0.81}} \qquad [16]$$

$$\tilde{a}_5 = \tilde{a}_6 \vee \tilde{a}_2 = \bar{0} \qquad [17]$$

$$\tilde{a}_5 \prec \tilde{a}_6 \vee \tilde{a}_6 \prec \tilde{a}_5 \vee \boxed{\tilde{a}_2 = \bar{1}} \qquad [18]$$

$$\boxed{\tilde{a}_5 = a} \qquad [19]$$

$$\left.\boxed{\tilde{a}_6 = \overline{0.9}}\right\} \qquad [20]$$

$$ordtcons(S^\phi) = \left\{\begin{array}{ll} \bar{0} \prec \overline{0.81} & [21] \\[2mm] \bar{0} \prec \overline{0.9} & [22] \\[2mm] \overline{0.81} \prec \bar{1} & [23] \\[2mm] \overline{0.9} \prec \bar{1} & [24] \\[2mm] \overline{0.81} \prec \overline{0.9} & [25] \end{array}\right\}$$

By a straightforward case analysis, we get that for $a = \bar{0}$, from [14], [15], [13], $\tilde{a}_9 = \bar{0}$, from [16], [21], [11], $\tilde{a}_4 = \bar{0}$, from [4], $\tilde{a}_1 = \bar{0}$, from [3], $\tilde{a}_0 = \bar{1}$, from [1] by **Rule (21)**, $\bar{1} \prec \bar{1}$, by **Rule (19)**, $\square$; for $a = \bar{1}$, from [14], [15], [13], $\tilde{a}_9 = \bar{1}$, from [16], [23], [11], $\tilde{a}_4 = \bar{0}$, which implies $\square$; for $\bar{0} \prec a \prec \bar{1}$, from [14], [15], [19], $\bar{0} \prec \tilde{a}_{13}, \tilde{a}_{14}, \tilde{a}_5 \prec \bar{1}$, from [13], $\bar{0} \prec \tilde{a}_9 \prec \bar{1}$, from [22], [24], [9], [10], [20], $\bar{0} \prec \tilde{a}_{11}, \tilde{a}_{12}, \tilde{a}_6 \prec \bar{1}$, from [8], $\bar{0} \prec \tilde{a}_8 \prec \bar{1}$, from [21], [23], [7], [16], $\bar{0} \prec \tilde{a}_7, \tilde{a}_{10} \prec \bar{1}$; for $\bar{0} \prec \tilde{a}_3 \prec \bar{1}$, from [5], [6], $\bar{0} \prec \tilde{a}_7, \tilde{a}_8 \prec \bar{1}$ by **Rule (18)** repeatedly, $\square$; for $\bar{0} \prec \tilde{a}_4 \prec \bar{1}$, from [11], [12], $\bar{0} \prec \tilde{a}_9, \tilde{a}_{10} \prec \bar{1}$ by **Rule (18)** repeatedly, $\square$;

Rule (23) : [1][3] :

$$\boxed{\tilde{a}_2 \prec \tilde{a}_1} \qquad [26]$$

for $\tilde{a}_3 = \bar{0}$ or $\tilde{a}_4 = \bar{0}$, from [4], $\tilde{a}_1 = \bar{0}$, from [26] by **Rule (20)**, $\tilde{a}_2 \prec \bar{0}$, by **Rule (19)**, $\square$; for $\tilde{a}_3 = \bar{1}, \tilde{a}_4 = \bar{1}$,

repeatedly **Rule (21)** : $[4][\tilde{a}_3 = \bar{1}][\tilde{a}_4 = \bar{1}]$:

$$\boxed{\tilde{a}_1 = \bar{1}} \qquad [27]$$

Rule (21) : [26][27] :

$$\boxed{\tilde{a}_2 \prec \bar{1}} \qquad [28]$$

Rule (23) : [18][28] :

$$\boxed{\tilde{a}_5 \prec \tilde{a}_6 \vee \tilde{a}_6 \prec \tilde{a}_5} \qquad [29]$$

Rule (21) : $[5][\tilde{a}_3 = \bar{1}]$:

$$\tilde{a}_7 = \tilde{a}_8 \vee \boxed{\bar{1} = \bar{0}} \tag{30}$$

Rule (19) : [30] :

$$\boxed{\tilde{a}_7 = \tilde{a}_8} \tag{31}$$

Rule (21) : [11][$\tilde{a}_4 = \bar{1}$] :

$$\tilde{a}_9 = \tilde{a}_{10} \vee \boxed{\bar{1} = \bar{0}} \tag{32}$$

Rule (19) : [32] :

$$\boxed{\tilde{a}_9 = \tilde{a}_{10}} \tag{33}$$

repeatedly **Rule (18)** : $[\bar{0} \prec a, \tilde{a}_5, \tilde{a}_6, \tilde{a}_7, \tilde{a}_8, \tilde{a}_9, \tilde{a}_{10}, \tilde{a}_{11}, \tilde{a}_{12}, \tilde{a}_{13}, \tilde{a}_{14} \prec \bar{1}]$;

$$[7][8][9][10][31]; [13][14][15][16][33]; [19][20][29] :$$

$$\square \tag{34}$$

We conclude that there exists a refutation of every guarded extension of S^ϕ; by Corollary 1 for S^ϕ, ϕ is a tautology.

5 Conclusions

In the paper, we have proposed a modification of the hyperresolution calculus from [14], which is suitable for automated deduction in the propositional product logic with explicit partial truth. The propositional product logic is expanded by a countable set of intermediate truth constants of the form $\bar{c}, c \in (0, 1)$. We have modified translation of a formula to an equivalent satisfiable finite order clausal theory. An order clause is a finite set of order literals of the form $\varepsilon_1 \diamond \varepsilon_2$ where ε_i is either a truth constant $\bar{0}$ or $\bar{1}$, or a conjunction of powers of propositional atoms or intermediate truth constants, and $\diamond$ is a connective $=$ or $\prec$. $=$ and $\prec$ are interpreted by the equality and standard strict linear order on $[0, 1]$, respectively. We have investigated the canonical standard completeness, where the semantics of the propositional product logic is given by the standard Π-algebra, and truth constants are interpreted by 'themselves'. The modified calculus is refutation sound and complete for finite guarded order clausal theories. A clausal theory is satisfiable if and only if there exists a satisfiable guarded extension of it. So, the *SAT* problem of a finite order clausal theory can be reduced to the *SAT* problem of a finite guarded order clausal theory. By means of the translation and calculus, we have solved the deduction problem $T \models \phi$ for a finite theory T and a formula ϕ in the expanded propositional product logic.

References

1. Biere, A., Heule, M.J., van Maaren, H., Walsh, T.: Handbook of Satisfiability. Frontiers in Artificial Intelligence and Applications, vol. 185. IOS Press, Amsterdam (2009)
2. Davis, M., Putnam, H.: A computing procedure for quantification theory. J. ACM **7**(3), 201–215 (1960)

3. Davis, M., Logemann, G., Loveland, D.: A machine program for theorem-proving. Commun. ACM **5**(7), 394–397 (1962)
4. Esteva, F., Godo, L.: Monoidal t-norm based logic: towards a logic for left-continuous t-norms. Fuzzy Sets Syst. **124**(3), 271–288 (2001)
5. Esteva, F., Godo, L., Montagna, F.: The ŁΠ and ŁΠ$\frac{1}{2}$ logics: two complete fuzzy systems joining Łukasiewicz and product logics. Arch. Math. Log. **40**(1), 39–67 (2001)
6. Esteva, F., Gispert, J., Godo, L., Noguera, C.: Adding truth-constants to logics of continuous t-norms: axiomatization and completeness results. Fuzzy Sets Syst. **158**(6), 597–618 (2007)
7. Esteva, F., Godo, L., Noguera, C.: On completeness results for the expansions with truth-constants of some predicate fuzzy logics. In: Stepnicka, M., Novák, V., Bodenhofer, U. (eds.) New Dimensions in Fuzzy Logic and Related Technologies. Proceedings of the 5th EUSFLAT Conference, Ostrava, Czech Republic, 11–14 September 2007, Volume 2: Regular Sessions. pp. 21–26. Universitas Ostraviensis (2007)
8. Esteva, F., Godo, L., Noguera, C.: First-order t-norm based fuzzy logics with truth-constants: distinguished semantics and completeness properties. Ann. Pure Appl. Logic **161**(2), 185–202 (2009)
9. Esteva, F., Godo, L., Noguera, C.: Expanding the propositional logic of a t-norm with truth-constants: completeness results for rational semantics. Soft Comput. **14**(3), 273–284 (2010)
10. Esteva, F., Godo, L., Noguera, C.: On expansions of WNM t-norm based logics with truth-constants. Fuzzy Sets Syst. **161**(3), 347–368 (2010)
11. Gallier, J.H.: Logic for Computer Science: Foundations of Automatic Theorem Proving. Harper & Row Publishers Inc., New York (1985)
12. Guller, D.: On the refutational completeness of signed binary resolution and hyperresolution. Fuzzy Sets Syst. **160**(8), 1162–1176 (2009)
13. Guller, D.: A DPLL procedure for the propositional product logic. In: Rosa, A.C., Dourado, A., Correia, K.M., Filipe, J., Kacprzyk, J. (eds.) IJCCI 2013 - Proceedings of the 5th International Joint Conference on Computational Intelligence, Vilamoura, Algarve, Portugal, 20–22 September 2013, pp. 213–224. SciTePress (2013). https://doi.org/10.5220/0004557402130224
14. Guller, D.: Hyperresolution for propositional Product logic. In: Guervós, J.J.M., Melício, F., Cadenas, J.M., Dourado, A., Madani, K., Ruano, A.E., Filipe, J. (eds.) Proceedings of the 8th International Joint Conference on Computational Intelligence, IJCCI 2016, Volume 2: FCTA, Porto, Portugal, 9–11 November 2016, pp. 30–41. SciTePress (2016). https://doi.org/10.5220/0006044300300041
15. Guller, D.: On the deduction problem in Gödel and Product logics. In: Computational Intelligence - IJCCI 2013. Studies in Computational Intelligence, vol. 613, pp. 299–321. Springer, Berlin (2016)
16. Hähnle, R.: Short conjunctive normal forms in finitely valued logics. J. Log. Comput. **4**(6), 905–927 (1994)
17. Hájek, P.: Metamathematics of Fuzzy Logic. Trends in Logic. Springer, Berlin (2001)
18. Klement, E., Mesiar, R.: Logical, Algebraic, Analytic and Probabilistic Aspects of Triangular Norms. Elsevier, Amsterdam (2005). https://books.google.sk/books?id=SB6motOxheQC
19. Klement, E., Mesiar, R., Pap, E.: Triangular Norms. Trends in Logic. Springer, Berlin (2013). https://books.google.sk/books?id=HXzvCAAAQBAJ
20. Marques-Silva, J.P., Sakallah, K.A.: GRASP: a search algorithm for propositional satisfiability. IEEE Trans. Comput. **48**(5), 506–521 (1999)
21. Montagna, F.: Notes on strong completeness in Łukasiewicz, product and BL logics and in their first-order extensions. In: Aguzzoli, S., Ciabattoni, A., Gerla, B., Manara, C., Marra, V. (eds.) Algebraic and Proof-Theoretic Aspects of Non-classical Logics, Papers in Honor of Daniele Mundici on the Occasion of His 60th birthday. Lecture Notes in Computer Science, vol. 4460, pp. 247–274. Springer, Berlin (2006). https://doi.org/10.1007/978-3-540-75939-3_15
22. Mostert, P.S., Shields, A.L.: On the structure of semigroups on a compact manifold with boundary. Ann. Math. 117–143 (1957)

23. Nonnengart, A., Rock, G., Weidenbach, C.: On generating small clause normal forms. In: Kirchner, C., Kirchner, H. (eds.) Automated Deduction - CADE-15, 15th International Conference on Automated Deduction, Lindau, Germany, 5–10 July 1998, Proceedings. Lecture Notes in Computer Science, vol. 1421, pp. 397–411. Springer, Berlin (1998)
24. Novák, V., Perfilieva, I., Močkoř, J.: Mathematical Principles of Fuzzy Logic. The Springer International Series in Engineering and Computer Science. Springer, Berlin (1999). http://books.google.sk/books?id=pJeu6Ue65S4C
25. Pavelka, J.: On fuzzy logic I, II, III. Semantical completeness of some many-valued propositional calculi. Math. Log. Q. 25(2529), 45–52, 119–134, 447–464 (1979)
26. Plaisted, D.A., Greenbaum, S.: A structure-preserving clause form translation. J. Symb. Comput. 2(3), 293–304 (1986)
27. Robinson, J.A.: Automatic deduction with hyper-resolution. Int. J. Comput. Math. 1(3), 227–234 (1965)
28. Robinson, J.A.: A machine-oriented logic based on the resolution principle. J. ACM 12(1), 23–41 (1965). https://doi.org/10.1145/321250.321253
29. Savický, P., Cignoli, R., Esteva, F., Godo, L., Noguera, C.: On product logic with truth-constants. J. Log. Comput. 16(2), 205–225 (2006)
30. Schöning, U., Torán, J.: The Satisfiability Problem: Algorithms and Analyses. Mathematik für Anwendungen, Lehmanns Media (2013)
31. Sheridan, D.: The optimality of a fast CNF conversion and its use with SAT. In: SAT (2004)
32. Silva, J.P.M., Sakallah, K.A.: GRASP - a new search algorithm for satisfiability. In: ICCAD, pp. 220–227 (1996)
33. de la Tour, T.B.: An optimality result for clause form translation. J. Symb. Comput. 14(4), 283–302 (1992)
34. Tseitin, G.S.: On the complexity of derivation in propositional calculus. In: Siekmann, J.H., Wrightson, G. (eds.) Automation of Reasoning 2: Classical Papers on Computational Logic 1967–1970, pp. 466–483. Springer, Berlin (1983). https://doi.org/10.1007/978-3-642-81955-1_28
35. Vidal, A., Bou, F., Esteva, F., Godo, L.: On strong standard completeness in some MTL$_\Delta$ expansions. Soft Comput. 21(1), 125–147 (2017). https://doi.org/10.1007/s00500-016-2338-0
36. Vidal, A., Esteva, F., Godo, L.: On modal extensions of product fuzzy logic. J. Log. Comput. 27(1), 299–336 (2015, 2017). https://doi.org/10.1093/logcom/exv046

Fuzzy Clustering of High Dimensional Data with Noise and Outliers

Ludmila Himmelspach and Stefan Conrad

Abstract Clustering high dimensional data is a challenging problem for fuzzy clustering algorithms because of so-called concentration of distance phenomenon. The most fuzzy clustering algorithms fail to work on high dimensional data producing cluster prototypes close to the center of gravity of the data set. The presence of noise and outliers in data is an additional problem for clustering algorithms because they might affect the computation of cluster centers. In this paper, we analyze and compare different promising fuzzy clustering algorithms in order to examine their ability to correctly determine cluster centers on high dimensional data with noise and outliers. We analyze the performance of clustering algorithms for different initializations of cluster centers: the original means of clusters and random data points in the data space.

Keywords Fuzzy clustering · C-means models · High dimensional data · Noise · Possibilistic clustering

1 Introduction

Clustering algorithms are used in many fields like bioinformatics, image processing, text mining, and many others. Data sets in these applications usually contain a large number of features. Therefore, there is a need for clustering algorithms that can handle high dimensional data. The hard *k-means* algorithm [1] is still mostly used for clustering high dimensional data, although it is comparatively unstable and sensitive to the initialization. It is not able to distinguish data items belonging to clusters from noise and outliers. This is another issue of the hard k-means algorithm because noise

L. Himmelspach (✉) · S. Conrad
Institute of Computer Science, Heinrich-Heine-Universität Düsseldorf,
40225 Düsseldorf, Germany
e-mail: himmelspach@cs.uni-duesseldorf.de

S. Conrad
e-mail: conrad@cs.uni-duesseldorf.de

© Springer Nature Switzerland AG 2019
J. J. Merelo et al. (eds.), *Computational Intelligence*,
Studies in Computational Intelligence 792,
https://doi.org/10.1007/978-3-319-99283-9_11

and outliers might influence the computation of cluster centers leading to inaccurate clustering results.

In the case of low dimensional data, the *fuzzy c-means* algorithm (FCM) [2, 3] which assigns data items to clusters with membership degrees might be a better choice because it is more stable and less sensitive to initialization [4]. The *possibilistic fuzzy c-means* algorithm (PFCM) [5] partitions data items in presence of noise and outliers. However, when FCM is applied on high dimensional data, it tends to produce cluster centers close to the center of gravity of the entire data set [6, 7]. In this work, we analyze different fuzzy clustering methods that are suitable for clustering high dimensional data. The first approach is the *attribute weighting fuzzy clustering algorithm* [8] that uses a new attribute weighting function to determine attributes that are important for each single cluster. This method was recommended in [7] for fuzzy clustering of high dimensional data. The second approach is the *multivariate fuzzy c-means* (MFCM) [9] that computes membership degrees of data items to each cluster in each feature. The third method is the *possibilistic multivariate fuzzy c-means* (PMFCM) [10] which is an extension of MFCM in a possibilistic clustering scheme. The last approach is PMFCM_HDD [11] which is another version for extending MFCM in the possibilistic clustering scheme. In PMFCM_HDD, the degree of typicality of data items to the clustering structure of the data set is computed for each single dimension. The main objective of this work is to analyze and compare fuzzy clustering algorithms in order to examine their ability to correctly determine cluster centers on high dimensional data in presence of noise and outliers. Additionally, we ascertain which fuzzy clustering algorithms are less sensitive to different initializations of cluster centers.

The rest of the paper is organized as follows: In the next section we give a short overview of fuzzy clustering methods for high-dimensional data. The evaluation results on artificial data sets containing different kinds of noise and outliers are presented in Sect. 3. Section 4 closes the paper with a short summary and the discussion of future research.

2 Fuzzy Clustering Algorithms

Fuzzy c-means (FCM) [2, 3] is a partitioning clustering algorithm that assigns data objects to clusters with membership degrees. The objective function of FCM is defined as:

$$J_m(U, V; X) = \sum_{k=1}^{n} \sum_{i=1}^{c} u_{ik}^m d^2(v_i, x_k),$$ (1)

where c is the number of clusters, $u_{ik} \in [0, 1]$ is the membership degree of data item x_k to cluster i, $m > 1$ is the fuzzification parameter, $d(v_i, x_k)$ is the distance between cluster prototype v_i and data item x_k. The objective function of FCM is usually minimized in an alternating optimization (AO) scheme [2] under constraint (2).

$$\sum_{i=1}^{c} u_{ik} = 1 \ \forall k \in \{1, \ldots, n\} \text{ and } \sum_{k=1}^{n} u_{ik} > 0 \ \forall i \in \{1, \ldots, c\}. \tag{2}$$

The algorithm begins with initialization of cluster prototypes. In each iteration of the algorithm, the membership degrees and the cluster prototypes are alternating updated according to Formulae (3) and (4).

$$u_{ik} = \frac{\left(d^2(v_i, x_k)\right)^{\frac{1}{1-m}}}{\sum_{l=1}^{c} \left(d^2(v_l, x_k)\right)^{\frac{1}{1-m}}}, \ 1 \leq i \leq c, \ 1 \leq k \leq n. \tag{3}$$

$$v_i = \frac{\sum_{k=1}^{n} u_{ik}^m x_k}{\sum_{k=1}^{n} u_{ik}^m}, \quad 1 \leq i \leq c. \tag{4}$$

The iterative process continues as long as the cluster prototypes change up to a chosen limit.

The FCM algorithm has several advantages over the hard k-means algorithm [1] in low dimensional data spaces. It is more stable, less sensitive to initialization, and is able to model soft transitions between clusters [4]. However, the hard k-means algorithm is still mostly used in real world applications for clustering high dimensional data because fuzzy c-means does not usually provide useful results on high dimensional data. Mostly it computes equal membership degrees of all data items to all clusters which results in the computation of final cluster prototypes close to the center of gravity of the entire data set. This is due to the *concentration of distance phenomenon* described in [12]. It says that the distance to the nearest data item approaches the distance to the farthest one with increasing number of dimensions.

2.1 Fuzzy Clustering Algorithms for High Dimensional Data

In [7], the author recommended to use the *attribute weighting fuzzy clustering algorithm* [8] for clustering high dimensional data. This method uses a distance function that weights single attributes for each cluster:

$$d^2(v_i, x_k) = \sum_{j=1}^{p} \alpha_{ij}^t (x_{kj} - v_{ij})^2, \quad 1 \leq i \leq c, \ 1 \leq k \leq n, \tag{5}$$

where p is the number of attributes, $t > 1$ is a fixed parameter that determines the strength of the attribute weighting, and

$$\sum_{j=1}^{p} \alpha_{ij} = 1 \ \forall i \in \{1, \ldots, c\}. \tag{6}$$

This approach works in the same way as FCM but it circumvents the concentration of distance phenomenon using a distance function that gives more weight to features that determine a particular cluster. The objective function of the attribute weighting fuzzy clustering algorithm is defined as:

$$J_{m,t}(U, V; X) = \sum_{k=1}^{n} \sum_{i=1}^{c} u_{ik}^{m} \sum_{j=1}^{p} \alpha_{ij}^{t} (v_{ij} - x_{kj})^2. \tag{7}$$

The attribute weights are updated in an additional iteration step according to Formula (8).

$$\alpha_{ij} = \left(\sum_{r=1}^{p} \left(\frac{\sum_{k=1}^{n} u_{ik}^{m}(x_{kj} - v_{ij})^2}{\sum_{k=1}^{n} u_{ik}^{m}(x_{kr} - v_{ir})^2} \right)^{\frac{1}{t-1}} \right)^{-1}, \ 1 \le i \le c, \ 1 \le k \le n. \tag{8}$$

The second approach that we analyze in this work is the *multivariate fuzzy c-means* (MFCM) algorithm. This fuzzy clustering method computes membership degrees of data items to clusters for each feature [9]. The objective function of MFCM is defined as follows:

$$J_m(U, V; X) = \sum_{k=1}^{n} \sum_{i=1}^{c} \sum_{j=1}^{p} u_{ikj}^{m}(v_{ij} - x_{kj})^2, \tag{9}$$

where $u_{ikj} \in [0, 1]$ is the membership degree of data object x_k to cluster i for feature j. The objective function of MFCM has to be minimized under constraint (10).

$$\sum_{i=1}^{c} \sum_{j=1}^{p} u_{ikj} = 1 \ \forall k \in \{1, \ldots, n\} \quad \text{and} \quad \sum_{j=1}^{p} \sum_{k=1}^{n} u_{ikj} > 0 \ \forall i \in \{1, \ldots, c\}. \tag{10}$$

The multivariate membership degrees and the cluster prototypes are updated in the iterative process of the algorithm according to (11) and (12).

$$u_{ikj} = \left(\sum_{h=1}^{c} \sum_{l=1}^{p} \left(\frac{(x_{kj} - v_{ij})^2}{(x_{kl} - v_{hl})^2} \right)^{\frac{1}{m-1}} \right)^{-1}, \ 1 \le i \le c, \ 1 \le k \le n, \ 1 \le j \le p. \tag{11}$$

$$v_{ij} = \frac{\sum\limits_{k=1}^{n} u_{ikj}^{m} x_{kj}}{\sum\limits_{k=1}^{n} u_{ikj}^{m}}, \quad 1 \le i \le c, \ 1 \le j \le p. \tag{12}$$

MFCM is not influenced by the concentration of distance phenomenon because it computes membership degrees in each feature depending on the partial distances in single dimensions. Therefore, this approach is suitable for clustering high dimensional data.

2.2 Possibilistic Clustering Algorithms for High Dimensional Data

The fuzzy clustering algorithms described before are not designed to cluster data in presence of noise and outliers. They assign such data items to clusters in the same way as data items within clusters. In this way, noise and outliers might affect the computation of cluster centers which leads to inaccurate partitioning results. There are different ways for avoiding this problem. The mostly used method is determining outliers before clustering. There are different methods for outlier detection but the most of them compare the distances from data points to their neighbors [13]. Another method called *noise clustering* introduces an additional cluster that contains all data items that are located far away from any of cluster centers [14, 15]. In this work, we extend the clustering algorithms described in the previous subsection using the *possibilistic fuzzy c-means* (PFCM) clustering model proposed in [5]. This approach extends the basic FCM by typicality values that express a degree of typicality of each data item to the overall clustering structure of the data set. The advantage of using typicality values is that outliers get less weight in the computation of cluster centers. The objective function of PFCM is defined as:

$$J_{m,\eta}(U, T, V; X) = \sum_{k=1}^{n} \sum_{i=1}^{c} (a u_{ik}^{m} + b t_{ik}^{\eta}) d^{2}(v_{i}, x_{k}) + \sum_{i=1}^{c} \gamma_{i} \sum_{k=1}^{n} (1 - t_{ik})^{\eta}, \tag{13}$$

where $t_{ik} \le 1$ is the typicality value of data item x_k to cluster i, $m > 1$ and $\eta > 1$ are the user defined constants. Similarly to FCM, the first term in (13) ensures that distances between data items and cluster centers are minimized, where constants $a > 0$ and $b > 0$ control the relative influence of fuzzy membership degrees and typicality values. The second term ensures that typicality values are determined as large as possible. The second summand is weighted by the parameter $\gamma_i > 0$. In [16], the authors recommended to run the basic FCM algorithm before PFCM and to choose γ_i by computing:

$$\gamma_i = K \frac{\sum\limits_{k=1}^{n} u_{ik}^m d^2(v_i, x_k)}{\sum\limits_{k=1}^{n} u_{ik}^m} \qquad 1 \le i \le c, \qquad (14)$$

where the $\{u_{ik}\}$ are the terminal membership degrees computed by FCM and $K > 0$ (usually $K = 1$). The objective function of PFCM has to be minimized under constraints (2) and (15).

$$\sum_{k=1}^{n} t_{ik} > 0, \ \forall i \in \{1, \ldots, c\} \qquad (15)$$

In [10], we extended the MFCM algorithm in the possibilistic clustering scheme to make it less sensitive to outliers. Since we considered outliers as data points that have a large overall distance to any cluster, we did not compute typicality values of data points to clusters for each feature. The objective function of the resulting approach that we refer here as *possibilistic multivariate fuzzy c-means* (PMFCM) is defined as

$$J_{m,\eta}(U, T, V; X) = \sum_{k=1}^{n} \sum_{i=1}^{c} \sum_{j=1}^{p} (a u_{ikj}^m + b t_{ik}^{\eta})(v_{ij} - x_{kj})^2$$
$$+ p \sum_{i=1}^{c} \gamma_i \sum_{k=1}^{n} (1 - t_{ik})^{\eta}. \qquad (16)$$

The objective function of PMFCM has to be minimized under constraint (17).

$$\sum_{i=1}^{c} u_{ikj} = 1 \ \forall k, j \ \text{ and } \ \sum_{k=1}^{n} u_{ikj} > 0 \ \forall i, j, \ \text{ and } \ \sum_{k=1}^{n} t_{ik} > 0 \ \forall i. \qquad (17)$$

In MFCM, the sum of membership degrees over all clusters and features to a particular data item was constrained to be 1. Since we want to retain equal influence of membership degrees and typicality values, we only constrain the sum of membership degrees over all clusters to a particular data item to be 1.

The membership degrees and the typicality values have to be updated according to Formulae (18) and (19). In PMFCM, the cluster centers are updated in a similar way as in PFCM according to Formula (20).

$$u_{ikj} = \left(\sum_{l=1}^{c} \left(\frac{(x_{kj} - v_{ij})^2}{(x_{kj} - v_{lj})^2} \right)^{\frac{1}{m-1}} \right)^{-1}, \ 1 \le i \le c, \ 1 \le k \le n, \ 1 \le j \le p. \quad (18)$$

$$t_{ik} = \left(1 + \left(\frac{b\sum_{j=1}^{p}(x_{kj} - v_{ij})^2}{\gamma_i\, p}\right)^{\frac{1}{\eta-1}}\right)^{-1} , \quad 1 \le i \le c,\ 1 \le k \le n. \tag{19}$$

$$v_{ij} = \frac{\sum_{k=1}^{n}(au_{ikj}^m + bt_{ik}^\eta)x_{kj}}{\sum_{k=1}^{n}(au_{ikj}^m + bt_{ik}^\eta)}, \quad 1 \le i \le c,\ 1 \le j \le p. \tag{20}$$

The membership degrees of data objects to clusters can be computed in this model as the average of the multivariate membership degrees over all variables, $u_{ik} = \frac{1}{p}\sum_{j=1}^{p} u_{ikj}$.

Due to the concentration of distance phenomenon, PMFCM might not produce meaningful typicality values because it uses the Euclidean distances between data items and cluster prototypes. Since distances between data items within clusters and cluster centers and distances between outliers and cluster centers might be similar in high dimensional data, the typicality values as they are computed in PMFCM might not be helpful for distinguishing between data items within clusters and outliers. Therefore, in [11], we proposed another version of PMFCM that computes typicality values of data items to clusters in each dimension. We refer this approach here as *possibilistic multivariate fuzzy c-means for high dimensional data* (PMFCM_HDD). The objective function of PMFCM_HDD is given in Formula (21).

$$J_{m,\eta}(U, T, V; X) = \sum_{k=1}^{n}\sum_{i=1}^{c}\sum_{j=1}^{p}(au_{ikj}^m + bt_{ikj}^\eta)(v_{ij} - x_{kj})^2$$
$$+ \sum_{i=1}^{c}\gamma_i \sum_{k=1}^{n}\sum_{j=1}^{p}(1 - t_{ikj})^\eta. \tag{21}$$

The objective function of PMFCM_HDD has to be minimized under the following constraint:

$$\sum_{i=1}^{c} u_{ikj} = 1\ \forall k, j \quad \text{and} \quad \sum_{k=1}^{n} u_{ikj} > 0\ \forall i, j, \quad \text{and} \quad \sum_{k=1}^{n} t_{ikj} > 0\ \forall i, j. \tag{22}$$

The membership degrees are updated in the same way as in PMFCM according to Formula (18). The update Formulae for typicality values and cluster centers are given in (23) and (24).

$$t_{ikj} = \left(1 + \left(\frac{b(x_{kj} - v_{ij})^2}{\gamma_i}\right)^{\frac{1}{\eta-1}}\right)^{-1} , \quad 1 \le i \le c,\ 1 \le k \le n,\ 1 \le j \le p. \tag{23}$$

$$v_{ij} = \frac{\sum\limits_{k=1}^{n} (au_{ikj}^m + bt_{ikj}^\eta)x_{kj}}{\sum\limits_{k=1}^{n} (au_{ikj}^m + bt_{ikj}^\eta)}, \quad 1 \le i \le c, \ 1 \le j \le p. \tag{24}$$

The typicality values of data objects to clusters can also be computed as average of the multivariate typicality values over all variables, $t_{ik} = \frac{1}{p} \sum_{j=1}^{p} t_{ikj}$.

3 Data Experiments

3.1 Test Data and Experimental Setup

We tested the four fuzzy clustering methods for high dimensional data described in Sect. 2 on artificial data sets containing different kinds of noise and outliers. The main data set was generated similarly to one that was used in [8]. It is a two-dimensional data set that consists of 1245 data points unequally distributed on one spherical cluster and three clusters that have a low variance in one of the dimensions. The data set is depicted in Fig. 1. We generated the second and the third data sets that are depicted in Figs. 2 and 3 by adding 150 and 300 noise points to the main data set.

In order to generate high dimensional data sets we added 18 additional dimensions containing feature values close to zero. Additionally, we generated the fourth data set by adding 300 noise points containing values different from zero in all 20 dimensions. In all data sets, we distributed clusters in the data space so that clusters were determined by different features. For example, data items of the first cluster had real values in the third and the fourth features, data items of the second cluster

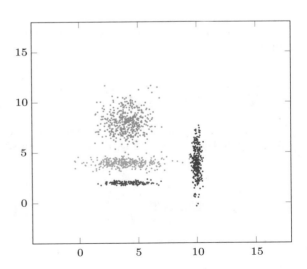

Fig. 1 Test data with four clusters (extracted from [11])

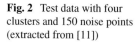

Fig. 2 Test data with four clusters and 150 noise points (extracted from [11])

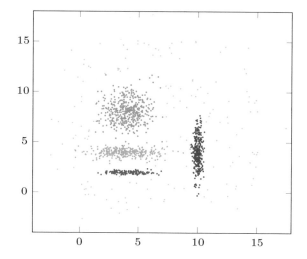

Fig. 3 Test data with four clusters and 300 noise points (extracted from [11])

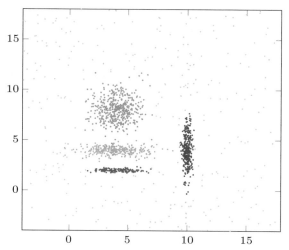

had real values in the sevenths and the eighth features etc. Furthermore, we completely distributed all noise points among the dimensions so that each dimension pair contained some noise points.

We had to modify the possibilistic clustering algorithms PMFCM and PMFCM_HDD. Instead of running FCM at the beginning, we only computed membership values to compute γ_i. If we ran FCM at the beginning, it output cluster centers close to the center of gravity of the data set which is a bad initialization for possibilistic clustering algorithms. In order to provide the best starting conditions for the clustering algorithms, in the first experiment, we initialized the cluster centers with the original means of clusters. Since the original cluster means are usually unknown, in the second experiment, we initialized the cluster centers with random points in the data space in order to analyze and compare the sensitivity of fuzzy clustering methods to initialization. In order to evaluate the experimental results, we computed the Frobenius distance d_{orig} between the original cluster means and the cluster centers obtained by the clustering algorithms. We also computed the sum of distances d_{means} between cluster centers produced by the clustering algorithms. Moreover, we computed the number of iterations #$iter$ that the clustering algorithms needed until the termination.

3.2 Experimental Results

Table 1 shows the experimental results for FCM with attribute weighting (indicated as FCM_AW in tables), MFCM and its possibilistic versions PMFCM and PMFCM_HDD for $a = 0.5$ and $b = 1000$ on the 20-dimensional data set without noise and outliers. The most accurate cluster centers were produced by MFCM. This is indicated by a low d_{orig} and high d_{means} (good separation between clusters). MFCM also needed much less iterations until termination than the other algorithms. The FCM algorithm with attribute weighting produced the least accurate cluster centers because it recognized the additional dimensions containing feature values close to zero as the important ones and gave small weights to important dimensions. This method was introduced by the authors as a clustering algorithm that performs the dimensionality reduction while clustering. Our results have shown that this method is not always able to distinguish between important and unimportant dimensions. The second best results were produced by the possibilistic clustering methods PMFCM and PMFCM_HDD. However, these methods needed much more iterations until termination than the other approaches.

Tables 2 and 3 show the experimental results on the second and the third data sets with 150 and 300 noise points. The clustering algorithms performed similarly as on the data set without noise and outliers. MFCM produced the most accurate final cluster prototypes and needed only few iterations until termination. Its possibilistic versions PMFCM and PMFCM_HDD also obtained final cluster centers that were close to the actual cluster means. Although FCM with attribute weighting produced

Table 1 Comparison between clustering methods on 20-dimensional data set with four clusters distributed in data space (initialization: original means of clusters)

	FCM_AW: $m = 2, t = 2$	MFCM: $m = 2$	PMFCM: $m = 2$, $\eta = 2$, $a = 0.5$, $b = 1000$	PMFCM_HDD: $m = 2$, $\eta = 2$, $a = 0.5$, $b = 1000$
d_{orig}	27.47	0.072	0.701	0.460
d_{means}	0.0017	65.97	65.79	66.39
#$iter$	38	5	85	85

Table 2 Comparison between clustering methods on 20-dimensional data set with four clusters and 150 noise points distributed in data space (initialization: original means of clusters)

	FCM_AW: $m = 2, t = 2$	MFCM: $m = 2$	PMFCM: $m = 2$, $\eta = 2, a = 0.5$, $b = 1000$	PMFCM_HDD: $m = 2, \eta = 2$, $a = 0.5$, $b = 1000$
d_{orig}	1.499	0.104	0.924	0.962
d_{means}	64.33	66.00	65.37	66.70
#iter	12	4	57	140

Table 3 Comparison between clustering methods on 20-dimensional data set with four clusters and 300 noise points distributed in data space (initialization: original means of clusters)

	FCM_AW: $m = 2, t = 2$	MFCM: $m = 2$	PMFCM: $m = 2$, $\eta = 2, a = 0.5$, $b = 1000$	PMFCM_HDD: $m = 2, \eta = 2$, $a = 0.5$, $b = 1000$
d_{orig}	3.733	0.072	0.787	0.784
d_{means}	61.34	65.97	65.47	65.74
#iter	27	5	53	76

the least accurate cluster prototypes, its performance was comparable to the performance of the other methods. Due to noise points, all features contained at least some values different from each other. Therefore, FCM_AW was able to correctly recognize features determining different clusters.

Table 4 shows the performance results of the algorithms on the fourth data set where noise points contained values different from zero in all features. MFCM produced cluster centers that were farther from each other than the real means of clusters. It also needed much more iterations than the other algorithms on this data set. FCM_AW produced cluster centers that were not close to the center of gravity of the data set anymore but they were too far from each other which is no surprise

Table 4 Comparison between clustering methods on 20-dimensional data set with four clusters distributed in data space and 300 noise points containing values in all features (initialization: original means of clusters)

	FCM_AW: $m = 2, t = 2$	MFCM: $m = 2$	PMFCM: $m = 2$, $\eta = 2, a = 0.5$, $b = 1000$	PMFCM_HDD: $m = 2, \eta = 2$, $a = 0.5$, $b = 1000$
d_{orig}	36.42	12.42	0.699	0.949
d_{means}	127.20	73.79	64.89	65.24
#iter	21	224	18	82

because both MFCM and FCM_AW are not designed to distinguish between data points within clusters and outliers. The most accurate cluster centers were obtained by PMFCM and PMFCM_HDD. PMFCM needed much less iterations until termination than the other clustering methods.

In the second experiment, we initialized the cluster prototypes with random values in the data space at the beginning of the algorithms. The following tables present the averaged performance results over 100 runs of the algorithms. Since MFCM needed a very large number of iterations until termination on the third data set, we averaged the results over 40 runs for this method (indicated with asterisks in the corresponding table). Table 5 shows the experimental results on the 20-dimensional data sets with four clusters distributed among dimensions without noise and outliers. We need both the distance between the original cluster means and the cluster centers obtained by the clustering algorithms d_{orig} and the sum of distances between the final cluster prototypes d_{means} for the evaluation of performance results of the algorithms. On the one hand, FCM_AW achieved the second best results for d_{orig}. On the other hand, according to its d_{means} value it produced almost the identical final cluster prototypes. The distance between the final cluster prototypes produced by MFCM was similar to the distance between the original means of clusters. On the other hand, the distance between the original and the obtained cluster centers was rather large which means that MFCM produced final cluster prototypes far from the original means of clusters. The final cluster prototypes obtained by the possibilistic versions of MFCM were much closer to each other than the original means of clusters, although PMFCM_HDD produced quite accurate final cluster prototypes in single runs.

Tables 6 and 7 show the performance results of the clustering algorithms on data sets with 150 and 300 noise points. While MFCM produced final clustering results farther from each other than the original means of clusters, the final cluster prototypes obtained by the other three clustering methods were closer to each other than the original cluster means.

The performance results of the clustering algorithms on the fourth data set are summarized in Table 8. Like in the case of the initialization of cluster prototypes with the original means of clusters (compare Table 4), FCM_AW and MFCM produced final cluster prototypes that were much farther from each other than the original means

Table 5 Comparison between clustering methods on 20-dimensional data set with four clusters distributed in data space (initialization: random)

	FCM_AW: $m = 2, t = 2$	MFCM: $m = 2$	PMFCM: $m = 2$, $\eta = 2, a = 0.5$, $b = 1000$	PMFCM_HDD: $m = 2, \eta = 2$, $a = 0.5$, $b = 1000$
$\bar{d}_{orig}$	13.96	21.52	16.65	13.41
$\bar{d}_{means}$	0.0019	64.25	35.31	33.65
$\overline{\#iter}$	39.45	21.15	30.67	210.51

Table 6 Comparison between clustering methods on 20-dimensional data set with four clusters and 150 noise points distributed in data space (initialization: random)

	FCM_AW: $m = 2, t = 2$	MFCM: $m = 2$	PMFCM: $m = 2$, $\eta = 2, a = 0.5$, $b = 1000$	PMFCM_HDD: $m = 2, \eta = 2$, $a = 0.5$, $b = 1000$
$\bar{d}_{orig}$	13.74	43.63	14.76	13.88
$\bar{d}_{means}$	19.93	149.88	32.82	21.53
$\overline{\#iter}$	86.69	472.15	31.17	101.42

Table 7 Comparison between clustering methods on 20-dimensional data set with four clusters and 300 noise points distributed in data space (initialization: random)

	FCM_AW: $m = 2, t = 2$	MFCM: $m = 2$	PMFCM: $m = 2$, $\eta = 2, a = 0.5$, $b = 1000$	PMFCM_HDD: $m = 2, \eta = 2$, $a = 0.5$, $b = 1000$
$\bar{d}_{orig}$	13.37	54.02*	14.64	15.11
$\bar{d}_{means}$	9.92	191.15*	27.30	5.01
$\overline{\#iter}$	130.98	2789.73*	29.72	17.67

Table 8 Comparison between clustering methods on 20-dimensional data set with four clusters distributed in data space and 300 noise points containing values in all features (initialization: random)

	FCM_AW: $m = 2, t = 2$	MFCM: $m = 2$	PMFCM: $m = 2$, $\eta = 2, a = 0.5$, $b = 1000$	PMFCM_HDD: $m = 2, \eta = 2$, $a = 0.5$, $b = 1000$
$\bar{d}_{orig}$	52.32	80.72	17.15	16.28
$\bar{d}_{means}$	92.07	240.63	30.86	41.29
$\overline{\#iter}$	21.28	131.52	19.31	121.1

of clusters. PMFCM and PMFCM_HDD obtained more accurate cluster prototypes, although they were much closer to each other than the original cluster means.

4   Conclusion and Future Work

Fuzzy clustering methods have several advantages over the hard clustering approaches in the case of low dimensional data. Clustering high dimensional data is still a challenging problem for fuzzy clustering algorithms because they are more affected by the concentration of distance phenomenon. Moreover, fuzzy clustering algorithms are more sensitive to initialization in high dimensional data spaces [6]. Noise and outliers in data sets additionally make the partitioning of data difficult because they affect the computation of cluster centers. In this paper, we analyzed different fuzzy clustering algorithms for high dimensional data in terms of correct determining final cluster prototypes in presence of noise and outliers. Our experiments showed that the performance results of FCM with attribute weighting strongly depend on the structure of the high dimensional data set even if the cluster prototypes are initialized with the original means of clusters. The multivariate FCM was able to produce accurate cluster centers as long as data items had real values in few features but it is very sensitive to initialization of cluster prototypes. The possibilistic fuzzy clustering methods PMFCM and PMFCM_HDD produced quite accurate final cluster centers on high dimensional data with noise and outliers as long as the initial cluster prototypes were close to the original means of clusters. Although their performance results for random initialization of cluster centers were promising in single runs, they are still too unstable to be used in the real world applications and need further development.

In our future work, we aim to extend the possibilistic fuzzy clustering methods PMFCM and PMFCM_HDD with a randomized seeding technique as proposed in [17] to make them less sensitive to initialization when applying on high dimensional data. Furthermore, we plan to apply fuzzy clustering algorithms for clustering text data and compare their performance with the LDA algorithm [18] and the common crisp clustering algorithms on small document collections.

References

1. MacQueen, J.: Some methods for classification and analysis of multivariate observations. In: Proceedings of the Fifth Berkeley Symposium on Mathematical Statistics and Probability, volume 1: Statistics, pp. 281–297. University of California Press, Berkeley (1967)
2. Bezdek, J.C.: Pattern Recognition with Fuzzy Objective Function Algorithms. Kluwer Academic Publishers, Norwell (1981)
3. Dunn, J.C.: A fuzzy relative of the isodata process and its use in detecting compact well-separated clusters. J. Cybern. **3**(3), 32–57 (1973)
4. Klawonn, F., Kruse, R., Winkler, R.: Fuzzy clustering: more than just fuzzification. Fuzzy Sets Syst. **281**, 272–279 (2015)
5. Pal, N.R., Pal, K., Keller, J.M., Bezdek, J.C.: A possibilistic fuzzy C-means clustering algorithm. IEEE Trans. Fuzzy Syst. **13**(4), 517–530 (2005)
6. Winkler, R., Klawonn, F., Kruse, R.: Fuzzy C-means in high dimensional spaces. Int. J. Fuzzy Syst. Appl. **1**(1), 1–16 (2011)
7. Klawonn, F.: What can fuzzy cluster analysis contribute to clustering of high-dimensional data?. In: Proceedings of the 10th International Workshop on Fuzzy Logic and Applications, pp. 1–14 (2013)
8. Keller, A., Klawonn, F.: Fuzzy clustering with weighting of data variables. Int. J. Uncertain. Fuzziness Knowl. Based Syst. **8**(6), 735–746 (2000)
9. Pimentel, B.A., de Souza, R.M.C.R.: A multivariate fuzzy C-means method. Appl. Soft Comput. **13**(4), 1592–1607 (2013)

10. Himmelspach, L., Conrad, S.: A possibilistic multivariate fuzzy C-means clustering algorithm. In: Proceedings of the 10th International Conference on Scalable Uncertainty Management, pp. 338–344 (2016)
11. Himmelspach, L., Conrad, S.: The effect of noise and outliers on fuzzy clustering of high dimensional data. In: Proceedings of the 8th International Joint Conference on Computational Intelligence, pp. 101–108 (2016)
12. Beyer, K.S., Goldstein, J., Ramakrishnan, R., Shaft, U.: When is nearest neighbor meaningful?. In: Proceedings of the 7th International Conference on Database Theory, pp. 217–235. Springer, Berlin (1999)
13. Kriegel, H., Kröger, P., Schubert, E., Zimek, A.: Outlier detection in axis-parallel subspaces of high dimensional data. In: Proceedings of the 13th Pacific-Asia Conference on Advances in Knowledge Discovery and Data Mining, pp. 831–838 (2009)
14. Dave, R.N., Krishnapuram, R.: Robust clustering methods: a unified view. IEEE Trans. Fuzzy Syst. **5**(2), 270–293 (1997)
15. Rehm, F., Klawonn, F., Kruse, R.: A novel approach to noise clustering for outlier detection. Soft Comput. **11**(5), 489–494 (2007)
16. Krishnapuram, R., Keller, J.M.: A possibilistic approach to clustering. IEEE Trans. Fuzzy Syst. **1**(2), 98–110 (1993)
17. Arthur, D., Vassilvitskii, S.: K-means++: the advantages of careful seeding. In: Proceedings of the Eighteenth Annual ACM-SIAM Symposium on Discrete Algorithms, pp. 1027–1035 (2007)
18. Blei, D.M., Ng, A.Y., Jordan, M.I.: Latent Dirichlet allocation. J. Mach. Learn. Res. **3**(4–5), 993–1022 (2003)

Part III
Neural Computation Theory
and Applications

E-Mail Spam Filter Based on Unsupervised Neural Architectures and Thematic Categories: Design and Analysis

Ylermi Cabrera-León, Patricio García Báez and Carmen Paz Suárez-Araujo

Abstract Spam, or unsolicited messages sent massively, is one of the threats that affects email and other media. Its huge quantity generates considerable economic and time losses. A solution to this issue is presented: a hybrid anti-spam filter based on unsupervised Artificial Neural Networks (ANNs). It consists of two steps, preprocessing and processing, both based on different computation models: programmed and neural (using Kohonen SOM). This system has been optimized by utilizing a dataset built with ham from "Enron Email" and spam from two different sources: traditional (user's inbox) and spamtrap-honeypot. The preprocessing was based on 13 thematic categories found in spams and hams, Term Frequency (TF) and three versions of Inverse Category Frequency (ICF). 1260 system configurations were analyzed with the most used performance measures, achieving $AUC > 0.95$ the optimal ones. Results were similar to other researchers' over the same corpus, although they utilize different Machine Learning (ML) methods and a number of attributes several orders of magnitude greater. The system was further tested with different datasets, characterized by heterogeneous origins, dates, users and types, including samples of image spam. In these new tests the filter obtained $0.75 < AUC < 0.96$. Degradation of the system performance can be explained by the differences in the characteristics of the datasets, particularly dates. This phenomenon is called "topic drift" and it commonly affects all classifiers and, to a larger extent, those that use offline learning, as is the case, especially in adversarial ML problems such as spam filtering.

Y. Cabrera-León · C. P. Suárez-Araujo (✉)
Instituto Universitario de Ciencias y Tecnologías Cibernéticas,
Universidad de Las Palmas de Gran Canaria, Las Palmas de Gran Canaria, Spain
e-mail: carmenpaz.suarez@ulpgc.es

Y. Cabrera-León
e-mail: ylermi.cabrera101@alu.ulpgc.es

P. García Báez
Departamento de Ingeniería Informática y de Sistemas,
Universidad de La Laguna, San Cristóbal de La Laguna, Spain
e-mail: pgarcia@ull.es

© Springer Nature Switzerland AG 2019
J. J. Merelo et al. (eds.), *Computational Intelligence*,
Studies in Computational Intelligence 792,
https://doi.org/10.1007/978-3-319-99283-9_12

Keywords Spam filtering · Artificial neural networks · Self-organizing maps ·
Thematic category · Term frequency · Inverse category frequency · Topic drift ·
Adversarial machine learning

1 Introduction and Background

Nowadays, the use and importance of telecommunication has increased, primarily due to the rise of Information and Communications Technology (ICT). Among the multiple ways to make such communication, email can be highlighted, mainly because it has been used extensively for decades. Unfortunately, this popularity has brought with it the appearance of threats such as hoaxes, cyber-attacks, computer viruses and, to a greater extent, spam.

Although there are many different ways of defining the word "spam" [1], in this document spam refers to any message, mostly email but other media are affected too [2], sent massively without the recipients having requested or desired it. The characteristic of "massive" must be highlighted because, for years, both spam volume and overall spam rate (in other words, the quantity of spam and the percentage of spam relative to all messages, respectively) have been extremely high: in 2008 about 62 trillion unwanted messages [3] and less than 1 out of 10 emails could be considered as ham (legitimate or desired messages), fortunately improving to 4 out of 10 in 2014 [4].

Considering that, in most cases, their content is offensive or questionable - e.g. scam, phishing, illegal drugs, pornography, replicas...[5] - it can be asserted that spam is a great scourge, creating quite substantial, both temporary and economic, losses: annually, American firms and consumers experience costs of near $20 billion due to spam whereas spammers (people or companies that send spam) and spam-advertised firms collect $200 million worldwide [6]. This occurs, mainly, due to sending spam being easy and having low cost, and that the recipient carries the bulk of the cost, in contrast to what happens with more traditional or off-line unsolicited marketing methods [7].

Throughout history, the goals of spam have changed from earning money to spreading malware, creating botnets, ransoming computers, and coaxing users to do illegal activities, which are almost always harmful for their interests but beneficial for the spammer [8]. In order to achieve them, spams must get to the recipient mail addresses and, in most cases, users should read them or proceed with other actions. Consequently, spams should surpass every anti-spam methods in between so they must evolve and adapt to the new defensive techniques that are continuously getting updated and being applied [9–11].

There have been a wide variety of proposals to solve the problem of email spam detection so far, and therefore there is a huge proliferation of papers in this regard, as it will be discussed in Sect. 1.1. In this document another solution to this problem is presented: a hybrid spam filtering system. It can be considered hybrid not only because its two main stages, preprocessing and processing, are based on different computing models - programmed and neural computation, respectively - but also because in the processing one the Self-Organizing Map (SOM), an unsupervised ANN, is followed by a non-neural supervised labeling part.

This system has been optimized using, as a data body, ham from "Enron Email" [12] and spam from two different sources: obtained through traditional ways (user's inbox), or through spamtraps and honeypots. Its quality and performance have been analyzed on several different datasets with both Kohonen maps' most used quality measures, Mean Quantization Error (MQE) and Topographic Error (TE), and the most common performance metrics for classifiers such as Receiver Operating Characteristic (ROC) curves and Area Under the Curve (AUC), among others.

As an extension of one of our previous works [13], in this one we wanted to demonstrate how the system performance was impacted by several characteristics of spam, hence evaluating the system's adaptability to varied incoming emails. The additional testing datasets were characterized by heterogeneous origins, dates, users and types, including samples of image spam. The "Matryoshka experiment" was conceived to reveal which characteristics of these datasets affected our system the most. Akin to the traditional Russian dolls, in the "Matryoshka experiment" each bigger dataset included all the smaller datasets, added one after the other. It should be noticed that the datasets were ordered a priori by the date of the emails in them, which means that a bigger dataset included more recent emails, apart from the older messages from the smaller datasets.

The remainder of this document is organized as follows. In Sect. 1.1 some of the numerous related works developed throughout the last decades are described. Through Sect. 2 the dataset and methods are explained. Section 3 shows the experimental results, followed by a discussion of them. Finally, the conclusions can be found in Sect. 4.

1.1 Related Works

As it is common to defensive and security systems in all areas (such as pathogenic diseases, armament, crime and predation), attackers (spammers in our case) are always one step ahead of defenders [14], therefore, in this area, the latter need to continuously face new threats and counter shortcomings, weaknesses and security flaws that the former have found, and later exploited, in anti-spam filters, other software or hardware [15]. Actually, this evolution explains the proliferation of multiple anti-spam techniques developed over the past decades [11].

Anti-spam methods may filter during any of the network hierarchical levels (mostly in the Application, Transport and Network layers of the TCP/IP model), i.e. in any of the steps involved with sending emails: in the sender device, *en route* and in the recipient. Alongside this manner, spam classifiers can also be grouped by these two ways: based on the design method, and based on the source of information. This section's scope has been reduced by just choosing user-level and administrator-level filters, whose techniques belong to any of the two previous groups, due to the fact that a user-level filtering system was developed here.

Based on the Design Method. There are two subgroups in this group, the latter being the most popular one:

Manual: easier to implement and more useful for email administrators. Despite their slow adaptation to changes in spam, whitelists and blacklists (respectively, lists of good and bad mail servers and ISPs) achieve just 1% of False Positives (FP) and False Negatives (FN) [16]. Furthermore, greylisting blocks email delivery temporarily with unrecognized senders, forcing resending, something not usually done by spammers [17]. It reduces bandwidth waste at the expense of delaying ham too [18].

Based on ML: it is the largest subgroup [19], where filters can be classified, at the same time, depending on the kind of architecture used (neural or not), or the quantity of human interaction required (supervised, semi-supervised or unsupervised). Making use of the latter, we could classify some anti-spam techniques as follows:

- Supervised: the most popular non-neural one is the Bayesian [20, 21], and, therefore, the most attacked by spammers through Bayesian poisoning [22–24].

  Metsis' article [25] must be described separately from other Bayesian filters because some of the datasets utilized in this document were built from theirs and later compared with, Sect. 2. Their system performs better when 3000 attributes, the greatest number they have tested with, are used, obtaining an average sensitivity of 0.9753 and specificity of 0.9726, and quite-near-perfection ROC curves.

  Nowadays, other popular non-neural one, due to its performance, is the Support Vector Machine (SVM) [26, 27], which is greatly kernel-dependent [28] and offers better results when several are combined with a voting strategy [29].

  On the other hand, for a long time, the perceptron, neural, has dominated as antispam [30, 31] but the raise of Bayesian and SVM changed this.

  Other supervised ANN is the Learning Vector Quantization (LVQ) with whom Chuan [32] built a good filter (96.20% F-measure, 98.97% precision and 93.58% sensitivity) if enough iterations, at least 1500, were made.

- Unsupervised: because no previous data labeling process is needed, emails should occupy less disk space and be more recent [33]. Among non-neural filters, we could find: the SpamCampaignAssassin [34] based on the detection of spam campaigns, one based on the alienness or searching of similarities in substrings [35], and other which uses suffix trees [36].

  On the other hand, there are also unsupervised neural techniques. The SOM Based Sequence Analysis [37] makes use of a double hierarchical-leveled SOM, where the second SOM is connected *a posteriori* with a k-Nearest Neighbors (k-NN) for categorization and sequence analysis.

  Two articles from Vrusias [38, 39] should be introduced separately from others due to their importance for the research shown in this document. They compare their SOM-based system with what they consider to be the best spam classifiers: the Multinomial Naïve Bayes Boolean, SVM and Boosted Decision Trees. It is a 10x10 SOM, sequentially trained for 1000 cycles, whose input vectors have 26 or 500 attributes, and where keywords were grouped (just when 26 attributes were used) and identified with Term Frequency·Inverse Document Frequency (TF·IDF) and weirdness. The main differences between their filters and this document's are: larger SOM, smaller input vectors, other learning algorithm and a similar way of identifying keywords were used here, Sect. 2. They also used datasets based on the "Enron-Spam" corpus from Metsis [25].

- Semi-supervised: not many labeled data, due to high costs [40], with a lot of unlabeled examples. Regularized Discriminant Expectation-Maximization [41] combine both transductive (for labeling unlabeled data) with inductive (to make a model in order to classify new data) methods, obtaining 91.66% detection rate and 2.96% FP.

  Learning with Local and Global Consistency, of which there is a variant proposed by [42], obtains better results than with k-NN and SVM [43, 44].

  Although SpamAssassin [45] is generally considered supervised, it can also utilize a semi-supervised learning rule [46].

  A semi-supervised version of SVM also exists, Transductive Support Vector Machine [47, 48], which sometimes performs worse than SVM as an anti-spam [49].

Based on the Source of Information. These methods, which use any part of an email i.e. envelope, header and body [50], can also be subdivided in the next three assortments:

Content of the email: the most prevalent way, either using the whole message [51] or just selecting parts with different methods: rules [52], detecting anchored parts [53] or spam campaigns [34], signatures of messages [54], or combining several techniques, such as in SpamAssassin [45, 55] and CRM114 [56] popular anti-spam filters.

User Feedback: in spite of users being considered the most reliable and robust anti-spam method, specially against content obfuscation made by spammers, [57] find out that they have up to 2% of classification errors, due to ham being very similar to spam - a.k.a. "hard ham" [58] - or the presence of "grey cases" [59], where categorization has an important subjective factor.

Information Relative to the System: they frequently take the advantage on the inherent difficulty for spammers to change the message headers and produce valid ones [60]. This can be detected by checking fields of some network protocols, specially the ones which contain the sender's IP and port, and the local sending time (countless emails sent during sender's sleeping time may indicate that sender's device belongs to a botnet). On the other hand, they can also detect the presence or absence of specific characteristics in the message, such as only images (usually no text at all, hence called "image spam") [41, 61], and attached files, prone to be malware-infected.

2 Dataset and Methods

2.1 Dataset

In order to work with ANNs effectively, it is crucial to have a broad and representative dataset, i.e. a set of emails where both spam and ham are widely represented [62]. Most email corpus are both restricted and costly in order to keep their users' privacy and obstruct spammers' countermeasures. In spite of this, several corpora are freely available nowadays [5, 19].

The proposed system has utilized a subset of emails from "Enron-Spam", a free and gratis corpus [25]. Our dataset was built using only "Enron1" and "Enron5" folders, Table 1. By doing this, we have worked with a balanced dataset, 5172 ham and 5175 spam, from the preprocessed version of "Enron-Spam", wherein:

- Ham belonged to the "Enron Email Corpus", which has been widely used with different preprocessing techniques applied on it [12, 63, 64]. In fact, Metsis [25] uses ham from 6 out of 7 Enron users' inboxes from the preprocessed version of Bekkerman's corpus [65], Table 1.
- Spam came from two different sources: received in a traditional way by one of the authors of the mentioned corpus, Georgios Paliouras (GP); and through spamtraps [66] and honeypots (SH), which are anti-spam techniques intended, respectively, to lure spam, and to bait, investigate and punish spammers. Unwanted messages from Bruce Guenter's "Spam Archive" (BG) [67] were not used in our case, unlike [25].

The E1 and E5 (E1E5) dataset was subsequently partitioned in the following balanced sets: 80% for training-validation the ANN and 20% for testing the system over data never seen before.

Apart from E1E5 used during the design of the system, other datasets were utilized to test more independently this design and, consequently, evaluate in a more realistic way the methodology of the system: both preprocessing and processing stages. Thus, the system was further tested on the following three datasets which came from different email corpora, Table 1:

- E2 and E4 (E2E4), as the "Enron2" (E2) and "Enron4" (E4) combination, is similar to E1E5 but with ham from other Enron users and a different quantity of spam from same sources.
- "SpamAssassin_2" (SA2), built choosing the most recent ham and spam folders from the SpamAssassin dataset, differs considerably from our previous datasets in terms of content, topics, origins and dates [66].
- "CSDMC2010_SPAM" (CS), which previously has been used in the data mining competition related with the ICONIP2010 conference, holds the newest emails that our system was tested with [68]. Moreover, the CS dataset contains several not preprocessed image spams, hence its content is the least similar to those of the rest.

Table 1 Original email corpora. In bold our datasets (based on [13])

Dataset	Ham–spam origins	No. of ham-spam	Ham dates	Spam dates
E1	**Farmer-d – GP**	**3672–1500**	**12/1999–01/2002**	**12/2003–09/2005**
E2	**Kaminski-v – SH**	**4361–1496**	**12/1999–05/2001**	**05/2001–07/2005**
E3	Kitchen-l – BG	4012–1500	02/2001–02/2002	08/2004–07/2005
E4	**Williams-w3 – GP**	**1500–4500**	**04/2001–02/2002**	**12/2003–09/2005**
E5	**Beck-s – SH**	**1500–3675**	**01/2000–05/2001**	**05/2001–07/2005**
E6	Lokay-m – BG	1500–4500	06/2000–03/2002	08/2004–07/2005
SA2	**Easy_ham_2 – spam_2**	1400–1397	**08/2002–12/2002**	**08/2002–12/2002**
CS	**TRAINING**	**2949–1378**	**09/2002–05/2010**	**09/2002–06/2009**

The experiments that were performed utilized the original data, which is not normalized, or normalized data. Hereinafter, each of these groups of experiments are indicated as Scenario 1 and Scenario 2, respectively.

2.2 Methods

The proposed intelligent anti-spam system consisted of two different computing stages or modules [13], Fig. 1, and it can be considered hybrid because each of them was based on a different computing scheme. The first one was the preprocessing stage, which was based on programmed computation, whereas the second one, the processing stage, made use of a neural computing scheme. The preprocessing module was responsible for obtaining a semantic and compact representation of the information environment, a set of feature vectors for emails to analyze. These vectors were the input data for the subsequent hybrid processing module, where the detection of spam by a SOM-based, unsupervised ANN, system was performed, followed by a non-neural supervised labeling method which worked with the outputs of the SOM.

Preprocessing Module. The preprocessing stage is quite important [69, 70], especially with unsupervised methods due to the fact that no kind of corrective signals nor correct outputs are provided. The main purposes of this stage were: reducing the dimensionality of the vocabulary, and only making use of the most relevant words. Three key concepts - thematic categories, ICF and Top $k\%$ of words - were used, which will be explained in the next paragraphs.

This preprocessing is founded on the premise that there are several spam thematic categories [5, 11], *ergo*, based on this, detection and differentiation from ham would be feasible. The most common thematic categories frequently found in spam and ham, which were expected to exist within our dataset, have been encountered and the most similar ones were lumped together in only 13 thematic categories, Table 2. Email words can belong to more than one category at the same time (e.g.: obfuscation

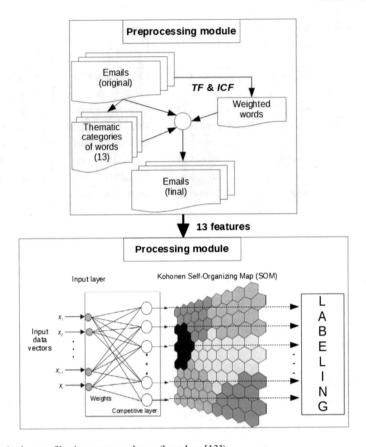

Fig. 1 Anti-spam filtering system scheme (based on [13])

and medicine, links and trading…). Initially, our words categorization in thematic categories was made manually, keeping in mind the word's use and context. Later and based on this, this process was automated. Manual words categorization brought with it two useful advantages:

- Lemmatization or stemming were not required [71], common in many anti-spam filters.
- Robust against words deliberately obfuscated, a countermeasure frequently applied by spammers to deceive or defeat anti-spam techniques [72, 73].

The existence of the aforementioned thematic categories directed this research to the usage of ICF, which is recommended by some authors when categories or classes exist in the data [74], rather than using Inverse Document Frequency (IDF). Both ICF and IDF are used for similar purposes: reduce the importance/weight given by TF to "stop words" i.e. extremely common words in categories or documents, respectively. Three different variants of ICF have been checked [75], (1), (2) and (3), where C

Table 2 Description of the 13 thematic categories within our spam and ham words (extracted from [13])

Thematic category	Description
Sex and Relationships	Mostly pornography, casual sex and dating websites
Medicine, Drugs and Body	Selling medicines or illegal drugs. Includes words related with body parts, and surgical procedures and tools
Betting, Gambling and Divination	Includes lotto, sports betting, casino, tarot, etc.
Banking, Investment, Insurance and Trading	Commerce, offers, funds, stock markets...
Links and Email addresses	Parts of links to websites and emails, mainly web extensions and domains
Other languages	Most of the emails were written in English but some were in Spanish, German, French, Dutch and Turkish, among others
Obfuscation	Very common. Words badly written *on purpose* by spammers to interfere with most content-based anti-spam filters
Business, Companies and Government	Name of firms, governmental agencies and analogous
Internet and Technology	ICT vocabulary
Documents and Paperwork	CV, diplomas, business documents, etc.
Names and Family	Names and surnames, also family members
Tourism and Regions	Countries, cities, holidays...
Attached files	Several file extensions and words related with "attach"

is the total number of categories, and f_t is the number of categories where token t happens:

$$ICF_{Log} = \log\left(\frac{|C|}{f_t}\right) \tag{1}$$

$$ICF_{Linear} = \frac{|C|}{f_t} \tag{2}$$

$$ICF_{Sqrt} = \sqrt{\left(\frac{|C|}{f_t}\right)} \tag{3}$$

Another interesting aspect in this process was to know if using all the words within a thematic category was better than using less, so the system was tested with different number of words in each one. This number is given by some percentage, k, which means that we chose the $k\%$ of the words with greater $TF_{category} \cdot ICF$, that is, the Top $k\%$ of words.

Our preprocessing stage could be divided in four phases, conducted in the indicated sequential order as they are interdependent:

Phase 0:   batch extraction of subject line and body of all emails-files in a path. Also, non alphanumeric characters were inserted between blank spaces to ease next phases. Optional as some datasets does not require it.

Phase 1:   only keeping *selected words*, that is, with length > 2, not very rare and also not too frequent as they might be "stop words" [76]. Each email was reduced to a text line, following the bag-of-words model i.e. several pairs of selected word next to its raw TF in this document: $TF_{document}$. At the end of each line two labels were added: spam/ham (not used during training) and the original, alphanumeric, email ID.

Phase 2:   previously making the described manual words categorization. Building a 13-dimensional, integer, array where each element, accumulator, represents the sum of the raw $TF_{document}$ of all the words belonging to a thematic category, by looking up the word in every category. At the end of each line the same two labels were inserted.

Phase 3:   automatizing words categorization (email words are counted using accumulators as in Phase 2 and associated to, by default 1, the winner thematic category) and weighting words with $TF_{category} \cdot ICF$ so extremely common words were given less importance. Ordering categories using those values permitted the obtainment of several Top $k\%$ of words to be tested. Building a 13-dimensional, floating point, array where each element represented the sum of the $TF_{document} \cdot ICF$, similar to Phase 2 but now weighted, of all words belonging to certain thematic category. Same two labels at the end of each line-email, Fig. 2.

Processing Module. The system's processing stage has as its information environment the feature vectors obtained in the preprocessing stage. It is hybrid as the first part was based on a type of ANNs, the well known Kohonen Self-Organizing Maps [77], whereas the second one was non-neural. Both parts will be explained below.

SOM, as an unsupervised neural architecture, is a very appropriate method for facing the problem to be solved. It quantifies the input space in different regions represented by a specific number of output neurons, a.k.a. detectors. In our case, there are two types of detectors: spam detectors and ham detectors.

Moreover, SOMs might be used as a visualization tool of high-dimensional data by projections over lower-dimensional maps [78]. During this projection process akin to multidimensional scaling, it is easily seen that SOMs try to extract the features of the input space in order to preserve its topological properties, similar to the idea of topographic maps that exist in the brains of highly developed animals [79].

Its structure is made of an input layer fully interconnected, by excitatory connections, with the neurons in the output layer. The latter is organized in a m-dimensional space, the most common being the 2D matrix. Within this matrix there is a neighborhood relationship between its nodes that is usually defined by an hexagonal or rectangular lattice. Also, the matrix shape can vary, the sheet one being the most common. All neurons within the output layer simultaneously present inhibitory lateral

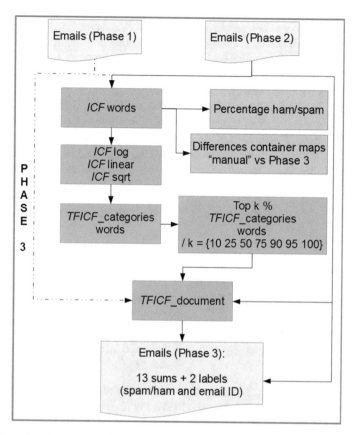

Fig. 2 Phase 3 of the preprocessing module and its relationship with the outputs of the Phases 1 and 2

connections among neural neighbors as well as excitatory self-connections. Their neurodynamic is simplified by computing the least (more frequently Euclidean and hence used in this work) distance between the inputs and a model [77], which is a parametric real vector, that can be considered as the weight vector in the neural architecture. The winning neuron, a.k.a. Best Matching Unit (BMU), will be the one with the minimum distance value, Eq. (4).

$$net_l(\mathbf{x}) = \|\mathbf{x} - \mathbf{w}_l\| = \sqrt{\sum_i (x_i - w_{li})^2}$$

$$u_l = \begin{cases} 1 & \text{if } l = \underset{k}{argmin} \{net_k(\mathbf{x})\} \\ 0 & \text{otherwise} \end{cases}$$

(4)

The learning process belongs to a winner-take-all, unsupervised and competitive training paradigm. The main variations are seen in the modification of the synaptic weights, which not only affects the winning neuron but also, to a lesser extent, the set of neurons in the winners' neighborhood N (thus, SOM training can be considered cooperative too), and consequently being able to generate topological relations, Eq. (5). During the training period, the neighborhood relationship between nodes h_{ji} decreases both in time and distance (commonly a Gaussian function), affecting only the BMU during the final phase. The learning rate α normally decreases with time, usually beginning near the unity and finishing close to zero during the fine tuning done in the last training cycles. However, a fixed value may be utilized but this is not recommended [80].

$$\Delta w_{li} = \begin{cases} \alpha(x_i - w_{li}) & \text{if } i \in N(\underset{k}{argmin} \{net_k(x)\}) \\ 0 & \text{otherwise} \end{cases} \tag{5}$$

SOMs can use two different learning methods: sequential and batch, Eq. (6), where $\overline{x}_j$ is the mean of the elements in a list of weights updates; h_{ji}, the neighborhood function; and n_i, the number of elements in that list. Batch learning method, which can be better and converges faster [81], has been employed in this article.

$$w_j(n+1) = \frac{\sum_i n_i \cdot h_{ji} \cdot \overline{x}_j}{\sum_i n_i \cdot h_{ji}} \tag{6}$$

The second part of the processing stage was a non-neural supervised classification method, which was appended after the SOM, Fig. 1. Its main aim was to label the results obtained by the SOM, that way classifying in spam or ham the emails inputted into the anti-spam filter. It was based on confidence thresholds, which are based on the minimum percentage from which consider an email as spam. These confidence thresholds, which were empirically chosen, will allow us to adjust the system relative to the FP, an important factor in this kind of filters. In a nutshell, an email was labeled as ham if the spam ratio $\frac{\#spam}{\#spam + \#ham}$ was lesser than this threshold, or as spam otherwise, Eq. (7).

$$label_{email} = \begin{cases} spam & \text{if } \frac{\#spam}{\#spam + \#ham} \geq \text{threshold} \\ ham & \text{otherwise} \end{cases} \tag{7}$$

On the other hand, the efficiency and quality of the proposed anti-spam system were determined through the usage of two different families of metrics:

- Quality of the SOM map [80]: MQE and TE, which measure the map resolution (how accurately the inputs are reflected in the output space) and the topology preservation (the order of the map), respectively.
- On-use performance measures such as F-score, accuracy, precision, specificity, sensitivity, ROC curve and AUC. All of them can be expressed in term of the elements of the confusion matrix: True Positives (TP), FP, True Negatives (TN)

and FN. Also, it might be included in this group one simple and low cost metric found out during this research that measures the least Euclidean distance between the ROC curve and the point of perfect classification in (0, 1).

For evaluation purposes and comparison with other researchers', the best considered performance measures are ROC curves[1] and AUC [82–84]. Indeed, AUC is a good measure of the fit of the classifiers when parameter adjustment is made.

The main development environment for our anti-spam filter was MATLAB, using the SOM Toolbox [85] for the SOM architecture and visualization tools, and the Parallel Computing Toolbox [86] to reduce the high computational costs of the experiments. Nevertheless, Python 3 was used to implement the Phase 0 of the preprocessing module due to its particularly useful characteristics when working with email files, especially with the EML format found on the additional testing datasets [50].

3 Results and Discussion

Several characteristics related to the information environment and the SOM structure were varied, Table 3. Considering all the possible combination of values, 1260 different system configurations were developed and their efficacy determined with the aforementioned metrics.

Relative to E1E5 dataset, results obtained with all metrics were quite positive, Table 4. All the 1260 analyzed configurations obtained AUC > 0.90, and even 204 got AUC > 0.95 which can be described as "excellent" classifiers in the anti-spam

Table 3 Tested values of the characteristics for all the 1260 configurations (based on [13])

Preprocessing characteristic	Tested values
ICF	Log, linear and sqrt
Top $k\%$ of words	100, 95, 90, 75, 50, 25 and 10
SOM characteristic	Tested values
SOM size	13x13, 20x20, 25x25, 40x40 and 50x50
SOM training algorithm	Batch
Number of epochs	100, 500, 1000, 3000, 5000 and 8000
Neighborhood function	Gaussian
SOM shape	Sheet
Lattice	Hexagonal
Weight initialization	Linear

---

[1]Each ROC curve was drawn with 102 specificity and sensitivity values, given by the same number of confidence thresholds, for enhanced ROC curve detail.

Table 4 Results of the anti-spam filter for each scenario (testing phase) with E1E5 (20%, 2069 emails) dataset

E1E5 (20%)		
Performance measurements	**Scenario 1**	Scenario 2
AUC	**0.971156**	0.968803
Discrete AUC	0.923584	**0.924011**
(Threshold)	(38)	(46)
Accuracy	0.924031	**0.924392**
Precision	0.898565	**0.900675**
F-score	0.927407	**0.927507**
Specificity	0.889005	**0.892034**
Sensitivity	**0.958163**	0.955988
TP	**939**	934
FP	106	**103**
TN	849	**851**
FN	**41**	43
%FP	11.1%	**10.8%**
%FN	**4.18%**	4.40%
Distance to (0, 1)	0.118618	**0.116593**
SOM map quality	**Scenario 1**	Scenario 2
MQE	39.592070	0.452661
TE	0.118567	**0.078204**

context. Additionally, results for most metrics were quite similar between configurations, even more when comparing the same scenario. It has been observed that all pairs Top 100% and Top 95% configurations, with identical rest of parameters, shared the same results. Consequently, Top 95% ones were preferred because of their faster learning due to using a smaller vocabulary. Furthermore, none of the best classifiers for each scenario used the biggest SOM sizes, 50x50, but smaller-sized ones. Obtained MQE and TE with normalized data are on the same range of values as other authors' [5].

Besides, normalized data behaved better with E1E5 in the majority of performance measures, which usually happens with Kohonen networks. However, as we consider AUC the preferred metric for comparisons, the optimal classifier with E1E5 dataset was the Scenario 1's configuration, that used the following characteristics: data without any kind of normalization applied to it, Top 25% of words, ICF log, Gaussian neighborhood, hexagonal lattice, 20x20 sheet-shaped map, and was trained for 100 epochs with the batch algorithm It overcame the Scenario 2's opponent that made use of normalized data, Top 95% of words, ICF sqrt, Gaussian neighborhood, hexagonal lattice, 20x20 sheet-shaped map, and trained for 8000 epochs. Both best candidates offered AUC $\approx$ 0.97, sensitivity > 0.95 and accuracy > 0.92.

Table 5 Results obtained by other researchers, indicating best ones in metrics that ours can be compared with (extracted from [13])

Research	Dataset	Method	Best results
Metsis [25]	Enron-Spam	Bayesian (several)	Sensitivity = [0.9232 − 0.9753]
Vrusias [39]	Enron-Spam	SOM	Precision = 0.992867
			Sensitivity = 0.920067
Chuan [32]	SpamAssassin	LVQ	Precision = 0.9897
			Sensitivity = 0.9358
Holden [87]	SpamAssassin	Bayesian (several, commercial)	Precision = [0.328 − 1]
			Sensitivity = [0.837 − 0.988]
Kufandirimbwa [30]	SpamAssassin	Perceptron algorithm	Precision = 0.97149
			Sensitivity = 0.77859
Luo [37]	Ling-Spam	Two-level SOMs + k-NN	Precision = [0.933 − 1]
			Sensitivity = [0.675 − 0.975]
Shunli [47]	ECML-PKDD 2006	Transductive SVM	AUC = 0.9321
Xie [27]	PU1 and PU2	SVM (several)	Accuracy (PU1) = [0.926 − 0.941]
			Accuracy (PU2) = [0.932–0.945]

Comparing our results in Table 4 with other researchers' in Table 5, the proposed system achieves worse than desired FP and FN (around 11 and 4%, respectively), which should be the correction priority in future works, while good and comparable values with performance metrics. Still, it should be noted that this comparison would have been more realistic if exactly the same emails and preprocessing methods had been tested with other processing techniques. This is expected to be done in future works, together with a more advanced system.

More specifically, comparing with the anti-spam filters proposed by Metsis [25], from whom most of the datasets used in this document were obtained, our results in Table 4 are comparable to theirs, Table 5, and our ROC curves in Fig. 4a are similar to their separated curves for "Enron1" and "Enron5" folders: specificity = 0.94433 and sensitivity = 0.96436 (average value and for only those two folders). The main differences are that their filters make use of a non-neural methodology with supervised learning strategy and up to 3000 attributes while our proposal is a SOM-based system which used only 13 for analogous results. Consequently, we could infer that we utilized both appropriate preprocessing methods, that let us obtain smaller yet more informative input vectors, and a powerful processing tool, which is able to work with such unlabeled vectors.

On the other side, another set of experiments were designed and conducted so that the adaptability to heterogeneous incoming emails was assessed. The proposed anti-spam filter was further tested with other datasets, which are different from E1E5, used during the design of the anti-spam. In our previous work [13], several datasets - E1E5, E2E4 and SA2 - were mixed in order to analyze both best configurations in a more realistic situation (i.e. with emails from diverse origins and in different proportions), discovering that "topic drift" happened and degraded the performance of our offline learning system, as expected. This term will be explained later.

In this document, a step forward is made and a new group of experiments is carried out. Its purpose is to reveal which of these characteristics affect our system the most:

- Users: the recipients and senders of the messages.
- Origin of the dataset: indicates from where the emails were obtained. It appears to be correlated with the dates of the emails and also implies variation of users.
- Type of the email is predominantly textual as our system is a textual and content-based filter that cannot handle image spam on its own. To prove how the system deal with this issue, some examples of not preprocessed image spams are utilized. Metsis' [25] dataset seems to contain image spams, in preprocessed form though, hence our system did not need to deal with that problem before [13].

In order to test them, the "Matryoshka experiment" was devised. Considering that the datasets are ordered by the date of the emails in them, the "Matryoshka experiment", similar to the typical Russian dolls, is built from layers of datasets, added one after the other and always including the previous smaller ones. As the datasets had different sizes from the beginning and in order to keep the same growth in the size of each dataset combination, each of these mixes got increased by the same quantity of emails, a quantity that was given by the size of the smallest used dataset, 2069 of E1E5. Henceforth, the following "dolls", ordered increasingly by size, were:

- Doll 1 is the smallest one and contains only 2069 emails from E1E5.
- Doll 2 is built by adding the same quantity of emails from E2E4 to the previous doll. Compared with E1E5, this dataset contains emails from different folders of the same email corpus, "Enron-Spam". So, this doll is used to analyze the influence of the characteristic called "users" by comparing with the results of the previous doll, Table 4.
- Doll 3, or E1E5 + E2E4 + SA2, is the first doll that includes emails from a dataset that has a completely different origin, Table 1.
- Doll 4 is the biggest one as it combines 2069 emails from each dataset, E1E5 + E2E4 + SA2 + CS. This doll is utilized to test how detrimental is the presence of image spams for our textual anti-spam.

Not only the individual dolls of the "Matryoshka experiment" are useful on their own, as stated, but also the whole experiment too because, when comparing each

Table 6 Results of the anti-spam filter for both scenarios in the rest of the "Matryoshka experiment"

	E1E5 + E2E4		E1E5 + E2E4 + SA2		E1E5 + E2E4 + SA2 + CS	
Performance Measurements	Scenario 1	Scenario 2	Scenario 1	Scenario 2	Scenario 1	Scenario 2
AUC	0.962637	0.960730	0.812940	0.811650	0.751110	0.733861
Discrete AUC	0.909253	0.909357	0.771565	0.768129	0.713050	0.693092
(Threshold)	(71)	(51)	(71)	(41)	(71)	(51)
Accuracy	0.909020	0.909987	0.770453	0.767319	0.692906	0.665606
Precision	0.924459	0.880378	0.715752	0.687967	0.604252	0.576021
F-score	0.908855	0.914763	0.793421	0.806730	0.720214	0.716137
Specificity	0.924737	0.866772	0.653130	0.561207	0.534846	0.439853
Sensitivity	0.893769	0.951943	0.890000	0.975051	0.891254	0.946332
TP	1750	1862	2581	2853	3098	3315
FP	143	253	1025	1294	2029	2440
TN	1757	1646	1930	1655	2333	1916
FN	208	94	319	73	378	188
% FP	7.53%	13.32%	34.69%	43.88%	46.52%	56.01%
% FN	10.62%	4.8%	11%	2.49%	10.87%	5.37%
Distance to (0, 1)	0.130190	0.141631	0.363894	0.439502	0.477696	0.562712

with the previous doll, the influence of origin-time can be checked. Considering the results of the "Matryoshka experiment" and their dolls independently, Tables 4 and 6 and Figs. 3 and 4, we can discuss these findings:

- Constraining to each doll, performance measures are quite similar between scenarios except specificity, where Scenario 1's configuration almost always outperformed Scenario 2's, and sensitivity, where the opposite happened. According to this, it may be interesting to further test a combination of both normalization approaches in order to negate each other's classification errors.
- "Users" was the characteristics that affects our system the least. As users that were tested belonged to the same corporation and, also, the emails came from their corporation's email addresses, they shared topics so our filter benefited from our preprocessing stage and the thematic categories.
- Consequences of "origin-time" changes:

  Specificity degraded at a faster pace than sensitivity, which stabilizes above 0.89.
  From doll 2 to 3, accuracy, AUC, precision and specificity lowered more than between doll 3 and 4.
  Precision and specificity raised from doll 1 to 2.
  False Negative Rate (FNR) stabilized around 11% for Scenario 1 whereas it is lower but unstable when using normalization.

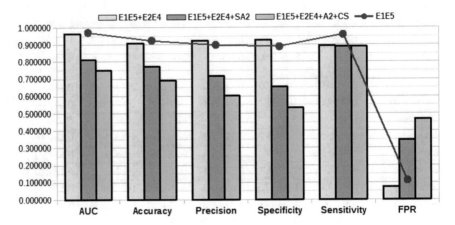

Fig. 3 Values of some performance measures for Scenario 1 in the "Matryoshka experiment"

False Positive Rate (FPR) kept raising, except between doll 1 and 2 when the opposite happened.

- Related with variations of "type of emails", image spams affected our system negatively albeit not as much as expected considering the lack of special tools and capabilities of the system to handle them.

When the analyzed emails share similar characteristics (e.g. content, topics and origins) with the dataset used for design, training, performance is excellent. But the system is still able to have good and very good performance with new received emails, whose attributes are highly dissimilar. These differences fully justify the results in each case, Tables 4 and 6. At the same time, results are quite promising and indicative of the goodness of the proposed methods.

Still, a generalization of thematic categories, of the words inside them or a mix of both might be one of the potential solutions in order to improve even more the performance of the proposed system when dealing with any email type. Another solution might be updating the filter periodically with new messages (i.e. providing it with online-learning capabilities) to both counteract the evolution of spam and ham content and topics. The latter phenomenon is known as "topic drift" [11] and it is related with the general problem of "concept drift" in ML: unexpected variations of the statistical properties of the target variable over time, which usually imply increasingly inaccurate predictions over the course of time [88, 89]. Both phenomena commonly affect all classifiers, especially when dealing with adversarial ML problems and, to a larger extent, to classifiers that made use of off-line learning, as is the case.

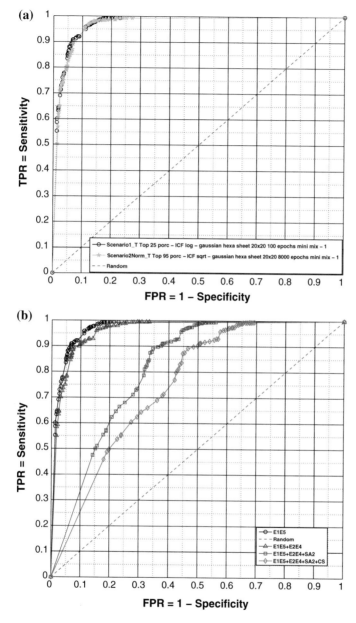

Fig. 4 **a** ROC curve of the anti-spam filter for each scenario (testing phase - "E1E5" dataset; based on [13]). **b** Comparison of the ROC curves obtained by the proposed system for Scenario 1 with several datasets

4 Conclusions

In this document a hybrid and modular anti-spam filtering system based on Kohonen Self-Organizing Maps and supervised non-neural computing has been presented. It has been proved that thematic categories can be found in spam and ham so, accordingly, both spam and ham words have been classified in the 13 categories found. The proposed system is robust to word obfuscation, commonly found in spam, and it is also independent of the need to use stemming or lemmatization algorithms, unlike other textual-based anti-spam filters.

All the studied configurations obtained good results with all metrics with E1E5, the dataset used during the design of the system. Results were identical when using the whole set, Top 100%, of keywords from each of the 13 categories or just the Top 95%, something that also brought lower runtime along. Results of other researchers [25] over the same corpus were comparable to ours, albeit their Bayesian methods made use of input vectors with up to 3000 attributes, a dimensionality several orders of magnitude greater than the one utilized by our Kohonen maps, 13.

As it happened in our previous work [13], the performance of the system got affected when newer and off-topic spams and hams were encountered. The origin-time of the emails was the characteristic related to "topic drift" that negatively impacted our system the most. Although the sensitivity of our system was robust enough against it, the rest of performance measures did not. On the other hand, the other characteristics - changes of the recipients and senders of emails, and the type of emails, including the presence of image spam - were less harmful. This performance decay is frequent in other offline-training anti-spams because topics drift along the time as both spam and spammers' techniques evolve [11].

Obtained results confirmed the high quality and goodness of our proposed system. The use of computational intelligence methodologies and hybrid schemes for designing anti-spam filtering systems were quite beneficial. Indeed, both evidences encourage us to keep researching over these topics. A big upgrading step might be the usage of some powerful hybrid neural architectures such as the Counterpropagation Network (CPN) [90, 91] or the Hybrid Unsupervised Modular Adaptive Neural Network (HUMANN) [92]. The performance decay due to "topic drift" can be solved by enhancing the generalization of the system or the thematic categories, or with regular retraining, as in online-training filters.

References

1. Subramaniam, T., Jalab, H.A., Taqa, A.Y.: Overview of textual anti-spam filtering techniques. **5**, 1869–1882
2. Cabrera-León, Y., García Báez, P., Suárez-Araujo, C.P.: Non-email spam and machine learning-based anti-spam filters: trends and some remarks. In: Moreno-Díaz, R., Pichler, F., Quesada-Arencibia, A. (eds.) Computer Aided Systems Theory–EUROCAST 2017. Lecture notes in computer science, vol. 10671, pp. 245–253. Springer, Cham
3. McAfee, ICF International: The Carbon Footprint of Email Spam Report

4. Statista: Global spam volume as percentage of total e-mail traffic from 2007–2015
5. Cabrera León, Y.: Análisis del uso de las redes neuronales artificiales en el diseño de filtros antispam: una propuesta basada en arquitecturas neuronales no supervisadas
6. Rao, J.M., Reiley, D.H.: The economics of spam. **26**, 87–110
7. Lieb, R.: Make Spammers Pay Before You Do
8. Alazab, M., Broadhurst, R.: Spam and criminal activity, 1–14
9. Calais Guerra, P.H., Guedes, D.O., Meira Jr., W., Hoepers, C., Chaves, M.H., Steding-Jessen, K.: Exploring the spam arms race to characterize spam evolution. In: Proceedings of the 7th Collaboration, Electronic Messaging, Anti-Abuse and Spam Conference
10. Pu, C., Webb, S.: Observed trends in spam construction techniques: a case study of spam evolution. In: Third Conference on Email and Anti-Spam (CEAS), pp. 1–9
11. Wang, D., Irani, D., Pu, C.: A study on evolution of email spam over fifteen years. In: 9th International Conference on Collaborative Computing: Networking, Applications and Worksharing (CollaborateCom 2013), pp. 1–10. IEEE
12. Cohen, W.W.: Enron Email Dataset
13. Cabrera-León, Y., García Báez, P., Suárez-Araujo, C.P.: Self-organizing maps in the design of anti-spam filters. a proposal based on thematic categories. In: Proceedings of the 8th IJCCI, vol. 3, pp. 21–32. NCTA, SCITEPRESS Digital Library
14. Postini, Inc: The shifting tactics of spammers: What you need to know about new email threats
15. Spammer-X, Posluns, J., Sjouwerman, S.: Inside the SPAM Cartel, 1st edn. Syngress, Elsevier
16. Erickson, D., Casado, M., McKeown, N.: The effectiveness of whitelisting: a user-study. In: Proceedings of Conference on Email and Anti-Spam, pp. 1–10
17. Kucherawy, M., Crocker, D.: RFC 6647 - Email Greylisting: An Applicability Statement for SMTP
18. Harris, E.: The Next Step in the Spam Control War: Greylisting
19. Guzella, T.S., Caminhas, W.M.: A review of machine learning approaches to Spam filtering. **36**, 10206–10222
20. Meyer, T.A., Whateley, B.: SpamBayes: Effective open-source, Bayesian based, email classification system. In: CEAS, (Citeseer)
21. Sahami, M., Dumais, S., Heckerman, D., Horvitz, E.: A bayesian approach to filtering junk E-mail
22. Lowd, D., Meek, C.: Good word attacks on statistical spam filters. In: Proceedings of the Second Conference on Email and Anti-Spam (CEAS), pp. 1–8
23. Sprengers, M., Heskes, T.T.: The effects of different bayesian poison methods on the quality of the bayesian spam filter spambayes
24. Wittel, G.L., Wu, S.F.: On attacking statistical spam filters. CEAS
25. Metsis, V., Androutsopoulos, I., Paliouras, G.: Spam filtering with naive bayes - which naive bayes? In: CEAS 2006 - Third Conference on Email and Anti-Spam
26. Drucker, H., Wu, D., Vapnik, V.N.: Support vector machines for spam categorization. **10**, 1048–1054
27. Xie, C., Ding, L., Du, X.: Anti-spam filters based on support vector machines. In: Advances in Computation and Intelligence. 4th International Symposium, ISICA 2009. Lecture notes in computer science, vol. 5821, pp. 349–357. Springer, Heidelberg
28. Chhabra, P., Wadhvani, R., Shukla, S.: Spam filtering using support vector machine. In: Special Issue of IJCCT Vol.1 Issue 2, 3, 4; 2010 for International Conference [ACCTA-2010], pp. 166–171
29. Blanco, N., Ricket, A.M., Martín-Merino, M.: Combining SVM classifiers for email anti-spam filtering. In: Sandoval, F., Prieto, A., Cabestany, J., Graña, M. (eds.) 9th International Work-Conference on Artificial Neural Networks, IWANN 2007. Computational and ambient intelligence of lecture notes in computer science, vol. 4507, pp. 903–910. Springer, Heidelberg
30. Kufandirimbwa, O., Gotora, R.: Spam detection using artificial neural networks (Perceptron Learning Rule). **1**, 22–29
31. Sculley, D., Wachman, G., Brodley, C.E.: Spam filtering using inexact string matching in explicit feature space with on-line linear classifiers. TREC

32. Chuan, Z., Xianliang, L., Mengshu, H., Xu, Z.: A LVQ-based neural network anti-spam email approach. **39**, 34–39 (6)
33. Cabrera León, Y., Acosta Padrón, O.: Spam: definition, statistics, anti-spam methods and legislation
34. Qian, F., Pathak, A., Hu, Y.C., Mao, Z.M., Xie, Y.: A case for unsupervised-learning-based spam filtering. ACM SIGMETRICS Perform. Eval. Rev. **38**, 367–368. ACM
35. Narisawa, K., Bannai, H., Hatano, K., Takeda, M.: Unsupervised spam detection based on string alienness measures. In: Discovery Science, pp. 161–172. Springer, Heidelberg
36. Uemura, T., Ikeda, D., Arimura, H.: Unsupervised spam detection by document complexity estimation. In: Discovery Science, pp. 319–331
37. Luo, X., Zincir-Heywood, N.: Comparison of a SOM based sequence analysis system and naive Bayesian classifier for spam filtering. In: Proceedings of the IEEE International Joint Conference On Neural Networks IJCNN'05, vol. 4, pp. 2571–2576
38. Vrusias, B.L., Golledge, I.: Adaptable text filters and unsupervised neural classifiers for spam detection. In: Proceedings of the International Workshop on Computational Intelligence in Security for Information Systems CISIS'08. Advances in soft computing, vol. 53, pp. 195–202. Springer, Heidelberg
39. Vrusias, B.L., Golledge, I.: Online self-organised map classifiers as text filters for spam email detection. **4**, 151–160
40. Chapelle, O., Schölkopf, B., Zien, A.: Semi-Supervised Learning, vol. 2. MIT Press
41. Gao, Y., Yan, M., Choudhary, A.: Semi supervised image spam hunter: a regularized discriminant EM approach. In: International Conference on Advanced Data Mining and Applications, pp. 152–164. Springer, Heidelberg
42. Pfahringer, B.: A semi-supervised spam mail detector, pp. 1–5
43. Santos, I., Sanz, B., Laorden, C., Brezo, F., Bringas, P.G.: (Computational Intelligence in Security for Information Systems: 4th International Conference, CISIS 2011, Held at IWANN 2011)
44. Zhou, D., Bousquet, O., Lal, T.N., Weston, J., Schölkopf, B.: Learning with local and global consistency. **16**, 321–328
45. Mason, J.: Filtering Spam with SpamAssassin (presentation)
46. Xu, J.M., Fumera, G., Roli, F., Zhou, Z.H.: Training spamassassin with active semi-supervised learning. In: Proceedings of the 6th Conference on Email and Anti-Spam (CEAS'09), pp. 1–8. (Citeseer)
47. Shunli, Z., Qingshuang, Y.: Personal spam filter by semi-supervised learning. In: Proceedings of the Third International Symposium on Com Puter Science and Computational Technology (ISCSCT'10), pp. 171–174
48. Zhou, D., Burges, C.J.C., Tao, T.: Transductive link spam detection. In: Proceedings of the 3rd International Workshop on Adversarial Information Retrieval on the Web, pp. 21–28
49. Mojdeh, M., Cormack, G.V.: Semi-supervised Spam Filtering: Does it Work? In: Proceedings of the 31st Annual International ACM SIGIR Conference on Research and Development in Information Retrieval, (ACM) 745–746
50. Resnick, P. (ed.) : RFC 5322 - Internet Message Format
51. Cormack, G.V., Mojdeh, M.: Machine learning for information retrieval: TREC 2009 web, relevance feedback and legal tracks. In: The Eighteenth Text REtrieval Conference Proceedings (TREC 2009), pp. 1–9
52. Malathi, R.: Email spam filter using supervised learning with bayesian neural network. **1**, 89–100
53. Pitsillidis, A., Levchenko, K., Kreibich, C., Kanich, C., Voelker, G.M., Paxson, V., Weaver, N., Savage, S.: Botnet judo: Fighting spam with itself. In: Symposium on Network and Distributed System Security (NDSS), pp. 1–19
54. Kolcz, A., Chowdhury, A., Alspector, J.: The impact of feature selection on signature-driven spam detection. In: Proceedings of the 1st Conference on Email and Anti-Spam (CEAS-2004), pp. 1–8

55. The Apache SpamAssassin Project: SpamAssassin v3.3.x: Tests Performed to Determine Spaminess and Haminess of a Message
56. Yerazunis, W., Kato, M., Kori, M., Shibata, H., Hackenberg, K.: Keeping the Good Stuff In: Confidential Information Firewalling with the CRM114 Spam Filter & Text Classifier, pp. 1–18
57. Graham-Cumming, J.: SpamOrHam, pp. 22–24
58. Feroze, M.A., Baig, Z.A., Johnstone, M.N.: A two-tiered user feedback-based approach for spam detection. In: Becker Westphall, C., Borcoci, E., Manoharan, S. (eds.) ICSNC 2015: The Tenth International Conference on Systems and Networks Communications, pp. 12–17. Curran Associates, Inc, Spain, 15–20 November 2015
59. Bruce, J.: Grey Mail: The New Email Nuisance To Hit Your Inbox
60. Ramachandran, A., Feamster, N.: Understanding the network-level behavior of spammers. ACM SIGCOMM Comput. Commun. Rev. **36**, 291–302
61. Fumera, G., Pillai, I., Roli, F.: Spam filtering based on the analysis of text information embedded into images. **7**, 2699–2720
62. Borovicka, T., Jirina Jr., M., Kordik, P., Jirina, M.: Selecting Representative Data Sets. In: Karahoca, A. (ed.) Advances in Data Mining Knowledge Discovery and Applications. (InTech)
63. Skillicorn, D.: Other Versions of the Enron Data (preprocessed)
64. Styler, W.: The EnronSent Corpus
65. Bekkerman, R., McCallum, A., Huang, G.: Automatic categorization of email into folders: Benchmark experiments on Enron and SRI corpora
66. The Apache SpamAssassin Project: Index of the SpamAssassin's Public Corpus
67. Guenter, B.: SPAM Archive: Email spam received yearly, since early 1998
68. CSMining Group: CSDMC2010 SPAM corpus
69. Hovold, J.: Naive Bayes Spam filtering using word-position-based attributes. In: CEAS
70. Zhang, Y.: Lecture for Chap. 2 - Data Preprocessing (course presentation)
71. Porter, M.F.: An algorithm for suffix stripping. **14**, 130–137
72. Freschi, V., Seraghiti, A., Bogliolo, A.: Filtering obfuscated email spam by means of phonetic string matching. Advances in Information Retrieval, pp. 505–509. Springer, Berlin
73. Liu, C., Stamm, S.: Fighting Unicode-obfuscated spam. In: Proceedings of the Anti-Phishing Working Groups 2nd Annual eCrime Researchers Summit, pp. 45–59. ACM
74. Wang, D., Zhang, H.: Inverse-category-frequency based supervised term weighting schemes for text categorization. **29**, 209–225
75. Lertnattee, V., Theeramunkong, T.: Analysis of inverse class frequency in centroid-based text classification. In: IEEE International Symposium on Communications and Information Technology (ISCIT 2004). vol. 2, pp. 1171–1176. IEEE
76. Zeimpekis, D., Kontopoulou, E.M., Gallopoulos, E.: Text to Matrix Generator (TMG)
77. Kohonen, T.: Self-Organizing Maps, 3rd edn. Springer, New York
78. Rojas, R.: Kohonen networks. In: Neural Networks: A Systematic Introduction, pp. 391–412. Springer, Berlin
79. Haykin, S.S.: Neural Networks. A Comprehensive Foundation, 2nd edn. Prentice-Hall International
80. Tan, H.S., George, S.E.: Investigating learning parameters in a standard 2-D SOM model to select good maps and avoid poor ones. In: Australasian Joint Conference on Artificial Intelligence, pp. 425–437, Springer, Berlin
81. Kohonen, T.: Essentials of the self-organizing map. **37**, 52–65
82. Fawcett, T.: ROC graphs: Notes and practical considerations for researchers. **31**, 1–38
83. Metz, C.E.: Basic principles of ROC analysis. In: Seminars in Nuclear Medicine. vol. 8, pp. 283–298. Elsevier
84. Slaby, A.: ROC analysis with Matlab. In: 29th International Conference On Information Technology Interfaces, 2007. ITI, pp. 191–196. IEEE
85. Vesanto, J., Himberg, J., Alhoniemi, E., Parhankangas, J.: SOM Toolbox for Matlab 5
86. MathWorks: Parallel Computing Toolbox for Matlab R2014a - User's Guide
87. Holden, S.: Spam Filtering II: Comparison of a number of Bayesian anti-spam filters over different email corpora

88. Gama, J., Žliobaitė, I., Bifet, A., Pechenizkiy, M., Bouchachia, A.: A survey on concept drift adaptation. **46**, 1–37

89. Žliobaitė, I., Pechenizkiy, M., Gama, J.: An overview of concept drift applications. In Japkowicz, N., Stefanowski, J., eds.: Big Data Analysis: New Algorithms for a New Society of Studies in Big Data, vol. 16, pp. 91–114. Springer International Publishing

90. Freeman, J.A., Skapura, D.M.: Neural Networks: Algorithms, Applications, and Programming Techniques. Computation and neural systems series. Addison-Wesley

91. Hecht-Nielsen, R.: Counterpropagation networks. **26**, 4979–4984

92. Suárez Araujo, C.P., García Báez, P., Hernández Trujillo, Y.: Neural computation methods in the determination of fungicides. In: Fungicides. Odile carisse edn. INTECH Open Access Publisher

A Noise Compensation Mechanism for an RGNG-Based Grid Cell Model

Jochen Kerdels and Gabriele Peters

Abstract Grid cells of the entorhinal cortex provide a rare view on the deep stages of information processing in the mammalian brain. Complementary to earlier grid cell models that interpret the behavior of grid cells as specialized parts within a system for navigation and orientation we developed a grid cell model that facilitates an abstract computational perspective on the behavior of these cells. Recently, we investigated the ability of our model to cope with increasing levels of input signal noise as it would be expected to occur in natural neurobiological circuits. Here we investigate these results further and introduce a new noise compensation mechanism to our model that normalizes the output activity of simulated grid cells irrespective of whether or not input noise is present. We present results from an extended series of simulation runs to characterize the involved parameters.

1 Introduction

The parahippocampal-hippocampal region takes part in the deep stages of information processing in the mammalian brain. It is generally assumed to play a vital role in the formation of declarative, in particular episodic, memory as well as navigation and orientation. The discovery of *grid cells*, whose activity correlates with the animal's location in a regular pattern, facilitates a rare view on the neuronal processing that occurs in this region of the brain [7, 9]. Complementary to earlier computational models of grid cells that interpret the behavior of grid cells as specialized parts within a system for navigation and orientation [1, 5, 8, 18, 19, 23] we introduced a new grid cell model that views the behavior of grid cells as just one instance of a general information processing scheme [10, 11]. The model relies on principles of self-organisation facilitated by the recursive growing neural gas (RGNG) algorithm.

J. Kerdels (✉) · G. Peters
University of Hagen, Universitätsstrasse 1, 58097 Hagen, Germany
e-mail: jochen.kerdels@fernuni-hagen.de

© Springer Nature Switzerland AG 2019
J. J. Merelo et al. (eds.), *Computational Intelligence*,
Studies in Computational Intelligence 792,
https://doi.org/10.1007/978-3-319-99283-9_13

We could demonstrate [10, 12] that our model can not only describe the basic properties of grid cell activity but also recently observed phenomena like *grid rescaling* [2, 3] as well as grid-like activity in primates that correlates with eye movements [14] instead of environmental location.

In addition, we recently investigated the ability of our model to cope with increasing levels of noise in it's input signal as it would be expected to occur in natural neurobiological circuits [13]. Even with noise levels up to 90% of the input signal amplitude the model was able to establish the expected activity patterns. However, with increasing levels of noise the average maximum activity of the model's output dropped by two orders of magnitude. Here we investigate this aspect further and present a noise compensation mechanism that we integrated in our grid cell model to normalize the average maximum output activity irrespective of whether or not input noise is present. The following section provides a short summary of our RGNG-based grid cell model, and Sect. 3 revisits our previous results [13] and analyzes the effects of input noise further. Subsequently, Sect. 4 introduces our proposed noise compensation mechanism and Sect. 5 presents the results obtained using this mechanism over various parameter ranges. Finally, Sect. 6 draws conclusions and outlines future work.

2   RGNG-Based Grid Cell Model

The RGNG-based grid cell model is a neuron-centric model in which neurons act in their "own interest" while being in competition with each other. A biological neuron receives thousands of inputs from other neurons and the entirety of these inputs and their possible values constitute the neuron's input space. We hypothesize that grid cells form a simple representation of their input space by learning a limited number of input patterns or *prototypes* that reflect the input space structure. Simultaneously, the competition among neurons in a local group of grid cells ensures that the simple representations learned by the individual cells are pairwise distinct and interleave in such a way that a complex representation of the input space emerges that is distributed over the entire group of neurons. We model this behavior by a two layer recursive growing neural gas that describes both the learning of prototypes within individual cells as well as the simultaneous competition among the cells in the group. Both processes are based on the same principles of self-organization utilizing a form of competitive Hebbian learning. For a formal description and an in-depth characterization of the model we refer to previous work [10, 13].

Here we focus on the operation of individual grid cells in the model. Their behavior is equivalent to that of a regular growing neural gas (GNG) as it was introduced by Fritzke [6]. A GNG is a network of units that is able to learn the topology of it's input space. Each unit is associated with a *reference vector* or *prototype* that represents a local region of the input space. The neighborhood relations of these local regions are reflected by the GNG network topology. In contrast to the original notion of GNG units as individual neurons our model interprets the GNG units as different

Fig. 1 Geometric
interpretation of ratio r,
which is used as a basis for
an approximation of the
modelled grid cell's activity.
Extracted from [13]

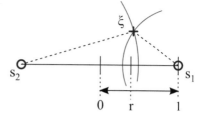

dendritic subsections of a *single* grid cell. Thus, we assume that competitive Hebbian learning can occur within the dendritic tree of a single neuron allowing the neuron to respond to multiple, different input patterns and to facilitate the formation of a simple prototype-based representation of the neuron's input space.

The basic learning mechanism selects for every input ξ the best and second best matching units (BMUs) s_1 and s_2 whose prototypes $s_1.w$ and $s_2.w$ are closest to the input ξ according to a distance function D. The GNG network is then updated by creating (or refreshing) an edge between s_1 and s_2 and the prototype of the BMU s_1 as well as the prototypes of all units connected to s_1 are adapted towards the input ξ. In addition, the output activity a_u of the modelled grid cell u in response to the input ξ is determined based on the relative distances of ξ towards $s_1.w$ and $s_2.w$:

$$a_u := e^{-\frac{(1-r)^2}{2\sigma^2}},$$

with $\sigma = 0.2$ and ratio r:

$$r := \frac{D(s_2.w, \xi) - D(s_1.w, \xi)}{D(s_1.w, s_2.w)},$$

using a distance function D. Figure 1 provides a geometric interpretation of the ratio r. If input ξ is close to BMU s_1 in relation to s_2, ratio r becomes 1. If on the other hand input ξ has about the same distance to s_1 as it has to s_2, ratio r becomes 0.

Based on this measure of activity it becomes possible to correlate the simulated grid cell's activity with further variables, e.g., the recorded location of an animal (Fig. 2a) in a typical experimental setup to study grid cells. Figure 2b shows such a correlation as a *firing rate map*, which is constructed according to the procedures described by Sargolini et al. [21] but using a 5 × 5 boxcar filter for smoothing instead of a Gaussian kernel as introduced by Stensola et al. [22]. This conforms to the de facto standard of rate map construction in the grid cell literature. Each rate map integrates position and activity data over 30,000 time steps corresponding to a single experimental trial with a duration of 10 min recorded at 50 Hz.

(a)   **(b)**

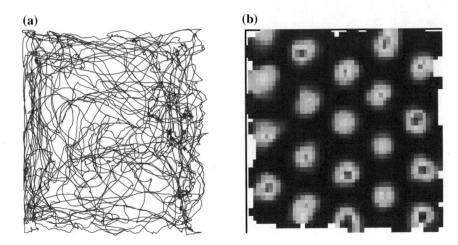

Fig. 2 **a** Example trace of rat movement within a rectangular, 1 m × 1 m environment recorded for a duration of 10 min. Movement data published by Sargolini et al. [21]. **b** Color-coded firing rate map of a simulated grid cell ranging from dark blue (no activity) to red (maximum activity). Extracted from [13]

3   Noise Resilience Revisited

Typical neurons in the the parahippocampal-hippocampal region of the brain have peak firing rates that range between 1 and 50 Hz [4, 9, 16, 21]. Some proportion of this firing rate is due to spontaneous activity of the corresponding neuron. According to Koch [15] this random activity can occur about once per second, i.e., at 1 Hz. Hence, the proportion of noise in a normalized firing rate resulting from this spontaneous firing can be expected to lie between 1.0 and 0.02 given the peak firing rates stated above.

We recently investigated the ability of the RGNG-based grid cell model to cope with noise in it's input signal that is caused by this spontaneous neural activity [13]. Since the model uses a vector of normalized neuronal activity as it's input signal, the proportion of noise in each input dimension depends on the assumed peak firing rate of the corresponding input neuron. Unfortunately, there is no empirical data on the distribution of peak firing rates in the input signal of biological grid cells. Thus, we assumed a uniform distribution and tested the model with increasing levels ξ_n of noise reflecting assumed minimal peak firing rates. For example, a maximum noise level of $\xi_n = 0.1$ corresponds to a minimal peak firing rate of 10 Hz, and a level of $\xi_n = 0.5$ corresponds to a minimal peak firing rate of 2 Hz in the input neurons.

The input signal used in the experiments was constructed by assuming that the animal location is encoded by two ensembles of input neurons that operate as one-dimensional ring attractor networks. In these networks a stable "bump" of activity encodes a linear position in a given direction. If the animal moves in that direction, the bump of activity is moved accordingly updating the encoded position. Similar types of input signals for grid cell models were proposed in the literature by, e.g., Mhatre et al. [17] as well as Pilly and Grossberg [20]. Formally, the input signal $\xi := (v^x, v^y)$ was implemented as two concatenated 50-dimensional vectors v^x and v^y. To generate an input signal a position $(x, y) \in [0, 1] \times [0, 1]$ was read from traces (Fig. 2a) of recorded rat movements that were published by Sargolini et al. [21] and mapped onto the corresponding elements of v^x and v^y as follows:

$$v_i^x := \max\left(1 - \left|\frac{i - \lfloor dx + 0.5 \rfloor}{s}\right|, \ 1 - \left|\frac{d + i - \lfloor dx + 0.5 \rfloor}{s}\right|, \ 0\right),$$

$$v_i^y := \max\left(1 - \left|\frac{i - \lfloor dy + 0.5 \rfloor}{s}\right|, \ 1 - \left|\frac{d + i - \lfloor dy + 0.5 \rfloor}{s}\right|, \ 0\right),$$

$$\forall i \in \{0 \ldots d - 1\},$$

with $d = 50$ and $s = 8$. The parameter s controls the slope of the activity bump with higher values of s resulting in a broader bump. Each input vector $\xi := (\tilde{v}^x, \tilde{v}^y)$ was then augmented by noise as follows:

$$\tilde{v}_i^x := \max\left[\min\left[v_i^x + \xi_n \left(2 U_{\mathrm{rnd}} - 1\right), 1 \right], 0 \right],$$

$$\tilde{v}_i^y := \max\left[\min\left[v_i^y + \xi_n \left(2 U_{\mathrm{rnd}} - 1\right), 1 \right], 0 \right],$$

$$\forall i \in \{0 \ldots d - 1\},$$

with maximum noise level ξ_n and uniform random values $U_{\mathrm{rnd}} \in [0, 1]$.

Using this type of input we ran a series of simulation runs with increasing levels ξ_n of noise. Each run simulated a group of 100 grid cells with 20 dendritic subsections per cell using a fixed set of model parameters.[1] Figure 3 summarizes the results of these simulations. Each column corresponds to a single simulation run and shows an exemplary rate map of a grid cell chosen randomly from the 100 simulated cells (top row), the average maximum activity (MX) and the average minimum activity (MN) present in the rate maps of all simulated grid cells (below the rate map), the distribution of *gridness* scores[2] (middle row), and an activity function plot that indicates which values of ratio r corresponds to the respective average maximum activity (bottom row). The exemplary rate maps as well as the gridness score distributions show that the RGNG-based grid cell model is able to sustain the expected grid-like activity patterns despite increasing levels of noise in it's input signal reflecting the robustness of the underlying principle of self-organisation. However, with increasing

---

[1]For a detailed description and motivation of all parameters we refer to [13].

[2]The *gridness score* ($[-2, 2]$) is a measure of how grid-like the firing pattern of a neuron is. Neurons with gridness scores greater 0.4 are commonly identified as grid cell.

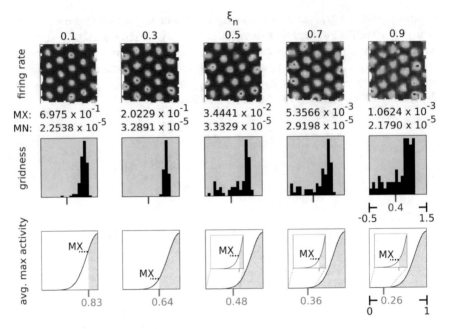

Fig. 3 Artificial rate maps (top row), gridness distributions (middle row), and activity function plots (bottom row) of simulation runs with varying levels ξ_n of noise (columns) added to the inputs. All simulation runs used a fixed set of parameters [13] and processed location inputs derived from movement data published by Sargolini et al. [21]. Each artificial rate map was chosen randomly from the particular set of rate maps. Average maximum activity (MX) and average minimum activity (MN) across all rate maps of a particular simulation given above gridness distributions. Gridness threshold of 0.4 indicated by red marks. Values of ratio r at average maximum activity (MX) given in blue. Insets show magnified regions of the activity function where MX values are low. Extracted from [13]

levels of noise the average maximum output activity of the simulated grid cells drops by two orders of magnitude, though the difference between average maximum and average minimum output activity is still at least two orders of magnitude for any tested noise level ξ_n.

The output activity a_u of a simulated grid cell u depends directly on the ratio r, which characterizes the relative distances between an input ξ and the prototypes of the best and second best matching units s_1 and s_2. Only if the input ξ is much closer to the prototype $s_1.w$ than $s_2.w$, the activity will approach a value of 1. Otherwise, if the distances between ξ and $s_1.w$ as well as ξ and $s_2.w$ are rather similar, the activity will be close to 0. This approximation of grid cell activity assumes that the input signals to the model originate from a sufficiently low-dimensional manifold in the high-dimensional input space. Only if this condition is met it is likely that some of the inputs will match the particular best matching prototype closely resulting in a strong activation of the corresponding grid cell. Adding noise to inputs from such a lower-dimensional manifold moves the inputs away from the manifold in random directions. As a consequence, each of the grid cell's prototypes becomes surrounded

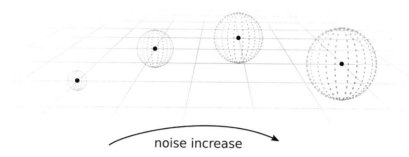

noise increase

Fig. 4 Illustration of high-dimensional "dead zones" (blue-dotted spheres) surrounding prototypes (black dots) that lie on a lower-dimensional manifold. The "dead zones" grow with increasing levels of noise

with a kind of "dead zone" for which it is unlikely that any input will originate from it (Fig. 4). This rather unintuitive property of randomness in high-dimensional space becomes more tangible if one considers the distribution of random points that lie within the unit sphere. For a point to lie close to the center of this sphere all of it's coordinates must be close to zero. If the absolute value of only one coordinate is large, i.e., close to one or negative one, the point will lie close to the surface of the sphere. Thus, with increasing dimension it becomes more and more unlikely for a random point that the absolute values of all of it's coordinates will be low. Likewise, it is equally unlikely that the high-dimensional noise added to an input will not move the input away from it's low dimensional manifold, and hence move it away from the grid cell's prototypes.

4 Noise Compensation

The results summarized above suggest that real grid cells should be able to process inputs with low peak firing rates, that they may show a similar reduction in activity when the proportion of noise in their inputs is high, and that they should not suffer a degradation of their firing field geometry in the presence of noise. To the best of our knowledge no experiments were conducted yet that investigated the behavior of grid cells in response to (controlled) noise in their input signals. Since grid cells do show a wide range of peak firing rates [4, 9, 16, 21], possible variations of noise in their input signals may provide an explanation for these observations.

However, grid cells may also employ strategies to directly compensate for noise, e.g., by changing electrotonic properties of their cell membranes [15]. To account for this possibility we added a noise compensation mechanism to our model that normalizes ratio r with respect to the level of noise. Like the RGNG-based grid cell model itself the implemented noise compensation is a computational mechanism that is an abstract representation of this potential ability of grid cells and does not relate to any specific neurobiological implementation. As described above, the addition of noise to the inputs of the model results in "dead zones" around the prototypes of each grid cell that effectively limit the maximum value of ratio r. To normalize r without apriori knowledge about the level of noise that is present, it is necessary to identify and track the border region of these "dead zones" around each prototype. For that purpose a buffer b_N of size N was added to each unit s of a grid cell's GNG containing the N largest values of ratio r encountered so far while s was the BMU. In addition, every entry of a buffer b_N has an *age* associated with it that is increased every time the simulated grid cell processes an input and the corresponding unit s is selected as BMU. Once the age of a buffer entry reaches a given age threshold A_{max}, the value is evicted from the buffer. This way, changes in the level of input noise that influence the size of the "dead zones" can be tracked. Using this additional information a normalized ratio $\hat{r}$ can then be defined as:

$$\hat{r} := \max\left[\, \min\left[\, \frac{r}{\widetilde{b_N}},\, 1 \right],\, 0 \right],$$

with $\widetilde{b_N}$ the median of all populated entries in buffer b_N.

5   Results

We characterized the normalized ratio $\hat{r}$ by conducting a series of simulation runs that covered combinations of varying age thresholds $A_{max} \in \{50, 250, 750, 1500, 3000\}$, varying buffer sizes $N \in \{5, 11, 21, 41, 81\}$, and varying levels of noise $\xi_n \in \{0.1, 0.3, 0.7, 0.9\}$. In addition, we reran corresponding simulations without normalizing ratio r to ensure that the input signal (including the random noise) was identical to the other simulations. All simulation runs used the same set of parameters as the previous simulations reported above and used the same location inputs derived from movement data published by Sargolini et al. [21]. However, the section of movement data that was analyzed here stems from a different experimental trial due to technical constraints of the simulation environment. This change resulted in slight variations in the gridness score distributions of the rerun results (compare Figs. 3 and 8).

Figures 5, 6, 7, and 8 summarize the simulation results of all simulation runs. Figure 5 shows one exemplary rate map for each simulation run. The rate maps were chosen randomly from one of the 100 grid cells per simulation. Figure 6 shows corresponding activity histograms that show the distribution of activity values present in all rate maps (100) within each simulation. Figure 7 provides for each simulation run a histogram of gridness scores calculated based on the corresponding firing rate maps. Lastly, Fig. 8 shows the results of the simulation reruns with non-normalized ratio r for comparison.

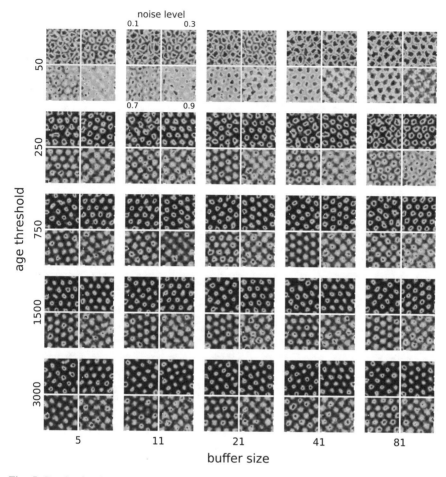

Fig. 5 Randomly chosen exemplary rate maps of simulation runs with varying age thresholds $A_{max} \in \{50, 250, 750, 1500, 3000\}$ (rows), varying buffer sizes $N \in \{5, 11, 21, 41, 81\}$ (columns), and varying levels of noise $\xi_n \in \{0.1, 0.3, 0.7, 0.9\}$ (quadrants of 2×2 blocks). All simulation runs used the same set of parameters as the simulation runs underlying the data presented in Fig. 3 and processed location inputs derived from movement data published by Sargolini et al. [21]

The firing rate maps shown in Fig. 5 reflect the influence of the two parameters A_{max} and N on the normalized ratio $\hat{r}$ in an intuitive way. The age threshold A_{max} determines the duration for which a recently encountered large value of non-normalized ratio r is kept in the buffer and used to determine the "dead zone" boundary surrounding the corresponding prototype. With 20 dendritic subsections, i.e., prototypes per grid cell and 30,000 time steps per 10 min trial, each prototype will be BMU for about 1500 time steps per trial or 2.5 time steps per second *on average*. Thus, with an age threshold of $A_{max} = 50$ recently encountered large values of r are

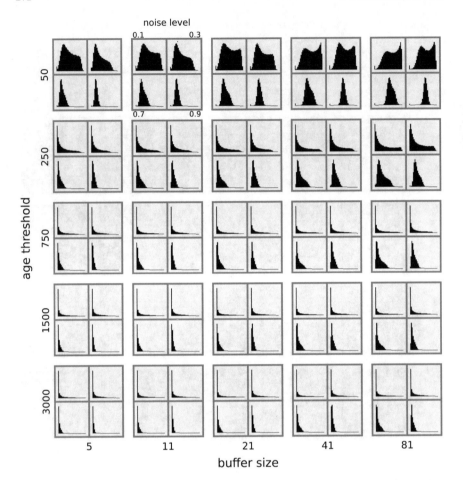

Fig. 6 Distributions of the activity values present in the firing rate maps of individual simulation runs. The shown data corresponds to the simulation runs shown in Fig. 5. The histograms range from 0 to 1

kept for about 20 seconds in the buffer before they are evicted. Larger values of A_{max} prolong this time:

$$
\begin{aligned}
A_{max} &= 50 \to 20\,\text{s} \quad\quad\quad\quad\;, \\
A_{max} &= 250 \to 100\,\text{s} \quad\quad\quad\quad\;, \\
A_{max} &= 750 \to 300\,\text{s}\ (5\,\text{min})\,, \\
A_{max} &= 1500 \to 600\,\text{s}\ (10\,\text{min})\,, \\
A_{max} &= 3000 \to 1200\,\text{s}\ (20\,\text{min})\,.
\end{aligned}
$$

Similarly, the buffer size N does not only define how many values of non-normalized ratio r are used to estimate a "dead zone" boundary, but it also implies how much time is needed to fill the buffer on average:

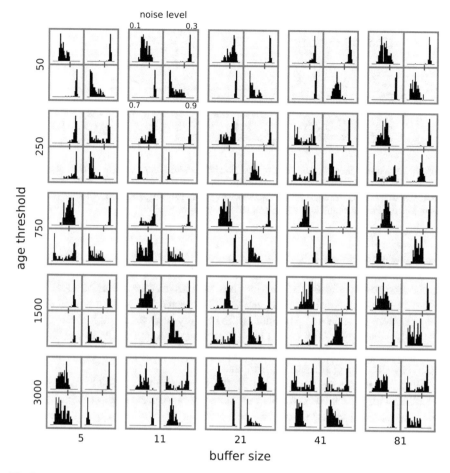

Fig. 7 Distributions of gridness scores calculated from the firing rate maps of individual simulation runs. The shown data corresponds to the simulation runs shown in Fig. 5. The histograms range from -1.5 to $+1.5$. Gridness score threshold of 0.4 indicated by red mark

$$N = 5 \rightarrow 2.0\,\text{s},$$
$$N = 11 \rightarrow 4.4\,\text{s},$$
$$N = 21 \rightarrow 8.4\,\text{s},$$
$$N = 41 \rightarrow 16.4\,\text{s},$$
$$N = 81 \rightarrow 32.4\,\text{s}.$$

In cases where the age threshold $A_{\max}$ is small and the buffer size N is large, the encountered values of ratio r are evicted faster than the buffer can be filled. As a consequence effectively all recent values of r are used to estimate the respective "dead zone" boundary. This effect can be observed in the first row of Fig. 5. With increasing buffer size the estimated "dead zone" boundary of each prototype moves towards the median of all encountered values of ratio r resulting in enlarged firing

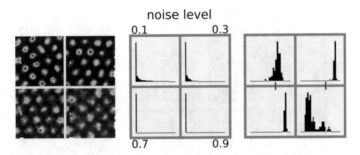

Fig. 8 Randomly chosen exemplary rate maps (left), distributions of activity values (middle), and distributions of gridness scores (right) from simulation runs with varying levels of noise $\xi_n \in \{0.1, 0.3, 0.7, 0.9\}$ (quadrants of 2×2 blocks) using **no noise compensation**. All other parameters were identical to the simulation runs presented in Figs. 5, 6, and 7

fields that are separated by only thin regions of lower activity. This overestimation of "dead zone" sizes is also reflected by the corresponding activity distributions shown in the first row of Fig. 6. The distributions are either unimodal or bimodal instead of being long-tail distributions as one would expected in case of a typical grid cell firing pattern (compare Fig. 8, middle). Especially in cases of high levels of noise ($\xi_n \geq 0.7$) and larger buffer sizes ($N \geq 21$) the minimum activity of the simulated cells increases to levels where the cell exhibits an unnatural continuous base level activity regardless of it's particular input.

With increasing age threshold A_{max} the time window in which the buffer can be filled with values of ratio r that are actually among the highest values that are likely to occur given the particular noise level increases. Consequently, the quality of the estimation of the particular "dead zone" boundary increases as well, which is reflected by decreasing sizes of firing fields (Fig. 5) and more long-tail distributions of activity values (Fig. 6). The activity distributions reveal that independent of a particular combination of age threshold and buffer size the tails of the distributions decrease with increasing levels of noise. Thus, the maximum activity of a simulated cell with normalized ratio $\hat{r}$ still decreases with noise, but the magnitude of this decrease is significantly lower compared to the decrease when the activity is based on a non-normalized ratio r (Figs. 3 and 8). In fact, the decrease in maximum activity due to noise using the normalized ratio $\hat{r}$ now matches the observed variability of peak firing rates in biological grid cells.

The influence of normalizing ratio r on the resulting gridness scores of the simulated grid cells is inconclusive (Fig. 7). In general, there appears to be a pattern where the gridness scores increase with low to medium levels of noise before they decrease again when the levels of noise increase further. Small amounts of noise may prevent the learning algorithm from getting stuck in local minima and thus may result in hexagonal firing patterns that are more regular. However, some simulation runs (e.g., $A_{max} = 750$, $N = 5$) deviate from this pattern and exhibit rather broad distributions of gridness score values at a noise level of $\xi_n = 0.7$, but a clear correlation with the parameters age threshold or buffer size is not recognizable. One likely explanation for the observed variability in gridness score distributions is the alignment of grid cell firing patterns in the RGNG-based grid cell model. In common with biological grid cells, a group of grid cells modeled by the RGNG-based model align the rotation and spacing of their firing patterns in a self-organizing manner. Depending on initial conditions and the input presented so far this alignment process can get stuck in a stable configuration that introduces irregularities in the otherwise hexagonal firing patterns of the simulated grid cells (e.g., $A_{max} = 1500$, $N = 41$, $\xi_n = 0.1$ in Fig. 5). As similar "defects" in the firing patterns of biological grid cells were experimentally observed and documented by Krupic et al. [16], it is inconclusive if this property of the RGNG-based model is a "bug" or a "feature". In the grid cell literature it is rather common to exclude the firing rate maps of cells with gridness scores lower than a given threshold (around 0.3–0.4) from publication resulting in a lack of knowledge about cells that are *almost* grid cells given their firing rate maps.

6 Conclusions

We presented a noise compensation mechanism for our RGNG-based grid cell model based on a dynamic normalization of the core measure used to derive the activity of a simulated grid cell. The normalization is controlled by two parameters, age threshold A_{max} and buffer size N, whose influence we characterized by an extended series of simulation runs. The results indicate that the age threshold parameter is more critical than the buffer size parameter. It should have a value that translates to a sliding time window of at least 5 min (in the context of the underlying experimental setup) in which unnormalized values are collected and used to dynamically estimate the normalization factor needed for the currently present noise level.

The proposed normalization procedure reduces the previously observed [13] drop in the output activity of simulated grid cells in the presence of high levels of input noise by one order of magnitude. The remaining reduction in output activity matches the observed variability of peak firing rates in biological grid cells. If there is a possible connection between the peak firing rates of grid cells and the level of noise present in their input signal is an open question and remains to be investigated.

References

1. Barry, C., Burgess, N.: Neural mechanisms of self-location. Curr. Biol. **24**(8), R330–R339 (2014)
2. Barry, C., Hayman, R., Burgess, N., Jeffery, K.J.: Experience-dependent rescaling of entorhinal grids. Nat. Neurosci. **10**(6), 682–684 (2007)
3. Barry, C., Ginzberg, L.L., OKeefe, J., Burgess, N.: Grid cell firing patterns signal environmental novelty by expansion. Proc. Natl. Acad. Sci. **109**(43), 17687–17692 (2012)
4. Boccara, C.N., Sargolini, F., Thoresen, V.H., Solstad, T., Witter, M.P., Moser, E.I., Moser, M.B.: Grid cells in pre-and parasubiculum. Nat. Neurosci. **13**(8), 987–994 (2010)

5. Burak, Y.: Spatial coding and attractor dynamics of grid cells in the entorhinal cortex. Curr. Opin. Neurobiol. **25**, 169–175 (2014), theoretical and computational neuroscience
6. Fritzke, B.: A growing neural gas network learns topologies. Advances in Neural Information Processing Systems, vol. 7, pp. 625–632. MIT Press, Cambridge (1995)
7. Fyhn, M., Molden, S., Witter, M.P., Moser, E.I., Moser, M.B.: Spatial representation in the entorhinal cortex. Science **305**(5688), 1258–1264 (2004)
8. Giocomo, L., Moser, M.B., Moser, E.: Computational models of grid cells. Neuron **71**(4), 589–603 (2011)
9. Hafting, T., Fyhn, M., Molden, S., Moser, M.B., Moser, E.I.: Microstructure of a spatial map in the entorhinal cortex. Nature **436**(7052), 801–806 (2005)
10. Kerdels, J.: A computational model of grid cells based on a recursive growing neural gas. Ph.D. thesis, FernUniversität in Hagen, Hagen (2016)
11. Kerdels, J., Peters, G.: A new view on grid cells beyond the cognitive map hypothesis. In: 8th Conference on Artificial General Intelligence (AGI 2015) (2015)
12. Kerdels, J., Peters, G.: Modelling the grid-like encoding of visual space in primates. In: Proceedings of the 8th International Joint Conference on Computational Intelligence, IJCCI 2016, vol. 3. NCTA, Porto, Portugal, 9–11 November 2016, pp. 42–49 (2016)
13. Kerdels, J., Peters, G.: Noise resilience of an RGNG-based grid cell model. In: Proceedings of the 8th International Joint Conference on Computational Intelligence, IJCCI 2016, vol. 3. NCTA, Porto, Portugal, 9–11 November 2016, pp. 33–41 (2016)
14. Killian, N.J., Jutras, M.J., Buffalo, E.A.: A map of visual space in the primate entorhinal cortex. Nature **491**(7426), 761–764 (2012)
15. Koch, C.: Biophysics of Computation: Information Processing in Single Neurons. Computational Neuroscience Series. Oxford University Press, Oxford (2004). http://books.google.de/books?id=J9juLkO7p80C
16. Krupic, J., Burgess, N., OKeefe, J.: Neural representations of location composed of spatially periodic bands. Science **337**(6096), 853–857 (2012)
17. Mhatre, H., Gorchetchnikov, A., Grossberg, S.: Grid cell hexagonal patterns formed by fast self-organized learning within entorhinal cortex (published online 2010). Hippocampus **22**(2), 320–334 (2010)
18. Moser, E.I., Moser, M.B.: A metric for space. Hippocampus **18**(12), 1142–1156 (2008)
19. Moser, E.I., Moser, M.B., Roudi, Y.: Network mechanisms of grid cells. Philos. Trans. R. Soc. B: Biol. Sci. **369**(1635) (2014)
20. Pilly, P.K., Grossberg, S.: How do spatial learning and memory occur in the brain? coordinated learning of entorhinal grid cells and hippocampal place cells. J. Cogn. Neurosci. **24**, 1031–1054 (2012)
21. Sargolini, F., Fyhn, M., Hafting, T., McNaughton, B.L., Witter, M.P., Moser, M.B., Moser, E.I.: Conjunctive representation of position, direction, and velocity in entorhinal cortex. Science **312**(5774), 758–762 (2006)
22. Stensola, H., Stensola, T., Solstad, T., Froland, K., Moser, M.B., Moser, E.I.: The entorhinal grid map is discretized. Nature **492**(7427), 72–78 (2012)
23. Welinder, P.E., Burak, Y., Fiete, I.R.: Grid cells: the position code, neural network models of activity, and the problem of learning. Hippocampus **18**(12), 1283–1300 (2008)

A Possible Encoding of 3D Visual Space in Primates

Jochen Kerdels and Gabriele Peters

Abstract Killian et al. were the first to report on entorhinal neurons in primates that show grid-like firing patterns in response to eye movements. We recently demonstrated that these *visual grid cells* can be modeled with our RGNG-based grid cell model. Here we revisit our previous approach and develop a more comprehensive encoding of the presumed input signal that incorporates binocular movement information and fixation points that originate from a three-dimensional environment. The resulting *volumetric firing rate maps* exhibit a peculiar structure of regularly spaced *activity columns* and provide the first model-based prediction on the expected activity patterns of *visual grid cells* in primates if their activity were to be correlated with fixation points from a three-dimensional environment.

1 Introduction

Investigating the information processing that occurs on a neuronal level in deep stages of the mammalian brain is a challenging task. Neurons in these regions of the brain commonly process information that is highly aggregated and difficult to correlate with external variables that are observable in an experimental setting. A rare exception in this regard are so-called *grid cells* [8, 10]. The activity of these neurons correlates in a regular and periodic fashion with the organism's location in it's environment and thus facilitates experimental inquiry. Since their discovery in the entorhinal cortex of rats cells with grid-like activity patterns were found in several mammalian species (rats, mice, bats, and humans) and multiple regions of the brain (entorhinal cortex, pre- and parasubiculum, hippocampus, parahippocampal gyrus, amygdala, cingulate cortex, and frontal cortex) [4, 6, 8, 10, 11, 25]. In all reported cases the activity of the observed neurons correlated with the organism's location supporting the early and predominant interpretation that grid cells are a functional component of a system

J. Kerdels (✉) · G. Peters
FernUniversität in Hagen – University of Hagen, Universitätsstrasse 1,
58097 Hagen, Germany
e-mail: Jochen.Kerdels@FernUni-Hagen.de

© Springer Nature Switzerland AG 2019
J. J. Merelo et al. (eds.), *Computational Intelligence*,
Studies in Computational Intelligence 792,
https://doi.org/10.1007/978-3-319-99283-9_14

for navigation and orientation. Particularly, the relatively stable and periodic firing patterns of grid cells are thought of to provide a kind of *metric for space* by means of path integration [19]. However, recent observations [2, 3, 17, 23] indicate that the firing patterns of grid cells are much more dynamic and adaptive than previously assumed questioning their utility as such a metric. Moreover, Killian et al. [16] observed neurons with grid-like firing patterns in the entorhinal cortex of primates whose activity does not correlate with the animal's location but instead with gaze-positions in the animal's field of view. These new observations challenge the original notion of grid cells as a specialized component for path integration [1, 5, 9, 19, 20, 24] and suggest that the grid-like activity patterns may reflect a more general form of information processing. To explore this alternative hypothesis we developed a new computational model of grid cells based on the *recursive growing neural gas* (RGNG algorithm), which describes the behavior of grid cells in terms of principles of self-organization that utilize a form of competitive Hebbian learning [12, 14]. We could demonstrate [12] that our RGNG-based model can not only describe the basic properties of grid cell activity but can also account for recently observed phenomena of dynamic grid pattern adaptation in response to environmental changes [2, 3, 17, 23].

Furthermore, we recently showed [15] that our model can describe the behavior of "visual" grid cells in primates as they were reported by Killian et al. [16]. However, the data presented by the latter is not as clean as comparable data from, e.g., experiments with rats (see Fig. 4). This difference may just be an artefact of challenging recording conditions in primates or caused by a limited amount of data, but it could also reflect a difference in the actual behavior of the observed cells. Here we investigate this question further by extending our previously used two-dimensional input model into a three-dimensional version. The following two sections summarize the RGNG-based grid cell model and revisit the results of our recent investigation into modelling and simulation of *visual grid cells* in primates [15]. Section 4 then introduces a three-dimensional input model that is based on the efference copy of motor neuron populations that putatively control the binocular gaze direction. Section 5 presents the simulation results we obtained using this new input model and shows how the results relate to the characteristics of the data published by Killian et al. [16]. Finally, Sect. 6 draws conclusions and provides a set of testable predictions for future neuroscientific research.

2   RGNG-Based Grid Cell Model

The predominant interpretation that grid cells are a functional component of a system for navigation and orientation is reflected by the majority of conventional grid cell models. Typically, they incorporate mechanisms that directly integrate information on the velocity and direction of the respective animal. Requiring this domain specific information as input renders these models incapable of describing the behavior of other neurons with similar grid-like firing patterns but different types of input signals like the *visual grid cells* observed by Killian et al. [16]. The RGNG-based grid cell model avoids such domain specific dependencies. It is a neuron-centric model in which neurons act in their "own interest" while being in local competition with each other. Biological neurons receive thousands of inputs from other neurons, and from the perspective of a single neuron these inputs are just electrochemical signals that carry no domain specific information. The entirety of these input signals and their possible values (i.e., states of activity) constitute the input space of a neuron. Besides additional sources of information such as neuromodulators the *Umwelt* of a neuron is primarily defined by the structure of this input space. We hypothesize that neurons "want" to maximize their activity in response to this Umwelt as recent findings indicate that neuronal activity increases the direct glucose uptake of neurons [18]. Hence, from the perspective of a "selfish" neuron being more active means getting more energy. To this end, we assume that neurons are able to learn a limited number of input patterns or *prototypes* within their dendritic tree such that encountering any of these patterns will cause the neuron to become active. In that case, maximizing activity translates into learning those input patterns that occur most often while simultaneously trying to avoid learning near duplicate patterns which would waste "dendritic memory" capacity. As a consequence, the set of learned input patterns form a simple prototype-based representation of the input space structure residing in the dendritic tree of the respective neuron. In addition, if multiple neurons compete against each other via local inhibition they will form input space representations that are pairwise distinct from each other given that each competing neuron still wants to maximize it's activity. In such a group of competing neurons the individual simple representations will interleave in such a way that a complex representation of the input space emerges that is distributed over the entire group. In our model we use a two layer recursive growing neural gas (RGNG) to describe both processes at once: the learning of prototypes within individual cells as well as the simultaneous competition among the cells in the group. Interestingly, in both cases the underlying dynamics follow the same principles of self-organization utilizing a form of competitive Hebbian learning.

Most importantly, the RGNG-based neuron model can operate on arbitrary input spaces. For any input space the modeled group of neurons will try to learn the structure of the particular input space as well as possible. If the activity of a single modeled neuron is then correlated with a suitable external variable, the individual firing fields that can be observed correspond to the individual prototypes or input space patterns that the neuron has learned. The "locations" of the firing fields in relation to the external variable are a pointwise mapping between the learned input space structure and the value range of that particular external variable. For instance, to observe the typical grid-like firing pattern of grid cells the input patterns must originate from a two-dimensional, uniformly distributed manifold in the input space and have to correspond to the location of the organism in it's environment. Notably, these basic requirements allow for a multitude of possible input spaces for grid cells, which can then be tested with respect to further observed properties like the dynamic adaptation of grid patterns to environmental changes. Likewise, by choosing a suitable input space it becomes possible to model observations of grid-like firing patterns in other domains like the *visual grid cells* reported by Killian et al. [16] (see next section). A formal description of the RGNG-based model is provided in the appendix. For an in-depth characterization of the model we refer to our prior work [12, 15].

3 Encoding of 2D Visual Space Revisited

Neurons with grid-like activity patterns were observed in several mammalian species and multiple regions of the brain [4, 6, 8, 10, 11, 16, 25]. Among these findings the observations of Killian et al. [16] stand out as they are the first to report grid-like activity patterns that are not correlated with the organism's location but with gaze-positions in the field of view. We recently investigated if and how these *visual grid cells* can be described by our RGNG-based neuron model [15]. In this section we revisit and summarize our results from this investigation.

We hypothesized that the observed neurons may receive an input signal that originates as a so-called *efference copy* from the populations of motor neurons that control the four main muscles attached to the eye (Fig. 1). In such a population signal the number of active neurons corresponds to the degree with which the particular muscle contracts. Hence, the signal would provide an indirect measure of the gaze-position of an eye. We constructed a corresponding input signal $\xi := (v^{x_0}, v^{x_1}, v^{y_0}, v^{y_1})$ for a given normalized gaze position (x, y) by using four concatenated d-dimensional vectors $v^{x_0}, v^{x_1}, v^{y_0}$ and v^{y_1}:

$$v_i^{x_0} := \max\left[\, \min\left[\, 1 - \delta\left(\tfrac{i+1}{d} - x\right),\ 1\right],\ 0\right],$$
$$v_i^{x_1} := \max\left[\, \min\left[\, 1 - \delta\left(\tfrac{i+1}{d} - (1-x)\right),\ 1\right],\ 0\right],$$
$$v_i^{y_0} := \max\left[\, \min\left[\, 1 - \delta\left(\tfrac{i+1}{d} - y\right),\ 1\right],\ 0\right],$$
$$v_i^{y_1} := \max\left[\, \min\left[\, 1 - \delta\left(\tfrac{i+1}{d} - (1-y)\right),\ 1\right],\ 0\right],$$
$$\forall i \in \{0 \ldots d-1\},$$

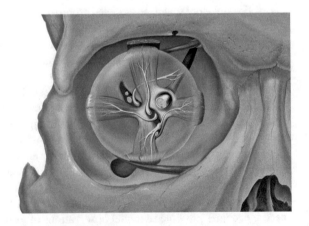

Fig. 1 Eye and orbit anatomy with motor nerves by Patrick J. Lynch, medical illustrator; C. Carl Jaffe, MD, cardiologist (CC BY 2.5). Extracted from [15]

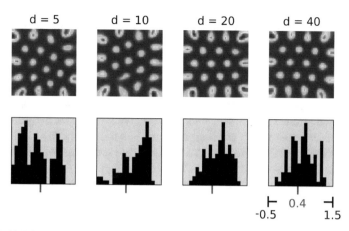

Fig. 2 Artificial rate maps and gridness distributions of simulation runs that processed input from a varying number d of presumed motor neurons per muscle (columns) that control the gaze-position. All simulation runs used a fixed set of parameters (Table 1) and processed two million random gaze locations encoded as the activity of corresponding motor neuron populations. Each artificial rate map was chosen randomly from the particular set of rate maps and displays all of the respective neuron's firing fields. The gridness distributions show the gridness values of all neurons from the particular neuron group. Gridness threshold of 0.4 indicated by red marks. Extracted from [15]

with $\delta = 4$ defining the "steepness" of the population signal and the size d of each motor neuron population.

Using this input space encoding we conducted a series of simulation runs with varying sizes d of the presumed motor neuron populations. Each run simulated a group of 100 neurons with each neuron using 20 prototypes to learn an input space representation.[1] The input consisted of two million, random gaze-positions per simulation run encoded as described above. Based on the resulting neuron activity and corresponding gaze-positions we constructed standard *firing rate maps* that represent a correlation of these variables. The maps were constructed according to the procedures described by Sargolini et al. [21] but using a 5×5 boxcar filter for smoothing instead of a Gaussian kernel as introduced by Stensola et al. [22]. This conforms to the de facto standard of rate map construction in the grid cell literature. The firing rate maps integrate activity and position information of 30,000 time steps sampled from the end of a simulation run, which corresponds to typical experimental trial durations in grid cell research, i.e., 10 min recorded at 50 Hz. In addition, we calculated the *gridness score* for each simulated neuron based on the constructed firing rate maps. The gridness score is an established measure to assess how grid-like an observed activity pattern is [21]. It ranges from -2 to 2 with scores greater zero indicating a grid-like activity pattern. To classify an observed neuron as grid cell, recent publications choose more conservative thresholds between 0.3 and 0.4.

Figure 2 summarizes the results of these simulation runs for motor neuron populations of size $d \in \{5, 10, 20, 40\}$ per eye muscle. Both the shown firing rate maps

---

[1]The full set of parameters used is given in the appendix (Table 1).

as well as the gridness score distributions show that a significant proportion of the simulated neurons form grid-like activity patterns in response to the input signal described above. Furthermore, the response of the neurons appears to be robust over the entire range of population sizes that were investigated. The fact that the activity of the simulated neurons forms a grid-like firing pattern when correlated with a two-dimensional, external variable (gaze-position) indicates, that the inputs themselves originate from a two-dimensional manifold lying within the high-dimensional $(d \times 4)$ input space – which is unsurprising in this case since we constructed the encoding of the input signal to have this property (but see Sect. 4 why this is only an *indication* in general).

The gaze-positions that were randomly chosen as input signals in the simulation runs cover the entire field of view and the resulting firing rate maps show the entire input space representation that was learned by a single neuron based on these inputs. As a consequence, a strong alignment of firing fields, i.e., prototypes at the borders of the firing rate maps can be observed since the borders of the firing rate maps coincide with the outer limits of the underlying, low-dimensional input space manifold in this case. As it is unlikely that experimental observations of natural neurons will cover the entire extent of an underlying input space, we investigated how the partial observation of firing fields may influence rate map appearance and gridness score distributions. To this end we conducted a second series of simulation runs using the same set of parameters as before with the exception of using 80 prototypes per neuron instead of 20. Furthermore, we constructed additional firing rate maps for each neuron that contain only one-quarter or one-sixteenth of the respective neuron's firing fields emulating a partial observation of natural neurons. Figure 3 shows the results of this second series of simulation runs. The additional firing rate maps that contain only a subset of firing fields display firing patterns that are much more regular and periodic compared to the firing patterns shown in Fig. 2. The corresponding gridness score distributions support this assessment by showing a larger proportion of simulated neurons with gridness scores above a threshold of 0.4. This indicates that the distortions introduced by the alignment of firing fields at the outer limits of the underlying input space manifold remain local with respect to their influence on the grid-like structure of other regions. Thus, natural neurons may receive signals from input space manifolds that are only partially two-dimensional and evenly distributed. In such a case grid-like firing patterns would only be observed if the experimental conditions happen to restrict the input signals to these regions of input space. Any shift towards regions with different properties would result in a distortion or loss of the grid-like structure.

4   Encoding of 3D Visual Space

In our recent investigation on modeling *visual grid cells* [15] we argued, as summarized above, that the firing rate map derived from the observation of a biological neuron would likely show only a subset of this neuron's firing fields since

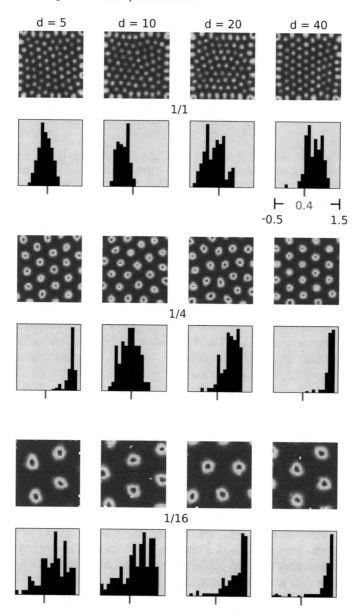

Fig. 3 Artificial rate maps and gridness distributions of simulation runs presented like in Fig. 2, but with 80 prototypes per neuron and containing either all, one-quarter, or one-sixteenth (rows) of the respective neuron's firing fields. Extracted from [15]

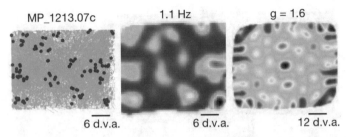

Fig. 4 Example of a neuron with a grid-like activity pattern in the monkey as observed by Killian et al. [16]. Left, plots of eye position (grey) and spikes (red) reveal non-uniform spatial density of spiking. For clarity, only spikes corresponding to locations of firing rate above half of the mean rate were plotted. The monkeys name and unit number are indicated at the top. Middle, spatial firing-rate maps show multiple distinct firing fields. Maps are colour coded from low (blue) to high (red) firing rates. The maximum firing rate of the map is indicated at the top. Right, the spatial periodicity of the firing fields shown against spatial autocorrelations. The colour scale limits are ±1 (blue to red), with green being 0 correlation. d.v.a., degrees of visual angle; g, gridness. Reprinted by permission from Macmillan Publishers Ltd.: Nature, Killian et al. [16], copyright 2012

experimental conditions would typically not allow to probe the entire input space of that neuron. Analyzing only a fraction of the firing fields of our simulated neurons showed an improvement in the overall regularity of the firing patterns and a reduction of the influence of alignment artifacts located at the outer limits of the underlying input space manifold. However, the activity patterns observed by Killian et al. [16] differ from these simulation results. They appear to be even less regular and "clean" in their grid-like structure (Fig. 4) as our simulated rate maps of the entire visual field (Fig. 2). It is possible that the distortions in the observed activity patterns are just artifacts that result from challenging recording conditions in primates, or potential inaccuracies in the measurement of the external variable (eye tracking); and in our previous investigation we assumed that this was indeed the case. Yet, it is also possible that the distortions are no artifacts and may reflect a more complex underlying input space structure.

To explore this alternative explanation we considered a more realistic input space model that includes both eyes and stereoscopic vision. Instead of encoding two-dimensional gaze-positions of a single eye we took random fixation points p_f in a three-dimensional view box (width: 400 cm × height: 250 cm × depth: 300 cm) and derived the corresponding horizontal and vertical viewing angles for both eyes at their respective locations. Allowed eye movement was restricted to a typical range of 50° adduction (towards the nose), 50° abduction (towards the temple), 60° infraduction (downwards), and 45° supraduction (upwards). Eyes were positioned 6 cm apart at 200 cm × 125 cm × 0 cm. Fixation points outside of the field of view of either eye as well as fixation points closer than 30 cm were rejected and not processed. After determining the horizontal and vertical viewing angles for each eye the angles were normalized to the interval [0, 1] and encoded as presumed population signal from the respective eight motor neuron populations (four per eye) as in our previous simulation experiments. A fixed population size of 20 motor neurons was used resulting in a

160-dimensional input space. We conducted a series of simulation runs with varying numbers of prototypes {20, 40, 80, 160} per neuron. Each run simulated a group of 100 neurons that processed 6 million input patterns derived from randomly chosen fixation points within the view box as described above. The full set of parameters used is provided in Table 2 in the Appendix.

5 Results

To assess the outcome of the simulation runs we constructed for each neuron a *volumetric firing rate map* that correlates the simulated neural activity with the corresponding fixation points within the view box. Figure 5 shows one exemplary firing rate map for each simulation run. The firing rate maps have a resolution of 3 cm × 3 cm × 3 cm, and a 5 × 5 × 5 boxcar filter adapted from the two-dimensional version introduced by Stensola et al. [22] was applied for smoothing. The normalized neuronal activity is color-coded from blue-transparent (low activity) to red-opaque (high activity). The maps are oriented such that the eyes are located at the center of the horizontal plane (width × height) looking upwards in the positive direction of the depth dimension.

The most prominent feature visible in these *volumetric firing rate maps* are columns of activity that radiate outwards from the location of the eyes. The columns appear evenly spaced, having similar diameters, and approaching a hexagonal arrangement with increasing depth. Columns at the outer limits of the input space manifold appear thicker and elongated outwards. The activity distribution within each column is non-homogeneous, but seems to never fall below a level that remains significantly larger than the activity outside of the column. In case of neurons with a number of 160 prototypes this continuity of activity appears to weaken as some columns show stretches of very low activity. This observation matches the expected increase in competition among individual neurons caused by the increasing number of prototypes per neuron, i.e., multiple neurons will compete for regions of input space within single columns.

In order to study the alignment of the activity columns with progressing depth we extracted a series of horizontal slices through the *volumetric firing rate maps* shown in Fig. 5. The resulting slices are shown in Fig. 6 with each slice integrating a depth range of 30 cm. The increase in regularity of the column alignment with increasing depth is clearly visible. The resulting patterns at a depth of 240 cm (Fig. 6, last row) vary between more diagonally aligned rectangular patterns in neurons with 20 and 40 prototypes to more hexagonal patterns in neurons with 80 and 160 prototypes. At depths of about 180 cm or less the outer limits of the underlying input space manifold determined by the maximum viewing angles of the modeled eyes become visible in the slices. Similarly to our previous results (Fig. 2) an alignment of the firing fields with respect to these limits can be observed. In addition, the outer firing fields appear larger and elongated outwards. This change in size and shape may in part be caused by cutting horizontally through columns that are angled outwards from the center.

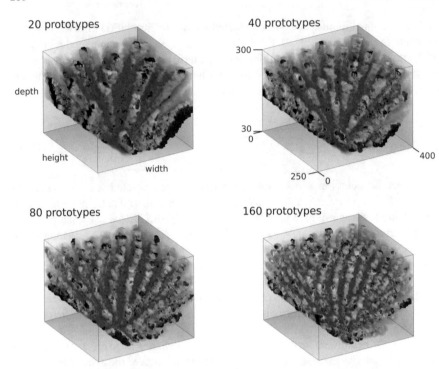

Fig. 5 Volumetric firing rate maps derived from simulation runs with varying numbers of prototypes {20, 40, 80, 160} per neuron. Each firing rate map was chosen randomly from the respective set of neurons. The rate maps correlate the simulated neuronal activity of a single neuron with the corresponding fixation point within the view box. Eyes are located at the center of the width and height dimensions looking upwards in the direction of the depth dimension. Normalized neuronal activity is color-coded from blue-transparent (low activity) to red-opaque (high activity). Resolution of the firing rate maps is 3 cm × 3 cm × 3 cm

However, if this would be the sole cause for the deformation we would expect to see a gradual increase in the deformation of firing fields with respect to their distance from the center, which is not observable in the slices.

The slices shown in Fig. 6 are also an approximation of the two-dimensional firing rate maps we would expect to observe if an animal with forward-facing binocular vision, like the macaque monkeys studied by Killian et al. [16], would be given the task of watching images presented on a two-dimensional monitor in front of them. In case of the experiments performed by Killian et al. the distance between the monkeys and the monitor that displayed the images was 60 cm, and the constructed firing rate maps covered 33° × 25° of visual angle. Values for adduction, abduction, infraduction, and supraduction as well as distance between the eyes in macaque monkeys was not provided by Killian et al. and we were unable to acquire this information reliably from other sources. Thus, we used common values for human eyes in our simulation experiments. Despite this shortcoming, the comparison of the

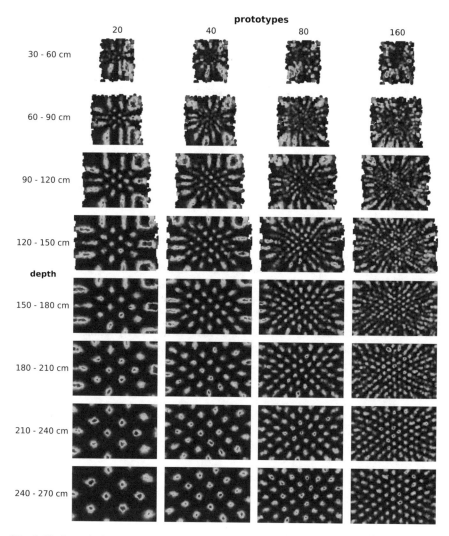

Fig. 6 Horizontal slices through the *volumetric firing rate maps* shown in Fig. 5. Each slice integrates a depth range of 30 cm. Normalized neuronal activity is color-coded from blue (low activity) to red (high activity). Regions with no data points are drawn in white

firing rate map provided by Killian et al. [16] (Fig. 4) and the slices of *volumetric firing rate maps* shown in the first rows of Fig. 6 shows some interesting structural similarities: In both cases the inner firing fields are weaker in activity, less regular, and smaller in size, whereas the outer firing fields are more pronounced, stronger in activity, and appear elongated outwards.

6 Conclusions

Killian et al. [16] were the first to report on entorhinal neurons in primates that show grid-like firing patterns in response to eye movements. We recently investigated how such a behavior could be modeled with our RGNG-based neuron model and we were able to demonstrate that it is feasible in principle [15]. In this paper we revisited our approach to modeling these neurons and provided a more comprehensive encoding of the presumed input signal that incorporates movement information from both eyes and fixation points that originate from a three-dimensional environment. The resulting *volumetric firing rate maps* of the simulated neurons exhibit a peculiar structure of regularly spaced *activity columns* that are angled outwards from the presumed eye locations. To the best of our knowledge these *volumetric firing rate maps* are the first model-based prediction on the expected activity patterns of *visual grid cells* in primates if their activity were to be correlated with fixation points from a three-dimensional environment.

In addition, horizontal slices through these *volumetric rate maps* show that the alignment of the activity columns increases in regularity with increasing distance from the eyes. Thus, for an experimental setup as it was used by Killian et al. [16] we predict that the distance between subject and the presented stimuli has a significant influence on the gridness of the observed activity patterns. A possible modification of the experimental design to test this prediction would be to use a video projector while placing the projection screen at various distances keeping the horizontal and vertical viewing angles constant.

Appendix

Recursive Growing Neural Gas

The recursive growing neural gas (RGNG) has essentially the same structure as the regular growing neural gas (GNG) proposed by Fritzke [7]. Like a GNG an RGNG g can be described by a tuple[2]:

$$g := (U, C, \theta) \in G,$$

with a set U of units, a set C of edges, and a set θ of parameters. Each unit u is described by a tuple:

$$u := (w, e) \in U, \quad w \in W := \mathbb{R}^n \cup G, \quad e \in \mathbb{R},$$

---

[2]The notation $g.\alpha$ is used to reference the element α within the tuple.

with the *prototype w*, and the *accumulated error e*. Note that in contrast to the regular GNG the prototype w of an RGNG unit can either be a n-dimensional vector or another RGNG. Each edge c is described by a tuple:

$$c := (V, t) \in C, \quad V \subseteq U \wedge |V| = 2, \quad t \in \mathbb{N},$$

with the units $v \in V$ connected by the edge and the *age t* of the edge. The *direct neighborhood* E_u of a unit $u \in U$ is defined as:

$$E_u := \{k | \exists (V, t) \in C, \quad V = \{u, k\}, \quad t \in \mathbb{N}\}.$$

The set θ of parameters consists of:

$$\theta := \{\epsilon_b, \epsilon_n, \epsilon_r, \lambda, \tau, \alpha, \beta, M\}.$$

Compared to the regular GNG the set of parameters has grown by $\theta.\epsilon_r$ and $\theta.M$. The former parameter is a third learning rate used in the adaptation function A (see below). The latter parameter is the maximum number of units in an RGNG. This number refers only to the number of "direct" units in a particular RGNG and does not include potential units present in RGNGs that are prototypes of these direct units.

Like its structure the behavior of the RGNG is basically identical to that of a regular GNG. However, since the prototypes of the units can either be vectors or RGNGs themselves, the behavior is now defined by four functions. The distance function

$$D(x, y) : W \times W \to \mathbb{R}$$

determines the distance either between two vectors, two RGNGs, or a vector and an RGNG. The interpolation function

$$I(x, y) : \left(\mathbb{R}^n \times \mathbb{R}^n\right) \cup (G \times G) \to W$$

generates a new vector or new RGNG by interpolating between two vectors or two RGNGs, respectively. The adaptation function

$$A(x, \xi, r) : W \times \mathbb{R}^n \times \mathbb{R} \to W$$

adapts either a vector or RGNG towards the input vector ξ by a given fraction r. Finally, the input function

$$F(g, \xi) : G \times \mathbb{R}^n \to G \times \mathbb{R}$$

feeds an input vector ξ into the RGNG g and returns the modified RGNG as well as the distance between ξ and the best matching unit (BMU, see below) of g. The input function F contains the core of the RGNG's behavior and utilizes the other three functions, but is also used, in turn, by those functions introducing several recursive paths to the program flow.

$F(g, \xi)$: The input function F is a generalized version of the original GNG algorithm that facilitates the use of prototypes other than vectors. In particular, it allows to use RGNGs themselves as prototypes resulting in a recursive structure. An input $\xi \in \mathbb{R}^n$ to the RGNG g is processed by the input function F as follows:

- Find the two units s_1 and s_2 with the smallest distance to the input ξ according to the distance function D:

$$s_1 := \arg\min_{u \in g.U} D(u.w, \xi),$$
$$s_2 := \arg\min_{u \in g.U \setminus \{s_1\}} D(u.w, \xi).$$

- Increment the age of all edges connected to s_1:

$$\Delta c.t = 1, \quad c \in g.C \wedge s_1 \in c.V.$$

- If no edge between s_1 and s_2 exists, create one:

$$g.C \Leftarrow g.C \cup \{(\{s_1, s_2\}, 0)\}.$$

- Reset the age of the edge between s_1 and s_2 to zero:

$$c.t \Leftarrow 0, \quad c \in g.C \wedge s_1, s_2 \in c.V.$$

- Add the squared distance between ξ and the prototype of s_1 to the accumulated error of s_1:

$$\Delta s_1.e = D(s_1.w, \xi)^2.$$

- Adapt the prototype of s_1 and all prototypes of its direct neighbors:

$$s_1.w \Leftarrow A(s_1.w, \xi, g.\theta.\epsilon_b),$$
$$s_n.w \Leftarrow A(s_n.w, \xi, g.\theta.\epsilon_n), \quad \forall s_n \in E_{s_1}.$$

- Remove all edges with an age above a given threshold τ and remove all units that no longer have any edges connected to them:

$$g.C \Leftarrow g.C \setminus \{c \mid c \in g.C \wedge c.t > g.\theta.\tau\},$$
$$g.U \Leftarrow g.U \setminus \{u \mid u \in g.U \wedge E_u = \emptyset\}.$$

- If an integer-multiple of $g.\theta.\lambda$ inputs was presented to the RGNG g and $|g.U| < g.\theta.M$, add a new unit u. The new unit is inserted "between" the unit j with the largest accumulated error and the unit k with the largest accumulated error among the direct neighbors of j. Thus, the prototype $u.w$ of the new unit is initialized as:

$$u.w := I(j.w, k.w), \quad j = \arg\max_{l \in g.U} (l.e),$$
$$k = \arg\max_{l \in E_j} (l.e).$$

The existing edge between units j and k is removed and edges between units j and u as well as units u and k are added:

$$g.C \Leftarrow g.C \setminus \{c \mid c \in g.C \wedge j, k \in c.V\},$$
$$g.C \Leftarrow g.C \cup \{(\{j, u\}, 0), (\{u, k\}, 0)\}.$$

The accumulated errors of units j and k are decreased and the accumulated error $u.e$ of the new unit is set to the decreased accumulated error of unit j:

$$\Delta j.e = -g.\theta.\alpha \cdot j.e, \quad \Delta k.e = -g.\theta.\alpha \cdot k.e,$$
$$u.e := j.e.$$

- Finally, decrease the accumulated error of all units:

$$\Delta u.e = -g.\theta.\beta \cdot u.e, \quad \forall u \in g.U.$$

The function F returns the tuple $(g, d_{\min})$ containing the now updated RGNG g and the distance $d_{\min} := D(s_1.w, \xi)$ between the prototype of unit s_1 and input ξ. Note that in contrast to the regular GNG there is no stopping criterion any more, i.e., the RGNG operates explicitly in an online fashion by continuously integrating new inputs. To prevent unbounded growth of the RGNG the maximum number of units $\theta.M$ was introduced to the set of parameters.

$D(x, y)$: The distance function D determines the distance between two prototypes x and y. The calculation of the actual distance depends on whether x and y are both vectors, a combination of vector and RGNG, or both RGNGs:

$$D(x, y) := \begin{cases} D_{RR}(x, y) & \text{if } x, y \in \mathbb{R}^n, \\ D_{GR}(x, y) & \text{if } x \in G \wedge y \in \mathbb{R}^n, \\ D_{RG}(x, y) & \text{if } x \in \mathbb{R}^n \wedge y \in G, \\ D_{GG}(x, y) & \text{if } x, y \in G. \end{cases}$$

In case the arguments of D are both vectors, the Minkowski distance is used:

$$D_{RR}(x, y) := \left(\sum_{i=1}^n |x_i - y_i|^p\right)^{\frac{1}{p}}, \quad x = (x_1, \dots, x_n),$$
$$y = (y_1, \dots, y_n),$$
$$p \in \mathbb{N}.$$

Using the Minkowski distance instead of the Euclidean distance allows to adjust the distance measure with respect to certain types of inputs via the parameter p. For example, setting p to higher values results in an emphasis of large changes in individual dimensions of the input vector versus changes that are distributed over many dimensions [13]. However, in the case of modeling the behavior of grid cells the parameter is set to a fixed value of 2 which makes the Minkowski distance equivalent to the Euclidean distance. The latter is required in this context as only the Euclidean distance allows the GNG to form an induced Delaunay triangulation of its input space.

In case the arguments of D are a combination of vector and RGNG, the vector is fed into the RGNG using function F and the returned minimum distance is taken as distance value:

$$D_{GR}(x, y) := F(x, y) \cdot d_{\min},$$
$$D_{RG}(x, y) := D_{GR}(y, x).$$

In case the arguments of D are both RGNGs, the distance is defined to be the pairwise minimum distance between the prototypes of the RGNGs' units, i.e., *single linkage* distance between the sets of units is used:

$$D_{GG}(x, y) := \min_{u \in x.U, k \in y.U} D(u.w, k.w).$$

The latter case is used by the interpolation function if the recursive depth of an RGNG is at least 2. As the RGNG-based grid cell model has only a recursive depth of 1 (see next section), the case is considered for reasons of completeness rather than necessity. Alternative measures to consider could be, e.g., *average* or *complete* linkage.

$I(x, y)$: The interpolation function I returns a new prototype as a result from interpolating between the prototypes x and y. The type of interpolation depends on whether the arguments are both vectors or both RGNGs:

$$I(x, y) := \begin{cases} I_{RR}(x, y) & \text{if } x, y \in \mathbb{R}^n, \\ I_{GG}(x, y) & \text{if } x, y \in G. \end{cases}$$

In case the arguments of I are both vectors, the resulting prototype is the arithmetic mean of the arguments:

$$I_{RR}(x, y) := \frac{x + y}{2}.$$

In case the arguments of I are both RGNGs, the resulting prototype is a new RGNG a. Assuming w.l.o.g. that $|x.U| \geq |y.U|$ the components of the interpolated RGNG a are defined as follows:

$$a := I(x, y),$$
$$a.U := \left\{ (w, 0) \; \middle| \; \begin{array}{l} w = I(u.w, k.w), \\ \forall u \in x.U, \\ k = \arg \min_{l \in y.U} D(u.w, l.w) \end{array} \right\},$$

$$a.C := \left\{ (\{l, m\}, 0) \; \middle| \; \begin{array}{l} \exists c \in x.C \\ \wedge \quad u, k \in c.V \\ \wedge \quad l.w = I(u.w, \cdot) \\ \wedge \quad m.w = I(k.w, \cdot) \end{array} \right\},$$

$$a.\theta := x.\theta .$$

The resulting RGNG a has the same number of units as RGNG x. Each unit of a has a prototype that was interpolated between the prototype of the corresponding unit in x and the nearest prototype found in the units of y. The edges and parameters of a correspond to the edges and parameters of x.

$A(x, \xi, r)$: The adaptation function A adapts a prototype x towards a vector ξ by a given fraction r. The type of adaptation depends on whether the given prototype is a vector or an RGNG:

$$A(x, \xi, r) := \begin{cases} A_R(x, \xi, r) \text{ if } x \in \mathbb{R}^n, \\ A_G(x, \xi, r) \text{ if } x \in G. \end{cases}$$

In case prototype x is a vector, the adaptation is performed as linear interpolation:

$$A_R(x, \xi, r) := (1 - r)x + r\xi.$$

In case prototype x is an RGNG, the adaptation is performed by feeding ξ into the RGNG. Importantly, the parameters ϵ_b and ϵ_n of the RGNG are temporarily changed to take the fraction r into account:

$$\theta^* := (\; r, \quad r \cdot x.\theta.\epsilon_r, \quad x.\theta.\epsilon_r, \quad x.\theta.\lambda, \quad x.\theta.\tau,$$
$$x.\theta.\alpha, \quad x.\theta.\beta, \quad x.\theta.M\;),$$
$$x^* := (x.U, \; x.C, \; \theta^*),$$
$$A_G(x, \xi, r) := F(x^*, \xi).x .$$

Note that in this case the new parameter $\theta.\epsilon_r$ is used to derive a temporary ϵ_n from the fraction r.

This concludes the formal definition of the RGNG algorithm.

Activity Approximation

The RGNG-based model describes a group of neurons for which we would like to derive their "activity" for any given input as a scalar that represents the momentary firing rate of the particular neuron. Yet, the RGNG algorithm itself does not provide a direct measure that could be used to this end. Therefore, we derive the activity a_u of a modelled neuron u based on the neuron's best and second best matching BL units s_1 and s_2 with respect to a given input ξ as:

$$a_u := e^{-\frac{(1-r)^2}{2\sigma^2}},$$

with $\sigma = 0.2$ and ratio r:

$$r := \frac{D(s_2.w, \xi) - D(s_1.w, \xi)}{D(s_1.w, s_2.w)}, \quad s_1, s_2 \in u.w.U,$$

using a distance function D. This measure of activity allows to correlate the response of a neuron to a given input with further variables.

Parameterization

Each layer of an RGNG requires its own set of parameters. In case of our two-layered grid cell model we use the sets of parameters θ_1 and θ_2, respectively. Parameter set θ_1 controls the main top layer RGNG while parameter set θ_2 controls all bottom layer RGNGs. Table 1 summarizes the parameter values used for the simulation runs presented in our previous work [15], while Table 2 contains the parameters of the simulation runs presented in this paper. For a detailed characterization of these parameters we refer to Kerdels [12].

Table 1 Parameters of the RGNG-based model used for the simulation runs in our previous work [15]. Parameters θ_1 control the top layer RGNG while parameters θ_2 control all bottom layer RGNGs of the model

θ_1	θ_2
$\epsilon_b = 0.04$	$\epsilon_b = 0.01$
$\epsilon_n = 0.04$	$\epsilon_n = 0.0001$
$\epsilon_r = 0.01$	$\epsilon_r = 0.01$
$\lambda = 1000$	$\lambda = 1000$
$\tau = 300$	$\tau = 300$
$\alpha = 0.5$	$\alpha = 0.5$
$\beta = 0.0005$	$\beta = 0.0005$
$M = 100$	$M = \{20, 80\}$

Table 2 Parameters of the RGNG-based model used for the simulation runs presented in this paper (Sect. 4)

θ_1	θ_2
$\epsilon_b = 0.004$	$\epsilon_b = 0.001$
$\epsilon_n = 0.004$	$\epsilon_n = 0.00001$
$\epsilon_r = 0.01$	$\epsilon_r = 0.01$
$\lambda = 1000$	$\lambda = 1000$
$\tau = 300$	$\tau = 300$
$\alpha = 0.5$	$\alpha = 0.5$
$\beta = 0.0005$	$\beta = 0.0005$
$M = 100$	$M = \{20, 40, 80, 160\}$

References

1. Barry, C., Burgess, N.: Neural mechanisms of self-location. Current Biol. **24**(8), R330–R339 (2014)
2. Barry, C., Ginzberg, L.L., OKeefe, J., Burgess, N.: Grid cell firing patterns signal environmental novelty by expansion. Proc. Nat. Acad. Sci. **109**(43), 17687–17692 (2012)
3. Barry, C., Hayman, R., Burgess, N., Jeffery, K.J.: Experience-dependent rescaling of entorhinal grids. Nat. Neurosci. **10**(6), 682–684 (2007)
4. Boccara, C.N., Sargolini, F., Thoresen, V.H., Solstad, T., Witter, M.P., Moser, E.I., Moser, M.B.: Grid cells in pre- and parasubiculum. Nat. Neurosci. **13**(8), 987–994 (2010)
5. Burak, Y.: Spatial coding and attractor dynamics of grid cells in the entorhinal cortex. Current Opin. Neurobiol. **25**(0), 169 – 175 (2014). Theoretical and computational neuroscience
6. Domnisoru, C., Kinkhabwala, A.A., Tank, D.W.: Membrane potential dynamics of grid cells. Nature **495**(7440), 199–204 (2013)
7. Fritzke, B.: A growing neural gas network learns topologies. In: Advances in Neural Information Processing Systems, vol. 7, pp. 625–632. MIT Press (1995)
8. Fyhn, M., Molden, S., Witter, M.P., Moser, E.I., Moser, M.B.: Spatial representation in the entorhinal cortex. Science **305**(5688), 1258–1264 (2004)
9. Giocomo, L., Moser, M.B., Moser, E.: Computational models of grid cells. Neuron **71**(4), 589–603 (2011)
10. Hafting, T., Fyhn, M., Molden, S., Moser, M.B., Moser, E.I.: Microstructure of a spatial map in the entorhinal cortex. Nature **436**(7052), 801–806 (2005)
11. Jacobs, J., Weidemann, C.T., Miller, J.F., Solway, A., Burke, J.F., Wei, X.X., Suthana, N., Sperling, M.R., Sharan, A.D., Fried, I., Kahana, M.J.: Direct recordings of grid-like neuronal activity in human spatial navigation. Nat. Neurosci. **16**(9), 1188–1190 (2013)
12. Kerdels, J.: A computational model of grid cells based on a recursive growing neural gas. Ph.D. thesis, FernUniversität in Hagen, Hagen (2016)
13. Kerdels, J., Peters, G.: Analysis of high-dimensional data using local input space histograms. Neurocomputing **169**, 272–280 (2015)
14. Kerdels, J., Peters, G.: A new view on grid cells beyond the cognitive map hypothesis. In: 8th Conference on Artificial General Intelligence (AGI 2015) (2015)
15. Kerdels, J., Peters, G.: Modelling the grid-like encoding of visual space in primates. In: Proceedings of the 8th International Joint Conference on Computational Intelligence, IJCCI 2016, Volume 3: NCTA, Porto, Portugal, 9–11 November 2016, pp. 42–49 (2016)
16. Killian, N.J., Jutras, M.J., Buffalo, E.A.: A map of visual space in the primate entorhinal cortex. Nature **491**(7426), 761–764 (2012)
17. Krupic, J., Bauza, M., Burton, S., Barry, C., O'Keefe, J.: Grid cell symmetry is shaped by environmental geometry. Nature **518**(7538), 232–235 (2015). https://doi.org/10.1038/nature14153
18. Lundgaard, I., Li, B., Xie, L., Kang, H., Sanggaard, S., Haswell, J.D.R., Sun, W., Goldman, S., Blekot, S., Nielsen, M., Takano, T., Deane, R., Nedergaard, M.: Direct neuronal glucose uptake heralds activity-dependent increases in cerebral metabolism. Nat. Commun. **6**, 6807 (2015)
19. Moser, E.I., Moser, M.B.: A metric for space. Hippocampus **18**(12), 1142–1156 (2008)
20. Moser, E.I., Moser, M.B., Roudi, Y.: Network mechanisms of grid cells. Philos. Trans. R. Soc. B Biol. Sci. **369**(1635) (2014)
21. Sargolini, F., Fyhn, M., Hafting, T., McNaughton, B.L., Witter, M.P., Moser, M.B., Moser, E.I.: Conjunctive representation of position, direction, and velocity in entorhinal cortex. Science **312**(5774), 758–762 (2006)
22. Stensola, H., Stensola, T., Solstad, T., Froland, K., Moser, M.B., Moser, E.I.: The entorhinal grid map is discretized. Nature **492**(7427), 72–78 (2012)
23. Stensola, T., Stensola, H., Moser, M.B., Moser, E.I.: Shearing-induced asymmetry in entorhinal grid cells. Nature **518**(7538), 207–212 (2015)
24. Welinder, P.E., Burak, Y., Fiete, I.R.: Grid cells: the position code, neural network models of activity, and the problem of learning. Hippocampus **18**(12), 1283–1300 (2008)
25. Yartsev, M.M., Witter, M.P., Ulanovsky, N.: Grid cells without theta oscillations in the entorhinal cortex of bats. Nature **479**(7371), 103–107 (2011)

Author Index

© Springer Nature Switzerland AG 2019
J. J. Merelo et al. (eds.), *Computational Intelligence*,
Studies in Computational Intelligence 792,
https://doi.org/10.1007/978-3-319-99283-9

Printed in the United States
By Bookmasters